# 15th International Conference on Environmental Engineering



Held as Part of the Canadian Society for Civil Engineering Annual Conference and General Meeting 2017

ISBN: 978-1-5108-7839-6

**Printed from e-media with permission by:**

Curran Associates, Inc.
57 Morehouse Lane
Red Hook, NY 12571

**Some format issues inherent in the e-media version may also appear in this print version.**



Printed by Curran Associates, Inc. (2018)

For permission requests, please contact Canadian Society for Civil Engineering at the address below.

Canadian Society for Civil Engineering
521-300 rue St-Sacrement
Montréal, QC
H2Y 1x4

Phone: (514) 933-2634
Fax: (514) 933-3504

info@csce.ca

**Additional copies of this publication are available from:**

Curran Associates, Inc.
57 Morehouse Lane
Red Hook, NY 12571 USA
Phone: 845-758-0400
Fax: 845-758-2633
Email: curran@proceedings.com
Web: www.proceedings.com

# TABLE OF CONTENTS

Leadership in Sustainable Infrastructure
*Leadership en Infrastructures Durables*

Vancouver, Canada
May 31 – June 3, 2017/ *Mai 31 – Juin 3, 2017*

# A REVIEW OF HISTORICAL EXTREME WIND SPEEDS IN A CHANGING CLIMATE AT SOME MAJOR CANADIAN CITIES

Li, Sihan[1,2], Irwin, Peter[1], Kilpatrick, John[1], Gibbons, Mike[1], Sifton Valerie[1]
[1] Rowan Williams Davies and Irwin Inc. (RWDI), Canada
[2] sihan.li@rwdi.com

**Abstract:** Some efforts have been made to consider the probable impacts of climate change on design wind speeds in tropical cyclone prone regions, such as Australia and the east coastal line of the United States. However, in Canada, the extreme wind speeds are mostly dominated by non-hurricane wind events. One recent study based on historical ground observations and re-analysis data indicates that stations over the Gulf of St. Lawrence of Canada mostly have no clear trend in terms of annual maximum wind speeds, but increasing variability was identified. A similar study shows that the historical monthly mean wind speed for the same region is decreasing. A recent study applies several weather forecast models for some possible future climate scenarios at various sites in Ontario, indicating an increase in the frequency of gust wind speeds between 28 km/h and 70 km/h. These studies point to three main tasks that need more research: obtaining a better understanding of any possible trends in historical data; the use of finer spatial scale regional climate model to predict future climate scenarios; and the development of more appropriate downscaling methods to relate the relationship between upper level and ground level wind speeds. The first task is critical to setting up the basis for measuring the quality of future climate modeling and to better understand how the current wind climate may translate to a probable future wind climate. From this perspective, this study explores the historical wind speed data in a number of major Canadian cities. Historical extreme wind observations are investigated in terms of the frequency of extreme wind events, seasonality and directionality. The probable impact on design wind speeds is discussed.

## 1 INTRODUCTION

Wind load is one of the most important parameters for building design, especially for high rise buildings. The design wind velocity is conventionally estimated from ground observations. Numerical simulation sometimes can be employed to extend the sample database used to evaluate the return period wind speed when the ground observations are insufficient, such as records for hurricane wind speeds (Vickery et al. 2009, Li and Hong 2014 & 2016). In Canada, the design wind speeds are dominated by extra-tropical wind storms or some local phenomena such as thunderstorms. The long term ground observations of wind speeds are typically used to estimate the return period wind speed and map the wind hazard for Canada (e.g. Hong et al. 2014). With increasing concern for the impact of a changing climate on infrastructure, some studies have applied numerical forecast models to predict the change of wind speed in future warmer climate scenarios. Among these studies, Cheng et al. (2013) indicate that in a warmer future climate, the frequency of gust wind speeds over some specific threshold values, such as 90 km/h, are most likely to increase for most regions of Canada. This gust wind speed is associated with a sudden, rapid and brief increase in wind speed. Conservatively, the corresponding hourly mean wind speed for this 90 km/h gust wind speed is no more than 54 km/h. This speed is less than the lowest value of design wind pressure velocity for 10-year (68.4 km/h) and 50-year (77.4 km/h) return period in the National Building Code of Canada (NBCC) 2015. The study of Cheng et al. (2013) indicates that for gust wind speeds greater than 90 km/h, the annual mean frequency of these speeds will increase in a future warmer climate. This may indicate that the frequency of wind speeds in the design speed range will increase as well. By definition, the return period wind speed is related to the probability of exceedance (occurrence of event over specific value). The change of annual occurrence of extreme wind speeds may have an impact on the estimated return period wind speeds. To deal with this changing occurrence rate of the extreme wind speeds, non-stationary probabilistic model needs to be employed. A non-stationary probabilistic model can be applied by assuming the probabilistic distribution parameters are varying with time. To fit such a non-stationary model, besides knowing the potential future change of the occurrence of extreme wind events, one has to know its historical characteristics as well. To the author's best knowledge, there are few studies publically available that investigate historical trends of the occurrence of extreme wind events. To better appreciate the historical characteristics of extreme wind events under a changing climate, one of the tasks in this study is to investigate trends in the occurrence rate of extreme wind events.

Besides the occurrence rate of the extreme wind speeds, the change in magnitude of the extreme wind speeds also has a direct impact on the estimated return period wind speeds. There are some studies exploring the trends of the historical wind speed data, such as Tuller (2004) and Wan et al. (2009). In these studies, the historical trend of long term (monthly in Wan et al. (2009) and annual in Tuller 2004) mean wind speeds is investigated. Tuller (2004) finds that four stations in west Canada have a decreasing trend for the annual mean wind speed, which is consistent with conclusions drawn by Wan et al. (2009). Broadly speaking, this decreasing trend identified by Wan et al. (2009) covers most of the south part of Canada, while only regions in the far North and east coast close to the Gulf of St. Lawrence have an increasing trend. A similar study by Hundecha et al. (2008) of the Gulf of St. Lawrence, using annual maximum wind speeds, found an increasing trend in part of the region, but a decreasing trend in the north part of this region. These studies generally agree that a decreasing trend appears in the long term mean wind speed for many meteorology stations of Canada. Only some limited effort was made to investigate the trend of extreme wind speeds, which are key to estimating design values. Due to the large uncertainty of current numerical weather forecast models, it is hard to reliably quantify the changes on annual maximum extreme wind speeds. Trends in the historical annual maximum wind speeds can be investigated as a basis for predicting future changes by assuming the same trends continue in future. This idea is one of the most popular approaches when non-stationary probabilistic models are employed to predict the percentile value of the extreme events in a future period (such as in Hundecha et al. 2008). The objective of this study is to conduct a trend analysis for the extreme wind speeds, which include the maximum wind speeds for an entire year, and for the summer, winter, and other seasons. The annual directional maximum wind speeds are considered as well in this study because the wind directionality often could be a key parameter for building design.

## 2 DATA AND STANDARDIZATION

This study uses meteorology stations at 20 cities across Canada. The geographical location and range of data used in this study are provided in Figure 1. It can be seen from Figure 1 that the selected stations are located in different regions across Canada, but are mostly located to the south. Note that the data for Niagara Falls is taken from the Niagara Falls International Airport on the US side of the border. Some of the data was removed in this study because of the bad data quality, such as insufficient data in a year, improper anemometer location (such as anemometer on building roof) etc. The data range used in this study for each station is listed in Table 1. Each station has more than 30 years of data. The data was standardized by conducting height corrections and exposure corrections. The height correction was conducted to convert the wind speed to standard 10m height by using the available anemometer height information provided for the meteorology station. For the exposure correction, the ESDU (2006) approach was adopted to consider the terrain change within tens of kilometers surrounding the site. The above processes are believed to homogenize the data reasonably well to standard height and terrain conditions. There might be some other factors, such as the change of anemometer type or human operation, which may also introduce some inhomogeneity into the data. These factors could not be accounted in this study due to lack of documentation and less knowledge on their impact. The statistics of the annual maximum wind speeds under stationarity assumptions, including mean value ($m_{Va}$), maximum value ($V_{a,max}$) and coefficient of variation (COV) of the annual maximum wind speeds, are summarized in Table 1. The return period wind speeds corresponding to design wind pressures for 50-year ($V_{50}$) return period given in NBCC 2015 are listed in Table 1 as well. For conversion of the design wind pressure to wind speed, the air density of 1.2929 kg/m$^3$ is used, consistent with the value assumed for the NBCC. It is believed that the length of records is also sufficient to carry out a trend analysis. From Table 1, it can be seen that some of the maximum values of annual maximum wind speeds are close to those 50-year return period wind speeds in NBCC. This is expected because by definition the 50-year return period wind speeds should be expected to be exceeded once in 50 years on average.

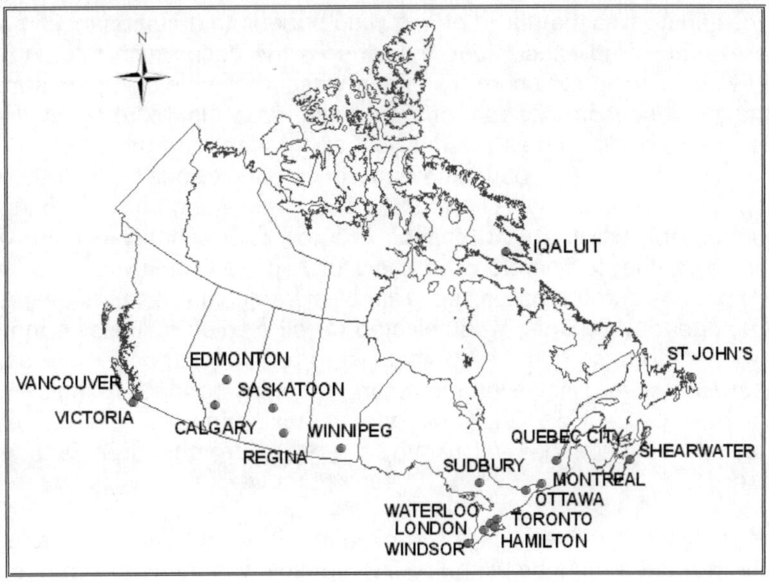


Figure 1: Geographic distribution of the meteorological station

Table 1: Statistics of the annual maximum wind speeds assuming stationarity for studied stations and their corresponding design wind speed.

| City | Number of Years | $m_{Va}$ (km/h) | $V_{a,max}$ (km/h) | COV | $V_{50}$ in NBCC 2015 (km/h) |
|---|---|---|---|---|---|
| Calgary | 30 | 75 | 94 | 0.11 | 98 |
| Edmonton | 43 | 65 | 83 | 0.09 | 95 |
| Shearwater (Halifax) | 34 | 70 | 96 | 0.14 | 108 |
| Hamilton | 46 | 74 | 98 | 0.11 | 96 |
| Iqaluit | 45 | 80 | 121 | 0.19 | 108 |
| London | 50 | 70 | 98 | 0.14 | 97 |
| Montreal | 52 | 69 | 86 | 0.12 | 92 |
| Niagara Falls | 44 | 72 | 94 | 0.12 | 93 |
| Ottawa | 55 | 64 | 83 | 0.13 | 91 |
| Quebec | 47 | 68 | 85 | 0.12 | 91 |
| Regina | 48 | 72 | 87 | 0.09 | 99 |
| Saskatoon | 46 | 66 | 95 | 0.15 | 93 |
| St. John's | 37 | 90 | 107 | 0.09 | 103 |
| Sudbury | 48 | 67 | 97 | 0.15 | 96 |
| Toronto | 53 | 75 | 91 | 0.09 | 94 |
| Vancouver | 54 | 62 | 87 | 0.14 | 98 |
| Victoria | 52 | 56 | 73 | 0.14 | 92 |
| Waterloo | 49 | 69 | 86 | 0.10 | 86 |
| Windsor | 48 | 74 | 99 | 0.12 | 97 |
| Winnipeg | 45 | 67 | 86 | 0.09 | 95 |

Although the data at the upper tail of the extreme value distribution has a direct impact on the estimated return period wind speed, the data at the lower tail also has an impact on the estimation of the slope parameter of the distribution. Therefore, the smaller values of annual maximum wind speed need to be considered as well. This perspective leads to the selection of threshold value for occurrence rate trend analysis in next section.

## 3 TREND ANALYSIS FOR EXTREME WIND SPEEDS

There are two major contributors to the return period wind speeds in a stationary climate. The first one is the magnitude of the extreme wind speeds and the other is the occurrence rate. A simple approach of estimating the return period wind speed under a stationary assumption is using the annual maximum wind speed. In this case the annual occurrence rate of the annual maximum extreme wind speeds is explicitly defined to be 1. Other probabilistic models or statistical approaches are available to estimate the return period wind speeds, such as the independent storm approach (Cook and Harris 2003, Harris 2009), peak over threshold method (POT, Simu and Heckert 1996) and a more general Poisson distribution method. Changes in the occurrence of extreme wind speeds due to non-stationary effects will have direct and explicit impacts on these methods. Therefore, in this study, a trend analysis was carried out for both annual maximum wind speeds and the occurrence rate of extreme wind speeds over a specific threshold value. A threshold wind speed of 55 km/h was selected for all studied stations. A more robust threshold wind speeds could be selected based on the 50-year return period wind speed or some percentile value of the annual maximum wind speed. However, the reason for applying this threshold wind speed is that a) this value is close to a lower percentile value of the annual maximum wind speed for each station; and b) by selecting other threshold wind speeds (+/-10 km/h), the results from the trend analysis for each station do not change.

For carrying out trend analyses, two methods are applied in this study. The first is to carry out a t-test for the best linear fit of the variable (annual maximum wind speeds or annual occurrence rate) as a function of time (i.e. number of years from start of the record). The null hypothesis is that the slope of the best linear fit is zero. If this hypothesis can be rejected, then the detected trend is considered to be statistically significant. Assuming the analyzed variable, Y, can be expressed as a linear function of time (T), $Y = aT + b$. The t-score for this test can be calculated by,

$$[1]\quad t_{score} = \hat{a} / \left( \sqrt{\left( \sum_{i=1}^{N} \left( y_i - \left( \hat{a} x_i + \hat{b} \right) \right)^2 \right) / (n-2)} / \sqrt{\sum_{i=1}^{N} (x_i - \bar{x})^2} \right)$$

where, $\hat{a}$ and $\hat{b}$ are the estimated slope and intercept, $x_i$ = time of observation from start of data, $y_i$ = observation variable (wind speed or occurrence rate), and $n$ = number of observations.

The analysis was assumed to indicate a statistically significant trend when the absolute value of $t_{score}$ is less than 0.05 (i.e., 5% significance level). Yue and Pilon (2004) suggest that the Mann-Kendall Tau trend analysis can be more effective in cases where the variable is non-normally distributed, which is the case for the annual maximum wind speed that typically can be fitted into a Gumbel distribution. Therefore, to further confirm the detected trend, if there is any, the Mann-Kendall trend analysis was also carried out for all studied stations. For the variables, *X*, ordered by time, such as the sequential annual maximum wind speed or occurrence rate of extreme storm event per year, the Mann-Kendall *S* score can be defined as,

$$[2]\quad S = \sum_{i<j} \operatorname{sgn}\left( X_j - X_i \right)$$

where *sng*(•) is the sign function.

The *S* score is asymptotically normal with zero mean and variance of $\sigma^2 = n(n-a)(2n+5)/18$. Again a 5% significance level was used to determine the null hypothesis.

As mentioned at the beginning of this section, two major variables, including annual maximum wind speeds and occurrence rate of specific extreme wind events, were used to carry out the trend analysis. These two major variables can be further classified by,

A1) annual maximum wind speeds of mixed wind climate;
A2) annual maximum wind speeds of synoptic (non-thunderstorm) wind climate;
A3) annual maximum wind speeds of thunderstorm only wind climate;
A4) annual maximum wind speeds of seasonal wind climate;
A5) annual maximum directional wind speeds (45-degree each bin);
A6) annual occurrence of hourly mean wind speeds exceeding specific threshold value (55 km/h);
A7) annual occurrence of thunderstorm.

Table 2: Summary of trend analysis for the annual maximum wind speed for studied stations

| Station Name | A1 | A2 | A3 | A6 | A7 |
|---|---|---|---|---|---|
| Calgary | Decreasing | Decreasing | - | - | - |
| Edmonton | Decreasing | Decreasing | - | - | - |
| Shearwater (Halifax) | Decreasing | Decreasing | Decreasing | Decreasing | - |
| Hamilton | - | - | - | - | Increasing |
| Iqaluit | Increasing | Increasing | N/A | Increasing | N/A |
| London | - | - | - | - | - |
| Montreal | - | - | N/A | N/A | - |
| Niagara Falls | - | - | - | - | - |
| Ottawa | Decreasing | Decreasing | - | Decreasing | - |
| Quebec City | - | - | Decreasing | Increasing | Decreasing |
| Regina | - | - | - | - | - |
| Saskatoon | - | - | - | - | - |
| St. John's | - | - | N/A | Decreasing | - |
| Sudbury | Decreasing | Decreasing | Decreasing | Decreasing | - |
| Toronto | - | - | - | - | - |
| Vancouver | Increasing | Increasing | N/A | Increasing | N/A |
| Victoria | - | - | N/A | - | N/A |
| Waterloo | - | - | - | Increasing | - |
| Windsor | Decreasing | - | - | Decreasing | Decreasing |
| Winnipeg | - | - | - | - | Decreasing |

Note: "-" denotes no statistically significant trends was detected, N/A denotes insufficient data to carry out trend analysis.

For a visual illustration, the best linear fits for the standardized annual maximum wind speeds for each studied station are plotted in Figure 2. The estimated slope and the calculated probability based on t-score are also given in each plot. For detection of the trend for annual maximum wind speeds, the Mann-Kendall trend analysis is more sensitive. Therefore, if the t-test indicated no trend but the Mann-Kendall test accepted trending, then the trend was accepted. For trend analysis for occurrence, when both tests reject the null hypothesis, the trend is confirmed. The results of the above trend analysis for tasks A1, A2, A3, A6 and A7 are listed in Table 2 for all studied stations. It can be seen from the Figure 2 and Table 2 that, for annual maximum wind speeds extracted from the mixed wind climate (A1), six of twenty stations are detected as having a decreasing trend, while only two stations are detected as having an increasing trend. The remaining 12 stations have no statistically significant trend. For stations that are detected having a trend for annual maximum wind speeds from the mixed wind climate, their synoptic annual maximum wind speeds (A2) have the same trend, except for Windsor, whose synoptic wind has no trend. There are three stations, Shearwater, Quebec City and Sudbury, which have a decreasing trend for thunderstorm extreme wind speeds (A3). Four stations were detected as having trend in the annual occurrence rate of thunderstorms (A7). These include Hamilton having increasing trend, while Quebec City, Windsor and Winnipeg having decreasing trend. The trend of annual occurrence for peak wind speed of independent storms exceeding 55 km/h (A6) for each station is generally similar to the analysis for case A1, except for Calgary and Edmonton detected as having no trend and Waterloo and Quebec City having an increasing trend, while St. John's having a decreasing trend.

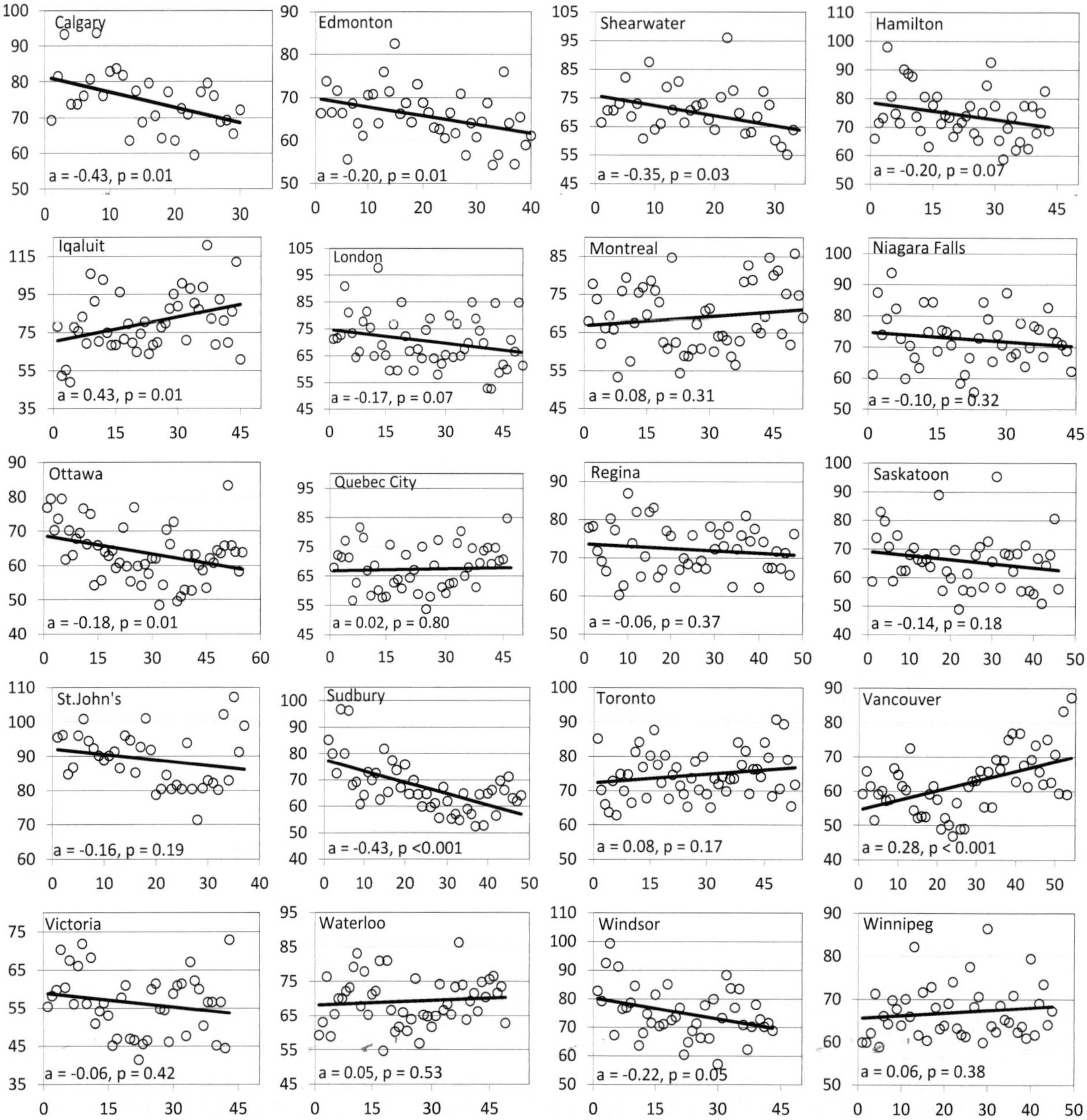


Figure 2: Trend analysis of annual maximum wind speed for each station applying T-test. X- axis represents time (year) from the start of the data and Y-axis represents the annual maximum wind speeds (km/h).

To carry out the trend analysis for seasonal extreme wind speeds, the data is first categorized based on recording time. The ranges of four seasons in this study are defined as, Spring being from March to May, Summer being from June to August, Autumn being from September to November, and Winter being from December to February of the following year. The frequencies of wind speed greater than 55 km/h for each season for each studied station are shown in Figure 3. It can be seen that for most stations, the extreme wind speeds (> 55 km/h) more frequently occur in winter or spring. The trend analysis is only carried out for the annual maximum wind speeds for each season. The results are illustrated in Figure 3 by showing a downwards arrow if there is a statistically significant decreasing trend, or an upwards arrow if an increasing trend was detected. Otherwise, no arrows are provided.

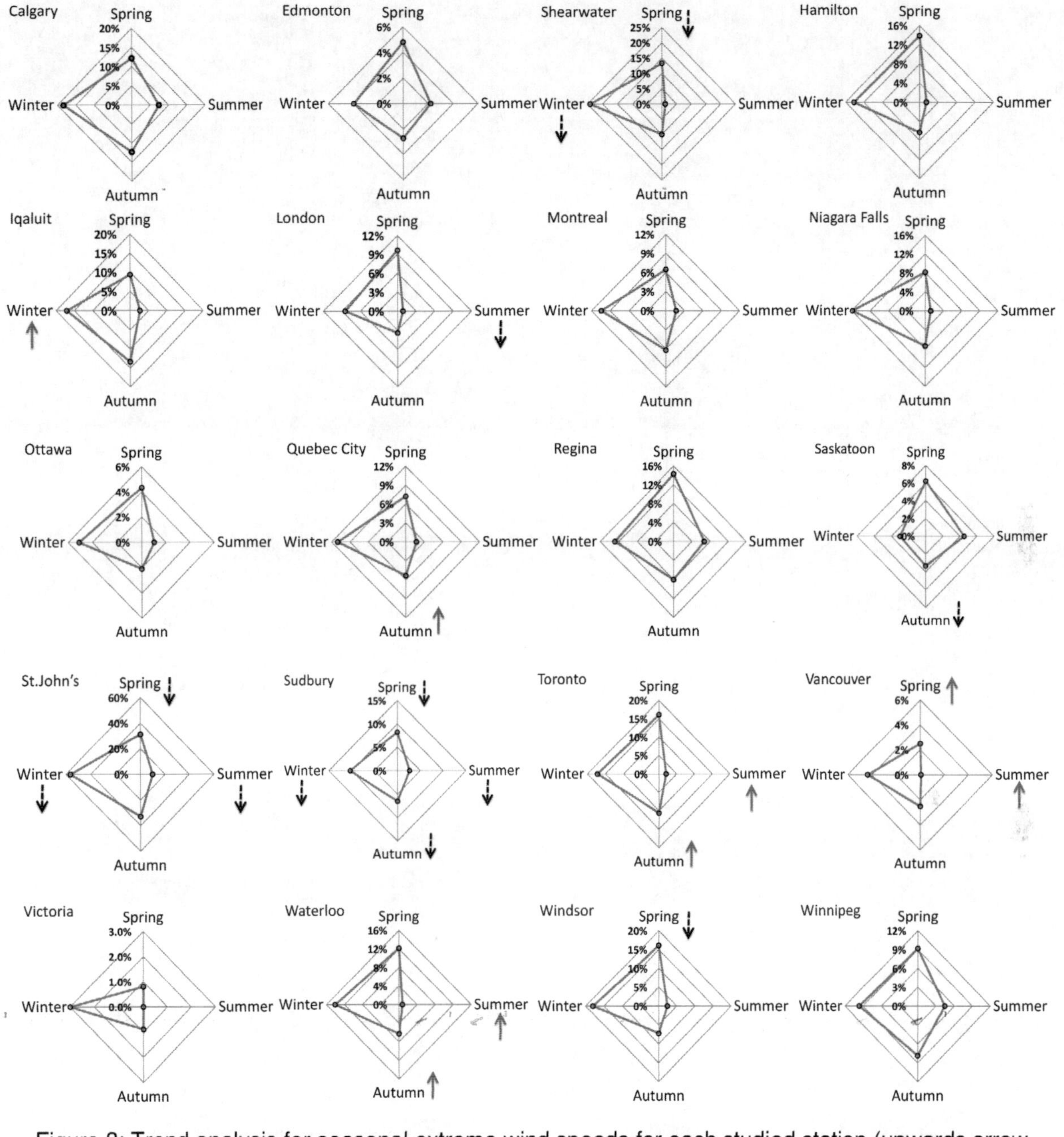


Figure 3: Trend analysis for seasonal extreme wind speeds for each studied station (upwards arrow denotes increasing trend and downwards arrow denotes decreasing trend)

It can be seen that the detected trends for the annual maximum wind speed in each season are not always consistent with the detected trend in the annual maximum wind speeds. For example, Calgary, Edmonton and Ottawa show no trends for the yearly maximum of seasonal wind speeds. However, their corresponding annual maximum wind speeds are detected as having trend. This may be because of the change of extreme wind climate is not concentrated in one season. Some of the detected trends support the trend analysis shown in Figure 2. Taking Shearwater station as an example, the winter and spring annual maximum value of seasonal extreme wind speeds are detected as having decreasing trends. The extreme wind speeds most frequently occur in winter and a bit lesser in spring. While extreme wind speeds in both seasons have a decreasing trend, the annual maximum wind speed for Shearwater station has a decreasing trend as well.

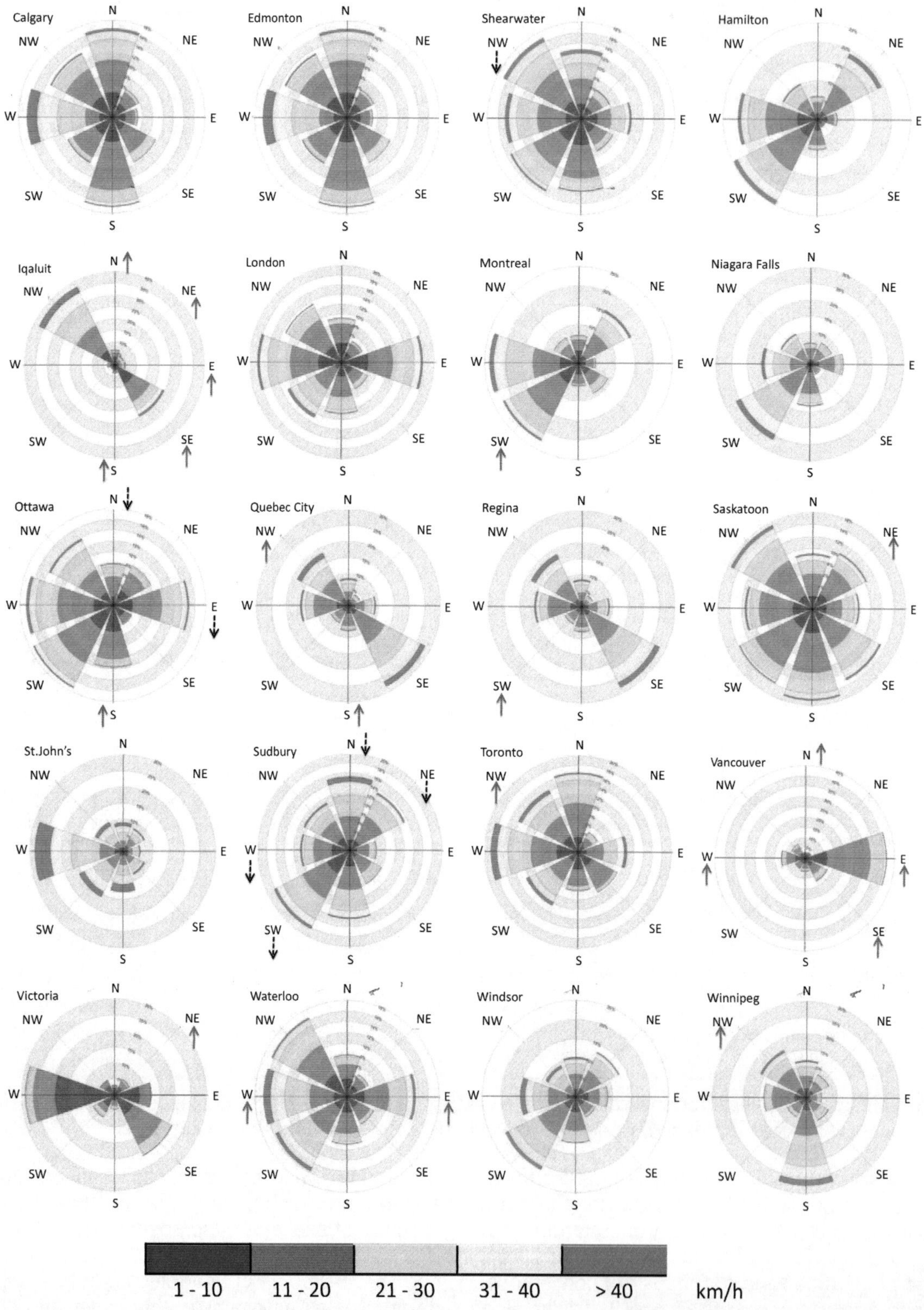


Figure 4: Trend analysis for directional extreme wind speeds for each studied station (upwards arrow denotes increasing trend and downwards arrow denotes decreasing trend)

For some stations, such as London, although the annual maximum value of seasonal extreme wind speeds had a trend, the most frequently occurring wind speed did not necessarily occur in that season. Consequently, the detected trend for seasonal extreme wind speeds is not found for annual maximum wind speed. A similar trend analysis was carried out for task A5, in which the threshold wind speed of 55 km/h was selected for reasons stated previously. To simplify the analysis, this study categorized the wind speed data into 8 wind angle bins. Each angle bin has a 45-degree range. To illustrate the frequency of wind speeds occurred in each wind angle bin, the calculated frequency for each station is plotted in Figure 4. It can be observed that the local wind climate is very diverse for these 20 studied stations. Some of the stations, such as Iqaluit and Regina, have only a few dominant wind angles where wind would most likely come from. Other stations may have more probable wind angles, e.g., Shearwater, Saskatoon and Waterloo. It should be noted that the wind rose of normal wind speeds does not necessarily indicate the dominant wind angle of the annual maximum wind speed. For example, Figure 3 shows that the most frequent wind direction for Vancouver is east, but, the dominant wind angle for annual maximum wind speeds is northwest (i.e., facing to the Strait of Georgia). The purpose of examining these frequency plots is to understand the relative frequency of the wind direction, which can be used to carry out wind directionality analysis in structural wind engineering design. While leaving any probable change/trend of relative frequency of wind directionality out of scope of this study, the extreme wind speeds in each wind angle bin are analyzed. The detected trends are given in Figure 4 by using the arrows with same meaning as those in Figure 3. Besides some similar observations as those indicated in Figure 3, some interesting findings can be also seen in Figure 4. From the above analysis, an increasing trend is detected for Iqaluit. Figure 4 indicates that this increasing trend is caused by wind from North to East, where the station faces the sea. In other words, it may indicate that during winter time, the storm which causes wind coming from around northeast may become more active and frequent. Such deduction needs further study from a point of view of a large scale macro-meteorology.

## 4 REMARKS AND DISCUSSION

This study uses ground observations of wind speeds at meteorology stations at various cities across Canada to carry out trend analyses. Both annual maximum wind speeds and occurrence rate of wind speeds exceeding 55 km/h are studied. T-test and Mann-Kendall trend analyses are used to detect possible trend of extreme wind speeds. It is found that for annual maximum wind speeds from mixed wind climates, there are eight of twenty stations that have a statistically significant trend. The detected trends are decreasing for six stations, while two stations are indicated having an increasing trend. Similar trends can be detected if the thunderstorm wind speeds are filtered out. However, trends were seldom detected in thunderstorm annual maximum wind speeds, or in annual occurrence rates of thunderstorms. There were three stations, including Shearwater, Quebec City and Sudbury that showed a decreasing trend of thunderstorm annul maximum wind speeds. Similar trend analyses are carried out for annual maximum values of seasonal extreme wind speeds and directional wind speeds. The results indicate that the trends detected for some stations mainly occurred in one or two seasons, those in which the wind speeds are most frequently occurring. For some other stations, there was no trend detected from the seasonal analysis, which indicates that the change of the wind speeds may be distributed more randomly among seasons. Results from the directional trend analysis indicate the possible wind direction where the extreme wind speeds having a trend. Such findings can be used for further study on the causes of that trend.

It is important to note that although changing trends in maximum wind speeds over time have been identified, this does not constitute evidence that these trends are directly caused by climate change, only that it is a possible indication. While reasonable efforts have been made to ensure that the data have been homogenized with respect to height and exposure, the influences from changes in instrumentation and exposure are much more complex than can be addressed through simplified adjustments. As such, there is still a degree of uncertainty in end results. In addition, there could be other larger scale influences that may produce impacts on climate over a longer period (e.g., inter-annual variations like El Niño\La Niña and inter-decadal variations like Pacific decadal oscillations) that have not been considered here. Some of these issues may have influenced some of the regional discontinuities of the detected trends that can be observed in this study. For example, the extreme wind speed for Vancouver international airport is detected as having an increasing trend. However, a station in the close area at Victoria international airport has no statistically significant trend. In terms of the directional extreme wind speeds, there is an increasing trend detected for both Vancouver and Victoria, but not for exactly the same wind directions. The physical mechanism causing this phenomenon is not clear yet. To better appreciate localized trends, research on the uncertainties of the trend analysis and regional trend analysis need to be carried out in future studies.

The final purpose of the trend analysis was to set up the basis for future studies of the impact of trends on the estimation of design wind speeds. For stations with a decreasing trend in annual maximum wind speeds, there might be no urgent action needed as a linear non-stationary probabilistic model will lead to a lower design wind speed. For those stations having an increasing trend, the estimated design wind speeds will be increased by using the non-stationary probabilistic model. The impact of such trends highly depends on the extent of the detected trends and the non-stationary model used. For example, one may simply use a linear non-stationary function to model the Gumbel distribution parameters. The detected trends in this study are mostly decreasing, while the trends in future scenarios found in some publications by using numerical forecast models are increasing. Therefore, before applying any non-stationary probabilistic model, further studies are needed to identify the causes of these differences.

## Acknowledgements

**The present studies were partly funded by RWDI and partly through funding provided by the National Research Council of Canada. The authors thank Mr. Juan Park for assisting with data processing.**

## References

Cheng, C. S., Lopes, E., Fu, C., & Huang, Z. 2014. Possible impacts of climate change on wind gusts under downscaled future climate conditions: Updated for Canada. *Journal of Climate*, *27*(3), 1255-1270.

Cook, N. J., Harris, R. I., & Whiting, R. 2003. Extreme wind speeds in mixed climates revisited. *Journal of Wind Engineering and Industrial Aerodynamics*, *91*(3), 403-422.

ESDU. 2006. *Strong Winds in the Atmospheric Boundary Layer. Part 1 : Mean-Hourly Wind Speeds. Data Item 82026.* ESDU International, London, UK.

Harris, R. I. 2009. XIMIS, a penultimate extreme value method suitable for all types of wind climate. *Journal of Wind Engineering and Industrial Aerodynamics*, *97*(5), 271-286.

Hong, H. P., Mara, T. G., Morris, R., Li, S. H., & Ye, W. 2014. Basis for recommending an update of wind velocity pressures in Canadian design codes. *Canadian Journal of Civil Engineering*, *41*(3), 206-221.

Hundecha, Y., St-Hilaire, A., Ouarda, T. B. M. J., El Adlouni, S., & Gachon, P. 2008. A nonstationary extreme value analysis for the assessment of changes in extreme annual wind speed over the Gulf of St. Lawrence, Canada. *Journal of applied meteorology and climatology*, *47*(11), 2745-2759.

Li, S. H., & Hong, H. P. 2014. Observations on a hurricane wind hazard model used to map extreme hurricane wind speed. *Journal of Structural Engineering*, *141*(10), 04014238.

Li, S. H., & Hong, H. P. 2016. Typhoon wind hazard estimation for China using an empirical track model. *Natural Hazards*, *82*(2), 1009-1029.

Simiu, E., & Heckert, N. A. 1996. Extreme wind distribution tails: a "peaks over threshold" approach. *Journal of Structural Engineering*, *122*(5), 539-547.

Tuller, S. E. 2004. Measured wind speed trends on the west coast of Canada. *International Journal of Climatology*, *24*(11), 1359-1374.

Vickery, P. J., Wadhera, D., Galsworthy, J., Peterka, J. A., Irwin, P. A., & Griffis, L. A. 2009. Ultimate wind load design gust wind speeds in the United States for use in ASCE-7. *Journal of structural engineering*, *136*(5), 613-625.

Wan, H., Wang, X. L., & Swail, V. R. 2010. Homogenization and trend analysis of Canadian near-surface wind speeds. *Journal of Climate*, *23*(5), 1209-1225.

Yue, S., & Pilon, P. 2004. A comparison of the power of the t test, Mann-Kendall and bootstrap tests for trend detection/Une comparaison de la puissance des tests t de Student, de Mann-Kendall et du bootstrap pour la détection de tendance. *Hydrological Sciences Journal*, *49*(1), 21-37.

Leadership in Sustainable Infrastructure
*Leadership en Infrastructures Durables*

Vancouver, Canada
May 31 – June 3, 2017/ *Mai 31 – Juin 3, 2017*

# ADVANCED OXIDATIVE PROCESSES FOR DEGRADATION OF CARBAMAZEPINE IN WASTEWATER EFFLUENTS

Purnima Somathilake[1], John Albino Dominic[2, 6], Gopal Achari[3], Cooper H Langford[4], and Joo-Hwa Tay[5]
[1, 2, 3, 4, 5] Department of Civil Engineering, University of Calgary, AB, Canada
[6] johnalbino.dominic@ucalgary.ca

## ABSTRACT

This paper presents a comprehensive evaluation of advanced oxidation based treatment processes for degradation of carbamazepine (CBZ) in wastewater effluents. CBZ is a prescribed antiepileptic drug and is one of the most frequently detected pharmaceutical in water bodies. Wastewater treatment plants (WWTPs) are considered to be the gate ways of CBZ into the environment as CBZ is resistant to biodegradation and mostly remains unaltered during the conventional treatment processes. Advanced Oxidation Processes such as ozonation, $H_2O_2$/UV and $TiO_2$/UV individually or in combinations have indicated possibility for degrading CBZ. In this study, various AOPs including UVC/$H_2O_2$, UVA/$H_2O_2$, Ozonation ($O_3$), and photocatalysis ($TiO_2$/UV) were evaluated for ability to degrade CBZ. Experiments were conducted in spiked pure water and postsecondary treated wastewater by varying doses of $H_2O_2$, $TiO_2$ and $O_3$. Among all the tested methods, fastest degradation of CBZ was observed in both, O3 (14.4 mg/h) and UV/$H_2O_2$ (100 mg/L), with a half-life period of 35 sec. Rate of degradation of CBZ in wastewater followed zero-order and first-order for $O_3$ and UV/$H_2O_2$ with rate constants of 4.3 mgL$^{-1}$min$^{-1}$ and 1.2 min$^{-1}$ respectively. UV/$H_2O_2$ was deduced to be most suitable method for degradation of CBZ as highest degree of mineralization with a total organic carbon reduction of 87% was observed after a reaction time of 8 hours while $O_3$ resulted with less than 10% for the same reaction time. These results indicate that degradation of CBZ in existing WWTPs could be enhanced by adding small amounts $H_2O_2$ to UV disinfection units.

Keywords: - carbamazepine; advanced oxidation processes; UV/$H_2O_2$; emerging substances of concern; wastewater treatment

## 1. INTRODUCTION

Carbamazepine (CBZ) is one of the most persistent pharmaceutical frequently detected in aquatic environments including surface water, groundwater, sewage and drinking water. It is prescribed for treating epileptic seizures, severe nerve pain (trigeminal neuralgia) and bipolar disorder. Presence of CBZ in environment is primarily attributed to human/animal excretion via urine/feces or from the disposal into sewers. Wastewater treatment plants (WWTP) are considered as the major gateways of CBZ into the environment as it is highly resistant to biological degradation and very low removal efficiencies (below 10%) are achieved in conventional wastewater treatment plants. CBZ is considered as an indicator of pharmaceutical pollution due to its resistance to its ubiquitous presence in wastewater effluent-impacted surface water and groundwater worldwide and its resistance to environmental degradation. Apart from being stable in the environment, CBZ bio-accumulates through food contamination in the aquatic environment and is possibly harmful for aquatic life since literature shows it to be toxic even at concentrations below 100 mg/L (Jelic et al., 2013). According to Zhan et al.,2008, potentially chronic toxicity rather than acute toxicity for aquatic organisms including bacteria, algae and fish such as feminizing effect on fish. Effects on human health is yet to be defined but precaution warrants developing efficient treatment methodologies (Cunningham et al., 2010, Kumar et al.,2010, Zhan et al.,2008, Jelic et al., 2013, Chen et al., 2006).

Advanced oxidation processes (AOP) have attracted considerable interest for treatment of pharmaceuticals in wastewater treatment as they generate hydroxyl radicals that can nonselectively degrade a wide range of pharmaceuticals including carbamazepine. Though many studies have investigated application of AOPs for degradation of CBZ, solid recommendations implementation of AOP based technology in existing WWTPs is still lacking. This study was focused on evaluation of AOPs, $TiO_2$/UV, UVC/ $H_2O_2$, UVA/ $H_2O_2$, and ozonation ($O_3$) from the view point of their suitability in implementing them in existing WWTPs with minimal modifications infrastructure. Initially, degradation of CBZ was studied though experiments conducted with pure water spiked with CBZ by varying doses of the oxidants/catalyst. Based on the performance of AOPs in pure water, selected AOPs with promising results were further evaluated in experiments conducted with wastewater spiked with CBZ.

## 2. MATERIALS AND METHODS

### 2.1 Chemicals

99.0% purity analytical grade CBZ was purchased from Sigma-Aldrich Co. (Canada). Degussa P25 $TiO_2$ powder (P25 D-60287 PI 52342) was purchased from Degussa AG (Frankfurt). Degussa P25 consists of 25:75 rutile : antase crystal form, with a purity of 99.9%. Post- secondary treated wastewater was collected from a WWTP in Calgary, Canada. High performance liquid chromatography (HPLC) grade methanol and water was purchased from VWR (Canada).

### 2.2 Method

Batch experiments were conducted in LZC-ORG photoreactor supplied by Luzchem Research Inc, Canada. The reaction vessel was a cylindrical quartz beaker (D = 7 cm, H = 13 cm) with a working volume of 200 mL. Initial concentration of CBZ in all experiments was maintained as 5 mg/L by spiking 10 mL of CBZ from a stock solution (100 mg/L) prepared by ultra-sonication for 8 hours. The rest of the reaction solution consisted of either pure water or secondary treated wastewater along with doses of $TiO_2$ or $H_2O_2$ or Ozone. Experimental conditions evaluated for each AOP method are outlined in Table 1. UV dose from UVA and UVC light source of 10 bulbs was estimated to be $1.321E^{18}$ and $8.95E^{17}$ photons/sec by using ferrioxalate actinometry. The reactor was equipped with five lamps on each side of the reaction vessel. UVA and UVC irradiations used light sources from Hitachi FL8BL-B (352 nm, 8W) fluorescent lamps and Sankyo Denki G8T5 (254 nm, 8W) low pressure mercury lamps respectively. In all experiments, a magnetic stirrer was used to homogenise the solution and 2ml samples were collected at specified time intervals. All experiments were conducted in duplicates, and the average values are reported. CBZ concentrations were quantified using Varian Prostar 210 HPLC instrument equipped with a UV detector. Errors in quantification of CBZ in most of the data points are estimated to be less than SD ±1%. CBZ concentrations were quantified using Varian Prostar 210 HPLC instrument equipped with a UV detector. Errors in quantification of CBZ in most of the data points are estimated to be less than SD ±1%.

## 3. RESULTS AND DISCUSSION

### 3.1 Degradation of CBZ in pure water

Degradation of CBZ by various AOPs in pure water are shown in Figure 1. More than 99% of 4.5 mg/L of initial CBZ was degraded by $O_3$ (14.4 mg/h), UVC/$H_2O_2$ (50 mg/L) and UVA/$TiO_2$ (0.5 g/L) in 0.5, 2 and 10 min respectively under optimal doses in pure water. Rate of degradation was the slowest under UVA/$H_2O_2$(500 mg/L) among all the AOPs.

Degradation of CBZ by photocatalysis was tested in the presence of various loadings of $TiO_2$ (P25 Degussa) under both UVA and UVC irradiations. Adsorption of CBZ on $TiO_2$ in the dark over a period of 30 min was less than 4% even for $TiO_2$ loading of 2 g/L. Though there was no considerable difference in the rate of CBZ degradation under various $TiO_2$ loadings, rate of degradation was faster under UVA irradiation. First order degradation rate coefficient for in case of UVC irradiation (*K*: 0.35 $min^{-1}$) for $TiO_2$ dose of 0.5 g/L was lower by 41.6% compared to UVA irradiation for the same dose (K: 0.6 $min^{-1}$). Strong absorption of $TiO_2$ in the near UV wavelength (λ: 385 nm) is the reason for faster degradation of CBZ observed in case of UVA/$TiO_2$.

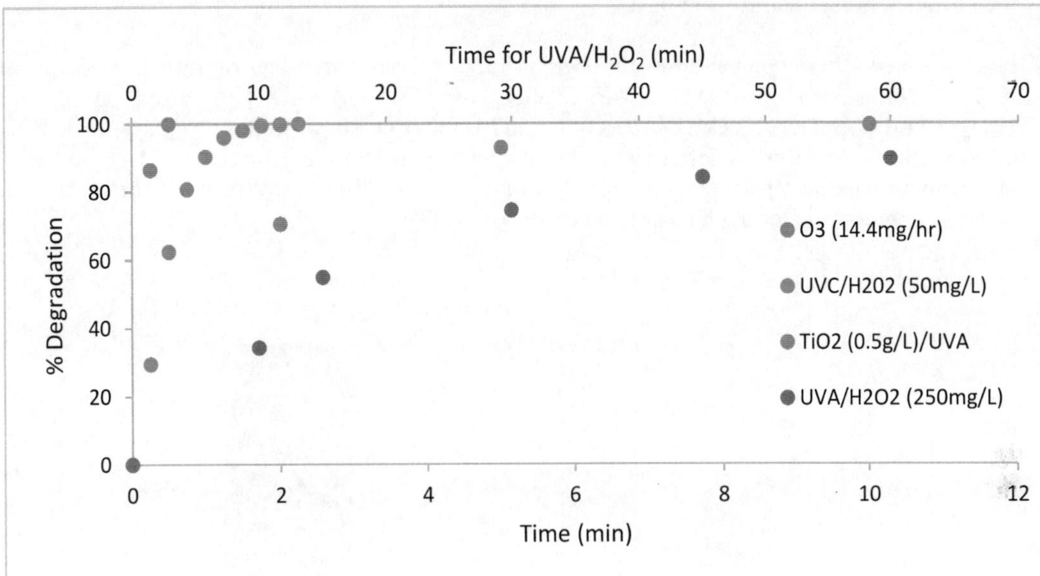


Figure 1: Degradation of CBZ by various AOPs in pure water

Table 1: Doses evaluated for each AOP method

| Treatment method | Doses | Water/wastewater |
|---|---|---|
| $TiO_2$/UVA | 0.5, 1, 1.5, 2 g/L | Water |
| UVA/$H_2O_2$ | 100, 250 & 500 mg/L | Water |
| UVA/$H_2O_2$ | 250, 500, 750 & 1000 mg/L | Wastewater |
| UVC/$H_2O_2$ | 10, 25, 50, 100 mg/L | Water |
| UVC/$H_2O_2$ | 10, 25, 50, 100 & 250 mg/L | Wastewater |
| Ozonation | 4.8, 9.6 & 14.4 mg/L | Water and wastewater |

The effect of $H_2O_2$ on CBZ degradation under both UVA and UVC irradiation was studied in the presence of by various doses of $H_2O_2$. Optimum dose of $H_2O_2$ for UVC/$H_2O_2$ and UVA/$H_2O_2$ treatment were estimated as 50 mg/L and 250 mg/L. Rate of degradation increased with increase in concentration of $H_2O_2$ dose under both UVA and UVC. However, significant improvement in rate of degradation is not observed above 50 and 250 mg/L of $H_2O_2$ under both UVA and UVC irradiations. Rate of CBZ degradation by UVA/$H_2O_2$ was slower and also required higher dose of $H_2O_2$ for optimal degradation compared to UVC irradiation. CBZ was highly recalcitrant to UVA photolysis as only 7.5 % of CBZ was degraded over period of 120 min while 31.7% was degraded under UVC photolysis. CBZ also exhibits strong absorption of UVC irradiation which peaks at 285 nm. Strong absorption of UVC by CBZ is assumed to sensitise the compound for faster degradation under UVC/$H_2O_2$ method.

Among all the tested AOPs, $O_3$ demonstrated the fastest rate of degradation of CBZ. More than 99.8% of CBZ was degraded within 2 min for the lowest dose of 4.8 mg/h of ozone tested in this study. For higher ozone doses of 9.6 and 14.4 mg/h, CBZ was degraded to below detection limits (>99.9%) by 0.75 and 0.5 min. Rate of CBZ degradation by ozone followed zero-order kinetics in contrast with other AOPs which followed first order kinetics. Zero-order kinetics during ozonation indicate that degradation of CBZ is not limited by the concentration of available ozone in the aqueous solution. High reactivity of CBZ with ozone at low ozone doses has been reported by Huber et al. (2003), however, in their studies they have used single spike of ozone dose in contrast to this present work in which ozone was continuously bubbled.

The half lives estimated for the optimal doses for degradation of CBZ in pure water by various AOPs are tabulated in Figure 2. According to the rate of degradation of CBZ the AOPs arranged in the order of $O_3$>UVC/$H_2O_2$>$TiO_2$/UVA>$TIO_2$/UVC>UVA/$H_2O_2$. $O_3$ and UVC/$H_2O_2$ show promise for easy implementation of these treatment methods as no major infrastructural units are required to remove residual $H_2O_2$ and $O_3$. Though $TiO_2$/UV also demonstrated faster degradation, removal of $TiO_2$ slurry from water requires energy and equipment and hence is not a viable option for implementing in existing WWTPs.

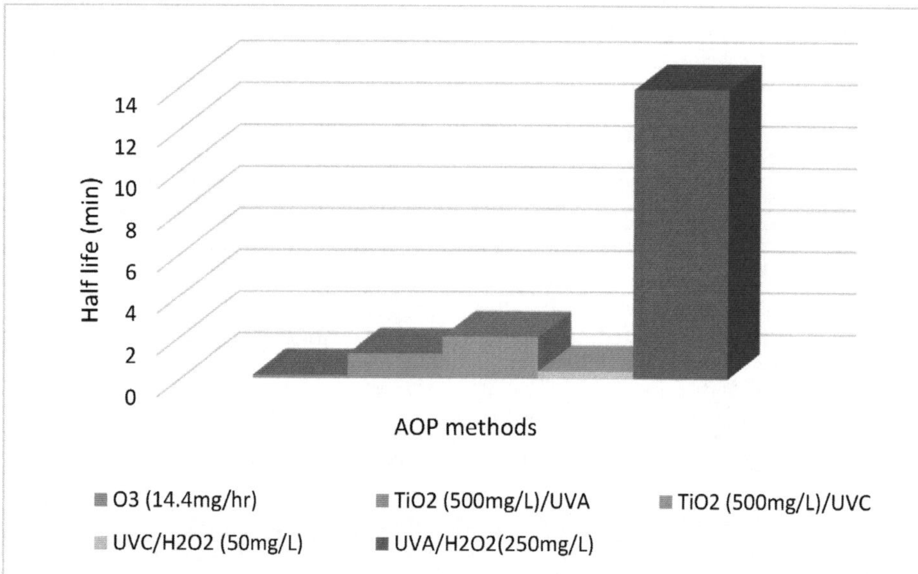


Figure 2: Half-life of CBZ for various AOPs in pure water

### 3.2 Degradation of CBZ in wastewater

Degradation of CBZ spiked in wastewater by AOPs $O_3$, UVA/ $H_2O_2$ and UVC/ $H_2O_2$ are shown in Figure 3. Rate of degradation of CBZ decreased in WW and also required higher doses of $H_2O_2$ under UVA and UVC irradiations. More than 99% of CBZ was degraded in the presence of 500 mg/L only after 240 min which is about 1.3 times higher that the time period observed in pure water experiments for achieving the same level of degradation with 250 mg/L of $H_2O_2$ in 180 min. In case of UVC/$H_2O_2$, the optimum concentration of $H_2O_2$ was estimated as 100 mg/L as rate of degradation increased only by 36.8% for a 60% increase in dose from 100 to 250 mg/L. Similar to UVA/$H_2O_2$ experiments higher doses of $H_2O_2$ were required to achieve optimal degradation rates in the presence of wastewater. Dissolved organic matter in secondary treated waste water can cause light shielding effect which reduce the rate of degradation. Also, the inorganic ions such as chloride, sulfate, nitrate and phosphate in the secondary treated water could compete for hydroxyl radicals as quenchers resulting in lower rates of degradation of carbamazepine.

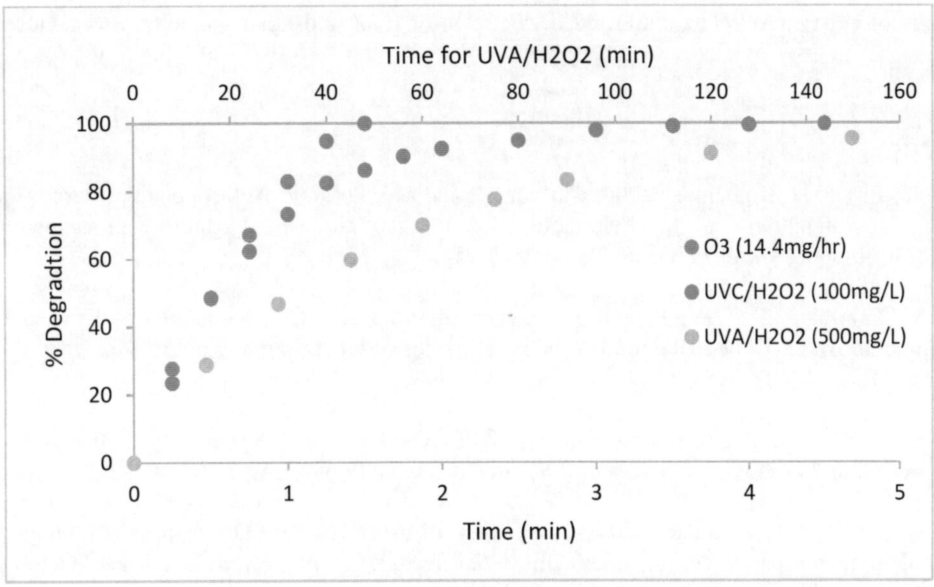


Figure 3: Percentage removal of CBZ by various AOPs in wastewater

Degradation of CBZ in wastewater by ozonation also followed zero order kinetics similar to experiments in pure water. However, in secondary treated wastewater, the rates of reaction were slower compared pure water spikes due to the presence of organic and inorganic matter that compete for ozone. For ozone dose of 14.4 mg/L, CBZ was degraded below detectable limits after a reaction time of 1.5 min which was 3 folds higher than reaction time observed in the presence of pure water.

Order of reaction and half-life of CBZ spiked in wastewater evaluated for various AOPs are tabulated in Table 2. The optimum dose of $H_2O_2$ required for UVC/$H_2O_2$ and UVA/$H_2O_2$ increased by 2 folds in the presence of WW as compared to pure water. However, experiments conducted in waste water show that $O_3$ and UVC/$H_2O_2$ are promising techniques for implementation in existing WWTPs as CBZ was rapidly degraded within 2 and 3 min respectively.

Table 2: Optimal doses and respective reaction rates in wastewater

| Treatment method | Dose | Order of Reaction | Half life |
|---|---|---|---|
| Ozone | 14.4 mg/h | Zero | 0.58 min |
| UVA/$H_2O_2$ | 500 mg/L | First | 34.65 min |
| UVC/$H_2O_2$ | 100 mg/L | First | 0.58 min |

## 4. CONCLUSIONS

This study investigated the application of advanced oxidation processes such as TiO2/UVA, H2O2/UV and O3 for the degradation of CBZ with perspective of implementing these technologies as tertiary treatment units in existing WWTPs. The study evaluated optimal conditions required for degradation of CBZ under each treatment method by conducting experiments in pure water and secondary treated wastewater. Ozonation yielded fastest rate of degradation of CBZ among all AOPs as 4.5mg/L of CBZ in wastewater was completely degraded within a period of 1.5 min (K: 4.3 mgL-1min-1). CBZ was also degraded by UVC/H2O2 photodegradation with a H2O2 dose of 100 mg/L within a period of 2.5 min (K:1.2 min-1). Findings from this study indicate that augmentation of H2O2 in UV disinfection units of existing WWTPs could aid degradation of CBZ and other recalcitrant organics in municipal wastewater.

## 5. REFERENCES


Clara, M., Strenn, B. and N. Kreuzinger 2004. Carbamazepine as a Possible Anthropogenic Marker in the Aquatic Environment: Investigations on the Behaviour of Carbamazepine in Wastewater Treatment and during Groundwater Infiltration. *Water Research* 38 (4): 947–954.

Domínguez, J.R., González, T., Palo, P. and J. Sánchez-Martín 2010. Electrochemical Advanced Oxidation of Carbamazepine on Boron-Doped Diamond Anodes. Influence of Operating Variables. *Industrial & Engineering Chemistry Research* 49 (18): 8353–8359.

Huber, M.M., Silvio C., Gun-Young P. and V.G. Urs 2003. Oxidation of Pharmaceuticals during Ozonation and Advanced Oxidation Processes. *Environmental Science & Technology* 37 (5): 1016–1024.

Jarvis, A.L., Melody J.B. and J.B. Randall 2014. The Effects of the Psychiatric Drug Carbamazepine on Freshwater Invertebrate Communities and Ecosystem Dynamics. *Science of The Total Environment* 496 (October): 461–470.

Jelic, A., Michael, I., Achilleos, A., Hapeshi, E., Lambropoulou, D., Perez, S., Petrovic, M., Fatta-Kassinos, D. and D. Barcelo 2013. Transformation Products and Reaction Pathways of Carbamazepine during Photocatalytic and Sonophotocatalytic Treatment. *Journal of Hazardous Materials* 263 (1): 177–186.

Klavarioti, M., Dionissios M. and K.Despo 2009. "Removal of Residual Pharmaceuticals from Aqueous Systems by Advanced Oxidation Processes." *Environment International* 35 (2): 402–417.

Mehrabani-Zeinabad, M., Achari, G. and C.H. Langford 2015 Advanced Oxidative Degradation of Bisphenol A and Bisphenol S. *Journal of Environmental Engineering and Science* 10 (4): 92–102.

Mohapatra, D.P., Brar, S. K., Tyagi, R.D., Picard, P. and R.Y. Surampalli 2014. Analysis and Advanced Oxidation Treatment of a Persistent Pharmaceutical Compound in Wastewater and Wastewater Sludge-Carbamazepine. *Science of The Total Environment* 470–471: 58–75.

Vogna, D, Raffaele M., Roberto A., Alessandra N. and Marco. D, 2004 Kinetic and Chemical Assessment of the UV/H2O2 Treatment of Antiepileptic Drug Carbamazepine. *Chemosphere* 54 (4): 497–505.

Leadership in Sustainable Infrastructure
*Leadership en Infrastructures Durables*

Vancouver, Canada
May 31 – June 3, 2017/ *Mai 31 – Juin 3, 2017*

# AKLAVIK, NWT – SOLID WASTE PLANNING STUDY

Johnson, Ken[1,4], Davignon, Jamie[2] and Behrens, Fred[3]
[1] Senior Environmental Planner and Engineer, Stantec, Edmonton, Alberta
[2] Environmental Engineer, Stantec, Whitehorse, Yukon
[3] Senior Administrative Officer, Hamlet of Aklavik, Aklavik, Northwest Territories
[4] kenneth.johnson@stantec.com

**Abstract:** The Hamlet of Aklavik is a 100 year old community within the Mackenzie Delta region of the Northwest Territories (NWT), and was originally was established as a fur trading post. The community's solid waste disposal facility has been experiencing operational problems in the recent years, related to the river floods in the spring, and drainage issues. In response to these issues, the Hamlet initiated a study to identify a new solid waste site. A scoping study identified 15 potential waste sites within 16 kilometers of the community, all situated along a seasonal access road. Waste generation for the community was estimated to be 200,000 m$^3$ for a 40 year horizon. This volume, with a 3 to 1 compaction ratio, produced a 66,000 m$^3$ volume requirement, and was rationalized to a 66,000 m$^2$ area requirement. Of the 15 original sites from the scoping study, 9 had sufficient area, which included a 50 m site buffer from the operating area to the adjacent pond areas. The estimated capital cost to develop these sites range from $11.7 million to $25.2 Million, based upon substantially long access roads, and the site development features. Based upon the planning analysis it was recommended to investigate 2 of the 9 sites further. Along with the planning analysis, a review of the existing site was completed, and it was recommended that the redevelopment of the existing site is a should be considered to obtain additional operating capacity for the facility (up to a 12 year period), improved regulatory compliance, and to achieve additional time for the implementation of a new site.

Keywords: - solid waste, planning, Northwest Territories, Aklavik

## 1 Introduction

### 1.1 Community Information

The Hamlet of Aklavik is located on the Peel Channel of the Mackenzie River Delta, 113 km south of the Arctic Coast, and 55 kilometers west of Inuvik. Its geographic coordinates are 68.219916 N latitude and -135.007788 W longitude. Aklavik has a population is 670 in 2015, and the majority of the residents are Inuvialuit and Gwich'in. The community falls into two different land claim settlement regions, the Inuvialuit Settlement Region (ISR) and the Gwich'in Settlement Region (GSR). The community has limited access, which is by air and barge only during the summer months, and by ice road from Inuvik during the winter months.

Aklavik is within the Boreal forest zone, with vegetation that includes white spruce (upper delta) or balsam, poplar and black spruce (lower delta) on high ground, and willows, alders, marshy vegetation and muskeg in the low lying areas. Aklavik is approximately seven meters above sea level, and is subject to periodic flooding. The lands which comprise the municipality fall within a Flood Risk Area under the Canada-NWT Flood Damage Reduction and Flood Risk Mapping Agreement.

Aklavik is within the delta region and is characterized by alluvial deposits of fine sand, and silt, and is situated in an area of discontinuous permafrost. The climate of Aklavik is characterized by cool summers, and long, cold winters. The daily average temperature of the warmest month in Aklavik is 13.9ºC and the coldest month is -26.3ºC. The annual precipitation as rainfall is 12.8 millimeters and 136.3 centimeters of snow.

### 1.2 Study Background

A report commissioned by the Hamlet of Aklavik in 2009 estimated that the existing solid waste site, immediately adjacent to the community, had a remaining service life of about five years. The disposal facility has also been experiencing operational problems in the recent years related to the river flooding in the spring, and runoff management issues. The 2009 report stated that "a new landfill outside of the floodplain in the area of the Richardson Mountains is the preferred option to be further evaluated by the community".

The Hamlet did not accept this recommendation, and identified their own future site within the delta; approximately 7 kilometers west of the community along the winter road to a gravel source. The identification of this alternate site prompted the Hamlet to advance a planning study to consider additional landfill sites within the delta area west of the community.

The Hamlet retained Stantec Consulting Ltd. based upon a proposal that included a scoping study that identified and presented preliminary assessments of 15 potential solid waste sites. The scope of work then advanced with a screening of these 15 waste sites, and analyses to develop a short list of sites for consideration by the community.

### 1.3 Ground Conditions

The alluvial deposits of fine sand and silt beneath Aklavik are stratified layers that extend to about 11 meters below the surface. The discontinuous permafrost has an active layer of 300 to 900 mm. Borehole information to a depth of 9.1 m below the surface identified soil stratigraphy of gravel and sand layers, layers of organics mixed with sand, over layers of silt. The gravel and sand layer is generally 600 mm thick and consists of gravel graded up to 100 mm in size and course- and fine-grained sand with some fines. The sand is brown to dark brown in color and contains small ice crystals.

The organic sand below the gravel and sand is a layer about 1.5 m thick, and contains various amounts of frozen and very moist organic materials. The moisture content near surface is about 10%, increasing with depth to about 36%. The silt below the organic sand extends to at least 9 meters, and consists of some clay with trace of fine grained sand and gravel.

## 2 Scoping Study for Initial Site Identification

### 2.1 Selection and Analysis for Scoping Study

Satellite imagery was used to identify the winter road access to the granular deposits to the west of the community. Along the winter road, a 2 kilometer corridor on each side of the road was identified as a reasonable limit to potential sites. Fifteen potential sites were identified (see Figure 1) for further analysis with each site having a 200 meter separation from the center of site to the adjacent open water. It was recognized from the Scoping Study that most of the sites were within the delta area, and potentially subject to flooding. Sites 1 and 2 are at the edge of the delta and would not have the same flood risk as the remainder of the sites.

## 2.2 Initial Analysis of Selected Sites from Scoping Study

An initial analysis of the 15 sites tabulated areas ranging from 90,000 $m^2$ to over 200,000 $m^2$. Access roads to the sites were identified off the existing winter road. The access to the sites would involve a combination of an all-weather access road along the existing winter road, and a new site access road. The site access roads varied in length from less than 100 meters to 2,000 meters. The existing winter road would require upgrading to an all-weather road to accommodate site access.

Figure 1: Fifteen sites identified as part of scoping study for new landfill in Aklavik

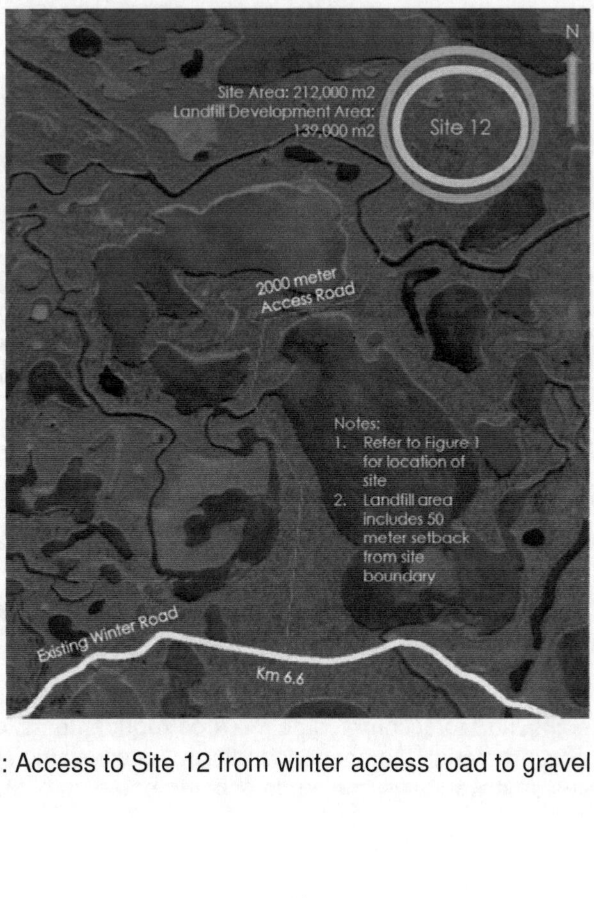


Figure 2: Access to Site 12 from winter access road to gravel sources

## 3 Waste Generation

### 3.1 Population Projection

The predicted population values were obtained from the Government of the Northwest Territories (GNWT) Bureau of Statistics. The design horizon for solid waste management in the NWT is 40 years, which recognizes the tremendous effort required in the implementation of a new landfill site. Population projections were available up to 2031, and values were extrapolated from this date estimated the population up to 2054, with a population of 710 people.

### 3.2 Capacity Requirements

To estimate the solid waste generation a standard GNWT waste generation rate was applied. The generation rate of 0.014 $m^3$ / (p*d), was applied with a one percent population growth rate and a starting population of 668. The total amount of solid waste generation over a 40 year planning horizon was estimated to be 200,000 $m^3$. A waste compaction rate of 3 to 1 was applied to the waste generated in the 40 year planning horizon to generate an anticipated waste volume of 66,000 $m^3$. This compaction ratio is appropriate to northern landfills in consideration of the frequency of compaction and the available equipment.

## 4 Site Development and Area Requirements

### 4.1 Site Development and Sustainability

The general features of the anticipated site development of a new site include areas for bulky waste, hazardous or special waste, recycled waste (waste diversion), and landfilled waste. There is no significant use of honey bags (bagged sewage), therefore a provision for a honey bag disposal area was not included. A honey bag disposal area would require special considerations because it would be a biohazardous waste. The site configuration may also include fencing, and drainage management features, for both on site drainage and off site drainage.

Although provisions for recycled waste and hazardous waste will be included in the site development, this aspect of sustainability is an ongoing challenge for small northern communities. Community capacity issues associated with operational funding, and human resources makes this activity impractical, particularly due to the transportation costs associated with waste materials either to a market for recycling or a facility for hazardous waste treatment. In most cases the recycled and hazardous waste are accumulated without any long term removal plan.

The operational site development for landfilled waste is cells that would have cross sections of 10 to 15 meters wide, and berms with a height of 1 to 1.5 meters on either side of the cell. The waste deposition in the working area of each cell would be accomplished in an active area that would be periodically consolidated and then compacted. Intermediate cover would be added to the compacted area, and compacted. Working areas and intermediate cover would provide the opportunity to isolate segments of the waste in order to reduce the opportunity for the contamination of on-site runoff, and fire protection to minimize the size of a potential landfill fire.

### 4.2 Area Requirements

The anticipated waste generation for a 40 year planning horizon is 200,000 $m^3$, and with a compaction ratio of 3 to 1, the needed capacity of the site would be 66,000 $m^3$. In applying the waste cell configuration with 66,000 $m^3$ for a 1.5 meter berm height, and developing a second phase of the landfill on top of the initial phase (3.0 meters in total height) it was determined that the landfilling area for household waste would be approximately 22,000 $m^2$.

For the purpose of site screening, and to accommodate the area requirements for bulky waste, hazardous waste, recycled waste, and cover material, the needed landfilling area was tripled for a total area requirement of 66,000 $m^2$. This area would also needs to accommodate an operator building, signage, miscellaneous storage, and internal roads on the site. The tripling of the was also based upon the general observation of the current landfill operating areas that have been developed by the Hamlet, and it also adds an additional conservative factor for the site selection process.

### 4.3 Earth Management

The soil requirements for the site will include material for an access road, material for internal roads on the site, material for berm construction, and material for intermediate cover, and ultimately final cover. Granular material is required for the site access road, and the internal roads, but granular material is not necessarily needed for berm construction and cover materials. The Hamlet has access to non-granular material from a borrow site at the east end of the community.

Berms may consist solely of non-granular material, as a less expensive alternative. A well planned earth management strategy for the new landfill will reduce the need for "higher quality" granular material, and create an opportunity to use poor quality borrow sources that are closer to the new landfill site.

### 4.4 Climate Change Considerations

Climate change should be considered during the development of the new solid waste facility. A report by the National Research Council (2011) confirms that the assumption of hydrological consistency and designing for this is no longer valid or practical. Deforestation, changes to wetlands, community growth, hydro projects, and other water diversions are a few examples of anthropogenic land cover changes that have a significant impact on the duration and intensity of floods and droughts. The changes highly impact downstream hydrology. This means that development within the Mackenzie River basin will have to be managed with a greater deal of uncertainty than in the past.

This same perspective applies to the Hamlet of Aklavik and any new landfill located within the delta area. There is a greater uncertainty in the river flooding with respect to frequency, duration, and extent of any flooding. Landfill facilities inherently apply berm structures as part of the facility development; therefore provisions for erosion protection may be a needed improvement in the construction. Since flooding is elevation dependent, building up the landfill area above the anticipated flood level may also be a mitigation opportunity.

## 5 Assessment of Solid Waste Sites

### 5.1 Assessment Criteria

The 15 sites identified in the scoping study were assessed with very limited information on terrain, geology, and the local environment to screen the sites to a short list for discussion with the community, and for future detailed evaluation based upon site reconnaissance. The evaluation criteria for the 15 sites included: available areas, geological features, topography, hydrology, and hydrogeology, setbacks, and accessibility. The available site information did not include any site specific topographic and hydrogeologic information, and therefore this information was excluded from the planning analyses.

### 5.2 Screening of Sites

Of the 15 potential sites considered for landfill development based upon the scoping study, Sites 5, 6, 11, 13, 14, and 15 were not considered further because they do not meet the active landfill area requirements of 66,000 m$^2$. This was determined by measuring the usable site area then adding a 50 m buffer to provide a considerable distance from the adjacent water bodies.

In addition, Site 15 was not considered for further analysis given its proximity to the airport of less than 3 km. A separation of less than 3 kilometers presents a bird hazard to aircraft from the potential movements of birds scavenging at the waste management area. Although the Canadian Manual of Airport Bird Hazard Control developed by Transport Canada recommends that activities such as landfills and lagoons pose a hazard to aircraft if located within 8 km, the GNWT adopted a 3 km setback requirement.

# 6 Cost Development for Solid Waste Sites

## 6.1 Basis of Opinion of Probable Cost (Capital Cost) for Landfill Development

The capital costs were developed from all the required components of an active landfill area. The components include: primary road cost (all season access road), secondary road cost (site access road from all season access road), bulky, recycled and landfilled waste area, hazardous waste storage, carcass, and burn pit cells, site operating structures, blowing debris control, on- and off-site drainage management, perimeter fencing, double swing gate, site signage, and engineering and contingency allowance (40% of capital cost).

The cost estimates were developed from an estimation of unit values for each of the elements presented above based upon the general site configuration for a 40 year development area of 66,000 $m^2$. Unit costs to apply to each of the unit values are based upon solid waste, and transportation related work from in house data base for work in the Mackenzie Delta, and the Kitikmeot region of Nunavut. The 40% Engineering and Contingency Allowance is an overall contingency allowance for construction and engineering applied to the conceptual level of costing.

## 6.2 Capital Costs and Life Cycle Cost of Solid Waste Sites (Table)

The cost estimates for the landfill sites remaining from the initial screening are summarized in Table 1. Sites 10 and 12 have the lowest capital costs by margins of $7 million because of the significantly lower cost of upgrading the winter access road to an all-weather road.

Table 1: Example table caption

| Site Number | Total Capital Cost | Annual O&M |
|---|---|---|
| 1 | $25,620,796 | $512,416 |
| 2 | $24,509,119 | $490,182 |
| 3 | $22,311,202 | $446,224 |
| 7 | $19,585,251 | $391,705 |
| 8 | $20,291,282 | $405,826 |
| 9 | $19,393,881 | $387,878 |
| 10 | $12,728,853 | $254,577 |
| 12 | $23,339,433 | $246,789 |

*Note: The O&M costs per year are estimated based on 2% of capital cost.

## 6.3 Life Cycle Cost Evaluation

A Life Cycle Cost Evaluation (40 year) was also prepared for the sites remaining from the screening process. As a sample, the life cycle operation and maintenance costs for the sites with the lowest capital (Sites 10 and 12), were the same with a value of $3.2 million.

# 7 Discussion

In the planning analysis context, a number of issues need to be considered in conjunction with the discussion, and ultimately the recommendation of potential waste management sites. For the development of the Hamlet of Aklavik solids waste facility, the following issues should be considered: access to site, proximity issues (human activities, natural features, and local receptors), site configuration, potential environmental or public health impacts, estimated capital cost to develop site, and estimate operation and maintenance costs.

Other issues including surface materials, snow accumulation, local hydrology, and vegetation should also be considered, although they are not taken into account at this stage due to the limited information available. The remaining sites from the screening process included sites 1, 2, 3, 7, 8, 9, 10 and 12. These sites provide an adequate site area, and active landfilling area.

An overriding issue associated with the development of a new solid waste for Aklavik is the site development in the delta area adjacent to the community instead of a development outside the delta area, which was recommended in a previous planning study. A risk of flooding exists for any facility within the delta area; however the risk is diminished to some degree with a location that is not immediately adjacent to the Peel Channel. As part of the subsequent phases of investigation, flood risk and flood proofing of any site must be considered in order to address anticipated regulatory concerns.

The process for the development of any landfill site, beyond the planning stage, takes a considerable amount of time and effort associated with biophysical studies, and regulatory reviews and approvals. In consideration of this time requirement, the redevelopment of the existing site is a reasonable step to obtain additional operating capacity for the facility. A concept for the redevelopment of the existing site is presented in Figure 3, which includes the provision for waste diversion and on-site drainage management. These provisions will also address existing compliance issues with the regulatory authorities. A key element in the redevelopment is the application of landfill cells instead of an open dumping area, which provides improved overall management, and may accommodate an ultimate operating height of 4 to 5 meters.

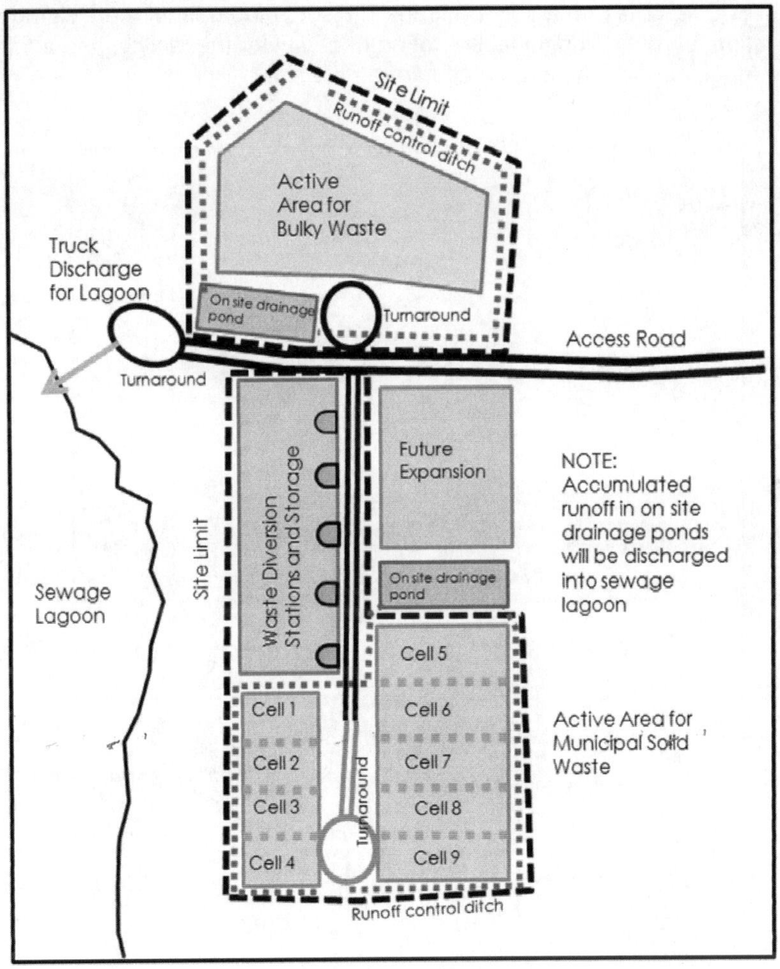


Figure 3: Redevelopment concept for existing landfill site

## 8 Conclusions

The waste generated over a 40 year planning horizon for the Hamlet of Aklavik is 200,000 $m^3$, and with a 3 to 1 compaction ratio, this results in 66,000 $m^3$ of solid waste. The area requirement for the waste generation is 22,000 $m^2$ for an ultimate site development with a 3 meter high site. In order to make sure there is adequate space waste diversion activities, site operating activities, and for cover material as part of proper waste disposal, the minimum area requirement was tripled, resulting in a minimum active landfill area of 66,000 m$^2$ for a 40 year operating horizon.

Limited information was available for a detailed consideration of the sites for topographic, hydrogeological, and biophysical characteristics as part of the planning analysis. The planning analysis was primarily based upon geographic characteristics of land areas, pond areas, and proximity to existing access.

Of the 15 original sites from the scoping study, 8 have sufficient area greater than 66,000 m$^2$, which included a 50 m site buffer from the operating area to the adjacent pond areas. These sites are 1, 2, 3, 4, 7, 8, 9, 10, and 12. The Opinion of the Probable Cost for the Capital Cost to develop these 8 sites range from $11.7 Million for Site 12, to $25.2 Million for Site 1. The operation and maintenance costs for the 8 sites range from $0.6 Million to $1.2 Million. The cost to each of the sites is the majority of the cost capital cost.

The least expensive sites to develop are Site 10 and 12 based upon Capital and O&M Costs. Development details for Site 12 is presented in Figure 4. Redevelopment of the existing site is a reasonable consideration to obtain additional operating capacity for the facility, and achieve additional time for the planning, engineering, and construction of a new site.

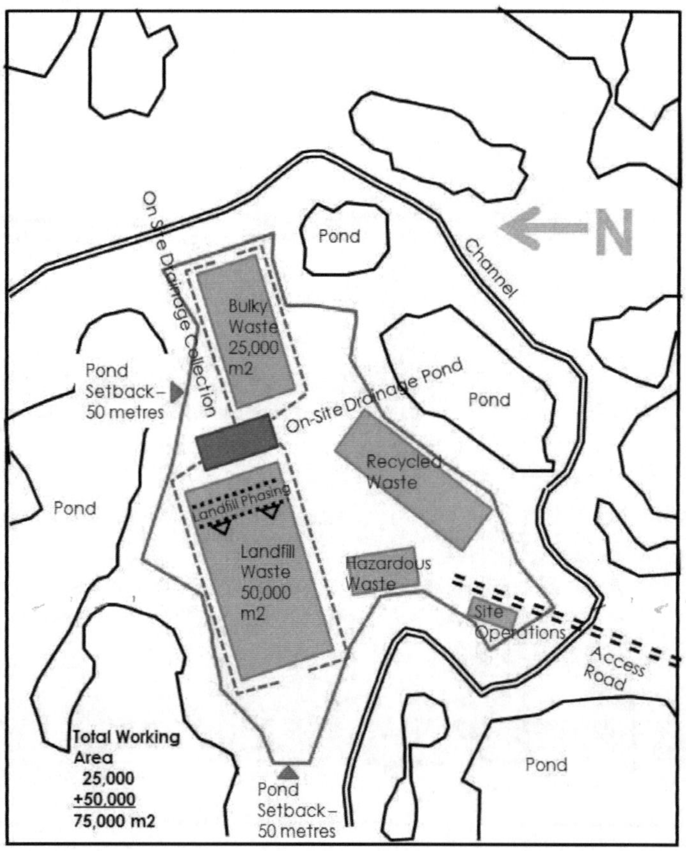


Figure 4: Site Development Concept for Site 12

## 9 Recommendations

Based upon the limited analysis, Sites 10 and 12 should be investigated further. These sites are recommended because they meet all of the selection criteria for site area, active landfilled area, and have the lowest capital and O&M costs. Further site investigations to be completed on Sites 10 and 12 should include topographic, hydrogeological, geotechnical biophysical analyses, and flood risk analyses.

The Hamlet of Aklavik should engage a discussion with the regulators based upon the planning study results in order to develop the potential scope of the detailed investigations. The Hamlet should anticipate a strong reluctance for the regulators to consider waste management sites in the delta area in consideration of the flooding issues that may develop.

Cost sharing with an all-weather access to the granular sites should be considered and advanced by the Hamlet. This would be a good opportunity to share the enormous costs associated with the development of an all-weather access road. Site 10, with a site area of 137,000 $m^2$, and an active landfill area of 79,000 $m^2$, has a total capital cost of $12.7 million, and annual operation and maintenance cost of $254,500. Site 12, with a site area of 211,600 $m^2$, and an active landfill area of 138,600 $m^2$ has a total capital cost of $12.3 million, and annual operation and maintenance cost of $246,800.

In consideration of the potential timeframe to develop a new solid site, Aklavik should advance the redevelopment of the existing solid waste site to increase the capacity and operating horizon, address regulatory compliance issues associated with drainage management, waste diversion, and operating areas.

Leadership in Sustainable Infrastructure
*Leadership en Infrastructures Durables*

Vancouver, Canada
May 31 – June 3, 2017/ *Mai 31 – Juin 3, 2017*

# AN EVALUATION MODEL FOR DEMOLITION WASTE IN THE RESIDENTIAL CONTEXT

Raghubar, Christopher[1,3], Oudjehane, Azzeddine[2]
[1] MASc Candidate (Ryerson University), Canada
[2] Associate Professor (SAIT), Canada
[3] christopher.raghubar@ryerson.ca

**Abstract:**

The waste generated during the demolition of residential homes consumes large volumes of limited space in municipal landfills. Unreserved disposal of demolition materials into landfills can be minimized through more methodical deconstruction methods. Sorted materials can then be reused or recycled whereas unsorted material can only be sent to a landfill. To create an incentive for individuals and companies to spend the additional time and money to sort waste, policies must be created by regional municipality. In Calgary for example, the incentive to approach demolition with an emphasis on future use of material is seen in the form of varying disposal rates; higher rate reserved for unsorted material and lower rates for materials that can be recycled. To understand how these different disposal rates affect the total demolition project cost for a typical Calgary home, the model and approach presented herein investigates the relationship between cost savings at disposal and labour investment during demolition. The ultimate goal of this study is to define a model to evaluate the substantial merit for tipping fees and waste management policies in terms of encouraging deconstruction in residential homes.

## 1 INTRODUCTION

To reduce the amount of waste created during the demolition process, the City of Calgary has introduced several initiatives to create incentives for individuals and companies to divert construction and demolition waste from traditional solid waste streams. The predominant method for reducing the amount of waste entering landfills is to recycle or reuse the waste generated during the demolition of buildings. In order to reuse or recycle materials, buildings need to be demolished in a fashion that facilitates source segregation of materials, allowing for future use. The method of demolition (selective vs conventional) is the predominant factor dictating how much material can be diverted from landfill. Neglecting salvage values and tipping fees, conventional demolition is typically the preferred method due to the high labour requirements inherent to selective demolition. However, the high tipping fees associated with mixed waste and the salvage value that recovered materials command have a profound effect on total demolition project costs.

### 1.1 The Conventional demolition

Conventional demolition is a preferred method for projects that have constrained timeframes, as the overall process of removing a home using primarily mechanical means takes very little time compared to deconstruction. Pun, Liu and Langston (2005) identify conventional demolition as mechanical demolition relying on equipment such as excavators and bulldozers to pull down a building. This method creates a heterogeneous stockpile of waste destined for a landfill, with very little material being reused or recycled. The reduced labour requirement greatly reduces the initial cost of demolition, but produces mixed waste that command a high tipping fee in comparison to segregated waste.

## 1.2 The Selective Demolition (Deconstruction)

Deconstruction is described as selective dismantling, or "a first in, last out process where components in a building assembly are removed in the opposite order they were installed" (Dantata, Touran, & Wang, 2004). Materials must be reused or recycled in order to remove them from the solid waste stream at the end of a buildings service life. Reused items retain some resale value but require a high degree of care when removing from the assembly. Examples of items that can be reused include fixtures (electrical or plumbing), aesthetic wood products (such as moldings or banisters), flooring and electrical equipment. The resale value of these items are highly variable as their value will be dependent on the condition and relative demand of the material. The International Panel on Climate Change (IPCC) broadly defines recycling as the recovery of materials (typically paper, plastics, glass, metals and plastics, sometimes wood and food waste) from the waste stream (IPCC, 2006). Reddy (2011) identifies the benefits of recycling as reducing the reliance on virgin materials, reducing pollution, reducing energy consumption, mitigating climate change and reducing the pressures on biodiversity. Recycling is essentially the reduction of a product into its base material which can be reprocessed into a usable item. Examples of items which can be recycled from a residential home include asphalt shingles, metal and plastic items, wood products, electrical cable and concrete. Deconstruction requires greater labour effort at the onset of the project to ensure that materials recovered retain high value, allowing for future use (Pun, Liu, & Langston, 2005). The attention to detail and the added steps associated with deconstruction contribute to a high up front cost; however, several other factors must be considered when employing selective demolition, as outline below:

- **Increased Planning** to safely and effectively deconstruct a building. The planning will be required to identify the safest way to sequence removal of building components. The planning will also be needed to identify which components will retain salvage value when reused and which items are limited to recycle or waste applications.
- **Larger Laydown** to allow for onsite sorting. For urban conditions the size of the available laydown may present a considerable challenge.
- **Greater Labour Skill** compared to conventional demolition. Labour involved in deconstruction will need to have relevant training to ensure that the goals of the project are maintained. Materials need to be disassembled in a correct order.
- **Increased Direct Supervision** to ensure that loads leaving site meet regulatory requirements for sorted material disposal rates.
- **Increased Transportation Costs** due to multiple material types transported separately or semi separately (Multiple bins on a single load).

Deconstruction can be broken up into four major stages: stripping interior components (doors, appliances and plumbing accessories), disassemble walls (remove sheathing, insulation, piping and wires), deconstruct roof and finally the removal of walls and floors (California EPA, 2001).

## 1.3 Current Municipal Policy

The City of Calgary has created specific policies to promote the more methodical and labour intensive practice of deconstruction. At municipal landfills in Calgary, there exists a tiered system for disposal rates; higher rates for mixed material destined for landfills and lower rates for sorted material that can be diverted from the solid waste stream (City of Calgary, 2016a). This system of categorized rates should make deconstruction a financially viable demolition method by leveling the difference in labour and disposal costs.

In 2015 the city of Calgary revised its 80/20 policy to the diversion of 70% of waste destined for landfill by 2025 (originally calling for 80% diversion by 2020). This reduction comes from the four sectors of waste producers: single family homes, multifamily homes, commercial, and construction/demolition (The City of Calgary, 2016b). To achieve this the City is relying primarily on the reuse/recycling of waste, rather than the reduction of waste production. Calgary, like other municipalities, experiences a large portion of waste entering the landfill system coming from construction and demolition activities. The City estimates that 20% of waste entering landfills comes from construction, demolition and renovation projects; and of this 20%, 31% is recyclable wood, 14% is drywall, 5% is asphalt shingles, 8% is other roofing materials, and 42% is other materials. (The City of Calgary, 2011).

In order to meet the waste reduction targets identified, the City of Calgary introduced a three-pronged approach consisting of economic incentives & disincentives, regulatory requirements and voluntary measures (The City of Calgary, 2011). The economic disincentive for dumping mixed loads is seen in the higher rates for designated material disposal at landfills - $170/tonne of mixed waste compared to $80/tonne of segregated construction waste and $113 for basic sanitary waste (The City of Calgary, 2016a). These materials are highly recyclable and common in conventional construction. The City of Calgary has the following materials classified as designated materials: Concrete (Whole or Crushed); Brick and Masonry Block (Whole or Crushed); Asphalt; Recyclable Wood; Drywall; Paper and Cardboard.

The purpose of this study is to quantify the relationship of total demolition project costs to gross floor area (GFA) of a residential home for both demolition methods, with the specific goal of determining if the current disposal rates stipulated by the City of Calgary promote selective demolition.

## 2 THE SCOPE

To measure the effectiveness of The City of Calgary's current waste diversion policy of tiered disposal rates, this study focuses on conventionally built single family residential buildings. Conventionally built refers to asbestos-free platform framed construction with 24" stud and joist spacing using modern building materials and codes; typical to that of homes built within the past few decades. The materials identified throughout this study were based on the materials typically used in residential home construction in the past 30 years. Multifamily residences such as town homes and condominiums were not included due to the changes in productivity and equipment requirements when demolishing attached buildings or greater than 2 story heights. Landscaping and earthworks were not considered as the change in lot size and end use will have a considerable impact on labour and disposal costs. Still, this should have minimal impact on the validity of this study as the increase in cost associated with earthworks will be similar in selective and conventional methods. Furthermore, the cost to transport the waste was neglected as each demolition method produced the same amount of waste and used the same waste operator. The selective method could result in slightly higher transport costs due to the separate loads required for each material; however, this increase in cost should be negligible compared to the total project costs. The developed functions and models can be used to determine the relationship of total project costs to GFA for buildings of different types of construction. The functions produced in this study to describe material quantities, material waste disposal fees, labour costs and total project demolition costs are applicable to conventionally built 2-story single family detached homes with GFAs between 150 $m^2$ and 250 $m^2$. Methods of demolition were limited to conventional demolition and selective demolition (deconstruction). There were no provisions made for hybrid techniques; referring to a demolition method including varying levels of conventional and selective demolition. Hybrid deconstruction can combine the benefits of the selective (reduced disposal fees) and conventional (low labour requirements) demolition in different assemblies of a home to minimize the total demolition project cost; thus optimizing benefit. Finally, labour rates were adjusted to the Calgary market; and current disposal rates stipulated by the City of Calgary and local waste handlers were used. Therefore, the results of the analysis presented in this report are valid only in the Calgary and must be adjusted to accurately reflect the current market and policies in other regions.

## 3 THE METHODOLOGY

A quantity takeoff was carried out on a complete set of construction drawings for a 186 m$^2$ Calgary home. The takeoff considered structural elements (exterior/bearing walls and floor systems); functional architectural systems (interior partitions); and building envelope materials (insulation and weather barriers). Utilities including electrical, HVAC, and plumbing were neglected due to the high metal content of these components. Due to the high salvage value as well as ease of extraction of the utility components, it was assumed that these materials would be removed for salvage prior to demolition in both selective and conventional methods. Architectural finishes such as trim and floor finishes were also neglected due to the high variability in unit weights among finishes.

The performed takeoff produced a building inventory that was used as a baseline for a typical 186 m$^2$ home. This building inventory was then extrapolated in order to describe the quantity of material in terms of GFA in homes with a GFA between 150 m$^2$ and 250 m$^2$, based on the rate that each material quantity increases with respect to GFA.

Cost functions to describe labour costs and disposal fees in selective and conventional demolition methods were then developed by applying labour rates and productivities to material quantity functions; and by applying material unit weights and disposal fees to material quantity functions. All labour rates and productivities were obtained from RS Means Building Construction Cost Data (2016), and material unit weights were obtained from various sources specified in the analysis section. Waste disposal fees were obtained from the City of Calgary (2016a) and Fish Creek Excavating (2016).

The total cost for each method was then determined by summing the labour costs and disposal fees for each method. This ultimately allowed for a comparison to be made on total demolition project costs based on GFA between 150 m$^2$ and 250 m$^2$ for selective and conventional demolition methods.

## 4 THE RESULTS

Quantity takeoff was performed on the structural, architectural and envelope components of the sample 186 m$^2$ home, neglecting all utilities as well as interior finishes. Considering the foundation, exterior and interior wall assemblies as well as flooring and roofing systems, the house yielded a total disposal weight of about 130 tonnes. The quantities taken off from the housing plans can be found in Table 1.0, and a complete breakdown of assemblies and material quantities can be found in Table 1.1.

**Table 1.0**: Takeoff Quantities by Level

| Basement | Main Floor | Second Floor |
|---|---|---|
| Found. Wall (Basement) = 40.8 m<br>Found. Wall (Garage) = 20.7 m | Ext. Wall = 40.2 m<br>Garage Ext. Wall = 20.7 m | Ext. Wall = 47 m |
| # Conc. Pads (Basement) = 3<br># Cont. Pads (Garage) = 5 | Int. Bearing Walls = 6.7 m<br>Arch. Walls = 21.5 m | Int. Plumbing Walls = 3.0 m<br>Arch. Walls = 46.2 m |
| Slab (Basement) = 87.7 m2<br>Slab (Garage) = 43.1 m2 | Floor Area = 87.7 m2 | Floor Area = 110.7 m2 |

The sample home yielded a building inventory consisting of:

- Cast-in-Place Concrete – 100.09 tonnes
    - Rebar – 4.0 tonnes
    - Concrete – 96.08 tonnes
- Dimensional Lumber – 5.33 tonnes
    - 2x4 lumber – 2.36 tonnes
    - 2x6 lumber – 2.96 tonnes
- Plastics – 0.66 tonnes
    - Vinyl Siding – 0.47 tonnes
    - Vapour Barrier – 0.18 tonnes

- Wood Products – 6.16 tonnes
- Asphalt Shingles – 1.53 tonnes
- Gypsum Board – 7.49 tonnes
- Fiberglass Insulation – 3.99 tonnes

**Table 1.1**: Breakdown of material weights by assembly

| Type of Assembly | Material | Material Class | Takeoff Quantity | Volume ($m^3$) | Weight (tonnes) |
|---|---|---|---|---|---|
| Foundation | Concrete | Concrete | 41.57 $m^3$ | 41.57 | 100.09 |
| | Rebar | Steel | *Included in Concrete | | 4.00 |
| Exterior Walls | 2x6 lumber | Wood | 992.43 m | 5.28 | 2.96 |
| | OSB Sheathing | Eng. Wood | 288.56 $m^2$ | 2.75 | 1.76 |
| | Insulation | Fiberglass | 234.49 $m^2$ | 32.75 | 3.68 |
| | Vapour Barrier | Plastic | 288.56 $m^2$ | - | 0.14 |
| | Vinyl Siding | Plastic | 288.56 $m^2$ | - | 0.47 |
| Interior Walls | 2x4 lumber | Wood | 217.76 $m^2$ | 2.42 | 1.35 |
| | Drywall | Gypsum | 567.92 $m^2$ | 7.21 | 5.55 |
| Floors | TJI Joists | Eng. Wood | 477.32 m | 1.89 | 1.21 |
| | OSB Sheathing | Eng. Wood | 116.13 $m^2$ | 1.11 | 0.71 |
| Roof | Shingles | Asphalt | 116.13 $m^2$ | - | 1.53 |
| | Trusses | Wood | 160.72 $m^2$ | 1.79 | 1.00 |
| | Insulation | Fiberglass | 28.32 $m^3$ | 35.40 | 0.31 |
| | | | | TOTAL | 129.27 |

The foundation considered a concrete density of 2400 kg/$m^3$, or 150 lb/$ft^3$ (McGraw-Hill, 2012) and a steel content (rebar) of 4% by weight. Walls considered 24" stud spacing plus 2 studs per corner with a density of 560 kg/$m^3$ for dimensional lumber (Engineers Edge, 2016). OBS sheathing quantities were based on the area of exterior wall with a density of 640 kg/$m^3$ for engineered wood (European Panel Federation, n.d.), while drywall was based on the area of both faces of interior wall with a density of 770 kg/$m^3$ (American Gypsum, 2010). Exterior walls also considered vapour barrier, building paper, and siding based on the area of exterior wall, with densities of 0.4555 kg/$m^2$ (3M, 2014), 0.183 kg/$m^2$ (University of Waterloo, 2001), and 1.95 kg/$m^2$ (NIST, 2007) respectively; as well as R20 fiberglass BATT insulation between wall studs with a density of 112 kg/$m^3$ (NIST 2007). Floors considered I-Joists spaced at 24" with a density of 580 kg/$m^3$ based on the densities of dimensional lumber and wood, as well as the area of web and flange within a 30cm deep cross section (BOCA International, 2002). OSB sheathing and drywall quantities were based on total area of floor. Roof covering material (OSB and shingles) were based on half the total floor area (considering a 2-story home) multiplied by a factor of 1.25 to consider roof slope, with a shingle density of 13.2 kg/$m^2$ (The Engineering Toolbox, n.d.). Trusses considered 27.4 m of 2x4 dimensional lumber for each 8-m span Pratt truss spaced at 60cm, and attic insulation considered 38cm of blown in fiberglass with a density of 8.81 kg/$m^3$ (NIST, 2007). A complete breakdown of the building inventory can be found in Table 1.1

## 5 THE ANALYSIS

Based on the quantities of material in each wall assembly, as well as the rate at which each assembly quantity increases with respect to floor area, functions were developed to describe the quantity of material in a home with a similar floor plan based on GFA in the domain of $\{x \in R \mid 150 \leq x \leq 250\}$. This domain was selected based on the 186 $m^2$ floor area and considering that layout and floor plan will change significantly as floor area increases or decreases. Relations of material to floor area can be found in below. Note that a relation to perimeter represents a proportionality to $\sqrt{A}$.

Basement Slab; Floors Systems; Roof System $\propto$ ***Area***

Foundation Walls; Exterior &Interior Wall Assemblies $\propto$ ***Perimeter***

## 5.1 Quantity of Material Based on Gross Floor Area

Using the quantities of materials in Table 2.1 and the relationships of material increase per assembly described above, the functions to describe the quantity of material per unit floor area in $m^2$: were developed. Total length in m, area measured in $m^2$, and volume in $m^3$. were then applied to every type of material identified in Table 2.1 and labour rates to obtain the total demolition costs of selective and conventional demolitions.

## 5.2 Material Weights Based on Gross Floor Area

The total weight of each material was found by applying material cross sections and thicknesses to quantities measured per unit area and length respectively, then applying material densities; and by applying material densities to quantities measured per unit volume. The weight of each material was then generated in the building inventory in tonnes.

## 5.3 Material Disposal Costs

The City of Calgary (2016a) and Fish Creek Excavating (2016) stipulate disposal costs for mixed and segregated waste as outlined in Table 2.0 The segregated waste produced in the selective demolition method will be disposed of at the rate of each respective material, while the waste produced in the conventional demolition method will be disposed of at the rate of $170/tonne for designated materials.

**Table 2.0** – Waste Disposal Rates

| Cost/tonne | Type of Waste |
|---|---|
| $0[1] | Hardened Concrete (concrete, rebar) |
| $80[2] | Sorted C&D Waste (Asphalt Roofing, Drywall, Wood) |
| $113[2] | Sanitary Waste (Insulation, Building Paper, Vapour Barrier, Siding) |
| $170[2] | Designated Materials |

[1]Disposal Rate of Fish Creek Excavating
[2]Disposal Rate of the City of Calgary

Note that the City of Calgary accepts concrete waste, however, it must be in small pieces free of rebar and commands a disposal rate of $0 – $5/tonne. Fish Creek was selected as the concrete waste handler as they accept reinforced or unreinforced hardened concrete free of charge.

### 5.3.1 Segregated Waste Disposal Cost

Concrete and rebar will be disposed of at the rate of $0/tonne for clean fill; asphalt roofing material, drywall, and wood will be disposed of at the rate of $80/tonne for sorted construction and demolition waste; and insulation, building paper, vapour barrier and vinyl siding will be disposed of at the rate of $113/tonne for sanitary waste.

$$\text{Disposal}_{\text{Segregated}} = (W_C + W_R)*(0) + (W_{AR} + W_{DW} + W_{DL} + W_{OSB} + W_{IJ})*(80) + (W_I + W_{EM} + W_s)*(113)$$
$$= 4.008548387A + 105.03233\sqrt{A}$$

Note that the various $W_i$ represent the material weights based on gross floor area as discussed in 5.2

### 5.3.2 Mixed Waste Disposal Cost

The mixed produced in the conventional demolition method will be disposed of at a rate of $170/tonne for designated materials. It should be noted that in the demolition sequence typically calls for the demolition of the superstructure followed by the substructure; allowing for the segregation of superstructure materials from substructure materials. For the sake of simplicity, segregation of material by demolition phase was neglected in the conventional demolition method. The waste disposal as a function of the gross floor area is shown below.

$$Disposal_{Mixed} = (W_C + W_R + W_{DW} + W_{DL} + W_{OSB} + W_{IJ} + W_{EM} + W_I + W_S)*(170)$$
$$=32.74129032A + 1113.597864\sqrt{A}$$

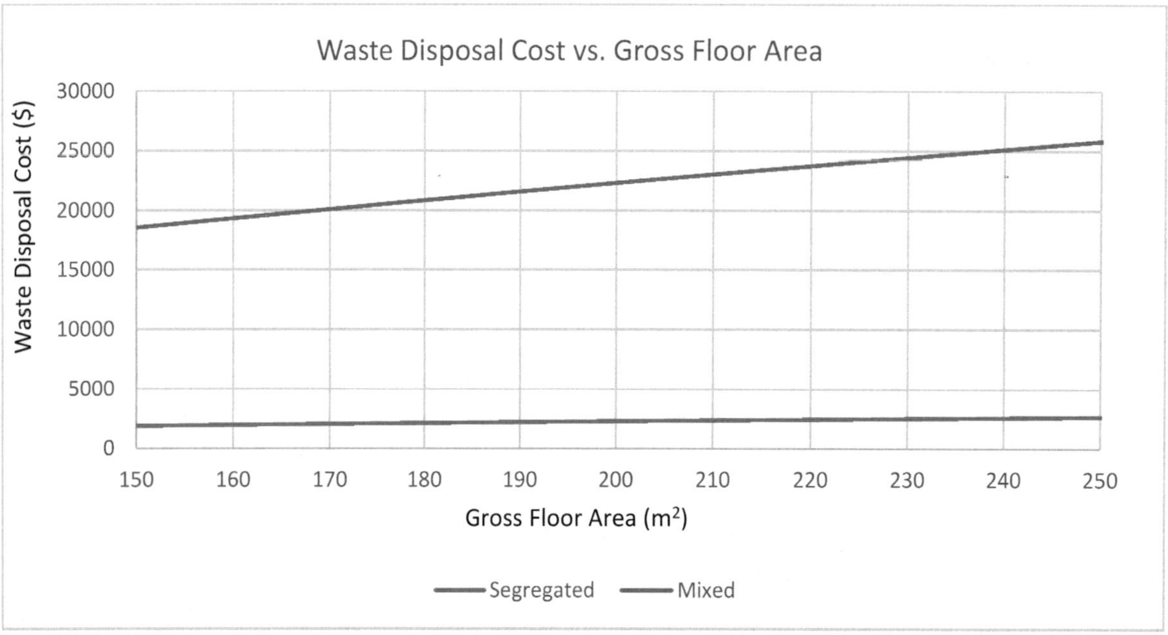


**Figure 3.0** – Gross Floor Area vs. Waste Disposal Costs

### 5.4 Labour Time based on Gross Floor Area

The total time to remove the total building stock in both selective and conventional methods was found by applying labour productivities to material quantities. The labour productivities were provided by RSMeans Building Construction Cost Data (2016); the functions relating the time required to remove each material in the selective method in days were tabulated. The analysis considered a labour productivity of 0.67 home/day for a 232 $m^2$ per home. This was normalized to a gross area-based productivity of 156 $m^2$/day. The time in days required to demolish this entire sample home was then defined as a function of the gross area.

### 5.5 Labour Costs

The segregated waste produced in the selective method used labour rates of \$303.20, \$606.40, \$1532.00, and \$1759.60 for each respective assembly, while the mixed waste produced in the conventional method used the rate of \$3345.20 for the entire building demolition. A multiplier of 1.048 was applied to labour to consider the Calgary Cost Index for demolition. RS Means (2016) stipulated crew sizes and daily costs for each assembly demolition as outlined below:

**Table 3.1**: Labour Rates by Assembly

| Cost/day | Crew | Assembly Applied to |
|---|---|---|
| $303.20 | 1 Labourer | Dimensional Lumber; Drywall; Insulation; Envelope Material; OBS-Roof; Siding |
| $606.40 | 2 Labourers | OSB-Floors/Walls; I-Joists; Trusses |
| $1532.00 | 1 Labour Foreman, 4 Labourers | Asphalt Roofing |
| $1759.60 | 1 Labourer, 1-250 cfm Air Compressor, 2-1.5" 50' Air Hoses, 2-60 lb Pavement Breakers | Concrete |
| $3345.20 | 1 Labour Foreman, 2 Laboureres, 1 Equipment Operator, 2 Truck Drivers | Whole Building |

### 5.5.1 Selective Demolition Cost (Producing Segregated Waste)

$Labour_S = [(t_{DL} + t_{DW} + t_I + t_{EM} + t_{OSB\text{-}R} + t_S)(303.20)] + [(t_{OSB} + t_{IJ} + t_{Tr})(606.40)] + [(t_R)(\$1532.00)] + [(t_C)(\$1759.60)] * 1.048$

$= 58.414472272382A + 1005.9513149606\sqrt{A}$

### 5.5.2 Conventional Demolition (Producing Mixed Waste)

$Labour_C = (t_{whole})(3345.20) * 1.048$

$=22.528794881201A$

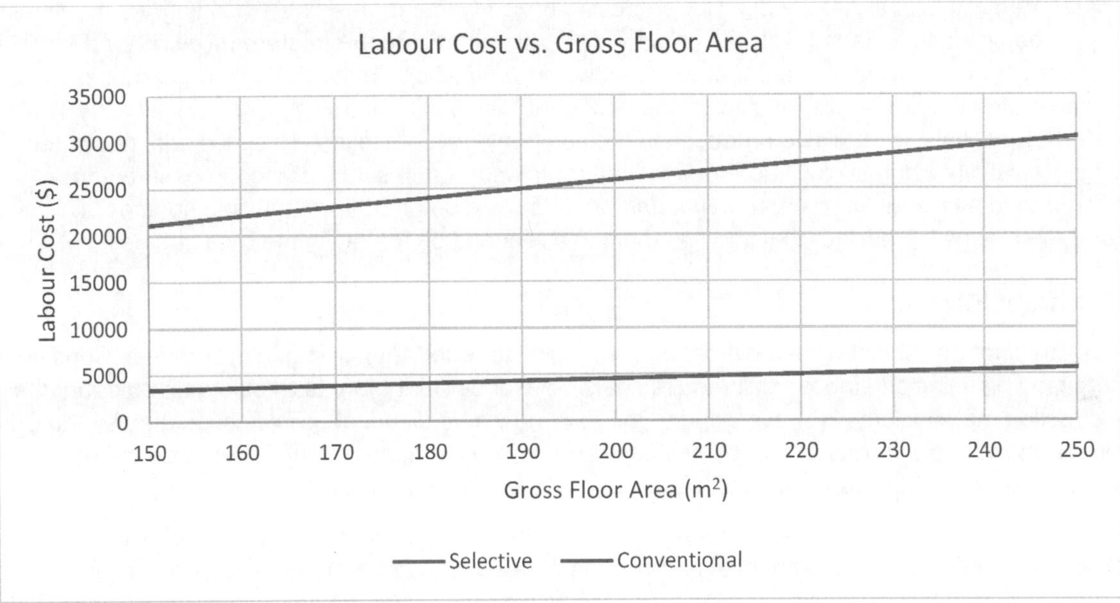


**Figure 3.1** – Labour Cost vs Gross Floor Area

## 5.6 Total Project Cost

Total project cost for selective and conventional demolitions were determined by adding the material disposal costs and labour costs for each method. Total costs are illustrated in Figure 3.2.

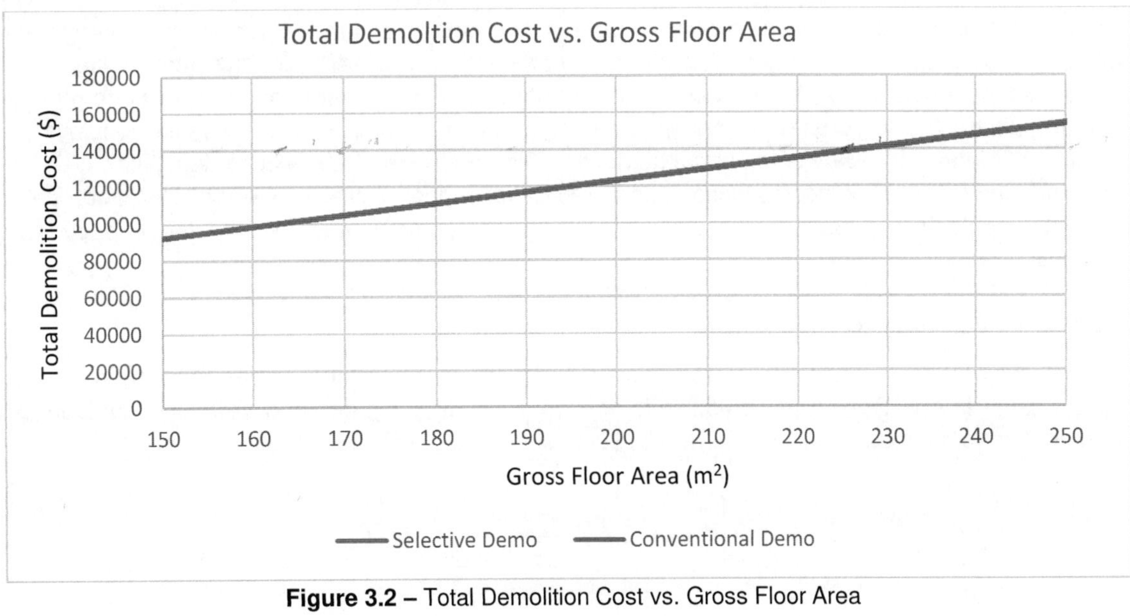


**Figure 3.2** – Total Demolition Cost vs. Gross Floor Area

It is thereby noted that with current disposal rate policies, the total demolition costs for selective and conventional methods are shockingly similar for the 186 $m^2$ sample home. The analysis revealed a cost increase of about 5% for selective over conventional, however, this cost is more than made up for through the social and environmental benefits associated with deconstruction and material reuse

## 6 LIMITATIONS OF THE STUDY

The analysis successfully identified total project costs for selective and conventional demolitions based on gross floor area in the domain of $\{x \in R \mid 150 \leq x \leq 250\}$. The selection of this domain presents its own shortcomings as floor plan and layout change significantly as area changes. The conventional demolition considered demolition of the entire building producing mixed waste for the entire building. As mentioned earlier, the demolition sequence typically calls for demolition of the superstructure followed by demolition of the substructure, allowing for fragmented mixed waste production. In the case of fragmentation, the foundation material could be disposed of at the cost of $0 at Fish Creek Excavating, significantly lower the material disposal cost of waste produced in the conventional demolition. This idea was neglected, reducing the validity of the study. Labour rates were obtained from a single source to ensure consistency, however, due to availability of data, material unit weights were obtained from multiple sources. This inherently increases the error of the study as there is a reduced uniformity in the data source.

## 7 CONCLUSION

There is no doubt that selective demolition requires a greater effort than that of conventional. Considering the increased planning and labour skill required, it is not wonder that there is a reluctance to adopt the deconstruction methodology. However, due to the existing waste disposal policies stipulated by the City of Calgary, the financial burden of adopting a selective demolition approach are countered by the immense mixed waste disposal fees associated with conventional demolition.

With confidence, it can be concluded through this study that current City of Calgary policies are promoting selective demolition within the residential sector. This conclusion can be made based on overlaying Figures 3.0 – Gross Floor Area vs. Waste Disposal Costs and 3.1 – Gross Area vs. Labour Costs. These figures demonstrate that although there is a much higher labour cost to performing selective demolition then conventional, there is a corresponding significantly higher cost to dispose of mixed conventional waste as opposed to sorted selective demolition waste when applying current City of Calgary tipping fees. Based on the analysis it was found that selective demolition costs about 5% more than conventional demolition.

This slight cost increase can be considered negligible, and is more than made up for in the environmental and social benefits stemming from deconstruction. The practice of disassembly allows for reduced demand on raw materials, increased service lives of landfills, and preservation of embodied energy. Furthermore, deconstruction creates a salvage material market for extracted materials, and employs a more people of higher skill than conventional demolition. The adoption of a selective demolition also results in reduced dust and noise production compared to conventional practices. From this, it can be seen that although selective demolition costs slightly more than conventional, the benefits produced far outweigh the costs.

## 8 ACKNOWLEDGMENTS

This project was part of a Capstone project as part of the BSc. in Construction Project Management, defended at SAIT in April 2016. We would like to acknowledge Riley Kynoch and Michael Wrightson for their contribution to the study.

# 9 REFERENCES


BOCA International Inc. (2002). *International Evaluation Report: Evaluation Report, Division 06 – Wood and Plastics, Section 06170 – Prefabricated Structural Wood.* Retrieved from http://richmond.ky.us/images/codes/files/references/Boise%20Cascade%20I-Joists.pdf

California EPA. (2001). *Deconstruction Training Manual: Waste Management Reuse and Recycling at Mather Field.* Sacramento: California Environemental Protection Agency.

Dantata, N., Touran, A., & Wang, J. (2004). *An analysis of cost and duration for deconstruction and demolition of residential buildings in Massachusetts.* Boston: Northeastern University Department of Civil and Environmental Engineering.2

Engineering Toolbox (n.d.). *Roofing Materials – Weights.* Retrieved from http://www.engineering toolbox.com/roofing-materials-weight-d_1498.html

Engineers Edge (2016). *Commercial Lumber Sizes Chart Table.* Retrieved from http://www.engineers edge.com/commercial_lumber_sizes.htm

European Panel Federation (n.d.). *Technical Information Sheet – OSB (Oriented Strand Board).* Retrieved from http://www.osb-info.org/technical.html

Fish Creek Excavating Ltd. (2016). *Concrete/Asphalt Disposal.* Retrieved from http://www.fishcreek ex.com/aggregate-recycling/concrete-asphalt-disposal

Gypsum Association (2010). *Gypsum Board Typical Mechanical and Physical Properties.* Retrieved from http://www.americangypsum.com/sites/default/files/documents/GA-235-10.pdf

IPCC. (2006). *2006 IPCC Guidelines for National Greenhouse Gas Inventories Chapter 2: Waste Generation, Composition and Management Data.* Hayama: IPCC.

McGraw-Hill (2012). *McGraw-Hill Encyclopedia of Science and Technology (20v).* New York, NY. McGraw-Hill Education

Pun, S. K., Liu, C., & Langston, C. (2005). *Case study of demolition costs of residential buildings.* Geelong, Australia: School of Architecture and Building, Deakin University.

RS Means. (2016). Building Construction Cost Data. Rockland Massachusetts, Construction Publishers & Consultants Press

Reddy, P. J. (2011). *Municipal Solid Waste Management.* Hyderabad: BS Publications.

The City of Calgary. (2011). *Construction and Demolition Waste Diversion Update.* Calgary: City of Calgary.

The City of Calgary. (2011). *Waste Diversion in the Construction and Demolition Sector in Calgary.* Calgary: City of Calgary. Retrieved from City of Calgary: http://www.calgary.ca/UEP/WRS/Pages/Commercial-Services/Construction-Demolition-Stakeholder-Engagement.aspx

The City of Calgary. (2016a). *Landfill Rates.* Retrieved from the City of Calgary: http://www.calgary.ca/UEP/WRS/Pages/Landfill-information/Landfill-Rates.aspx

The City of Calgary. (2016b). *Leading Calgary to zero waste.* Retrieved from City of Calgary: http://www.calgary.ca/UEP/WRS/Pages/About-WRS/Calgary-Waste-Goals.aspx

University of Waterloo (2001). *Wrapping it Up: Building Paper and Housewraps.* Retrieved from http://www.civil.uwaterloo.ca/beg/Downloads/Can_arch_Bldg_Paper-2001.PDF

3M (2014). *Air and Vapor Barrier 3015.* Retrieved from http://multimedia.3m.com/mws/media/849276O/3m-air-and-vapor-barrier-3015-technical-data-page.pdf

Leadership in Sustainable Infrastructure
*Leadership en Infrastructures Durables*

Vancouver, Canada
May 31 – June 3, 2017/ *Mai 31 – Juin 3, 2017*

# AN INNOVATIVE APPROACH TO TREAT AMMONIA-RICH WASTEWATER BY PARTIAL NITRIFICATION/ANAMMOX IN BIOCAST REACTOR

Saborimanesh, Nayereh[1], Castillo Arriagada, Esteban[1], Walsh, David[1], Yerushalmi, Laleh[1], Mulligan, Catherine. N[1, 2]
[1] Concordia University, Canada
[2] mulligan@civil.concordia.ca

**Abstract:** The release of untreated or improperly treated ammonium-rich wastewater to the environment has deleteriously impacted human health and aquatic life. Biologically-based technologies are extensively used to transform nitrogenous pollutants to non-polluting forms (e.g. $N_2$). These technologies have proven to be effective, but obtaining an efficient removal of nitrogenous pollutants in a one-vessel treatment system that can simultaneously provide a balance in the environmental conditions (e.g. dissolved oxygen) for the nitrifiers and anaerobic ammonia oxidizers ("Anammox"), that require aerobic and anaerobic conditions, is extremely challenging. The main objectives of this study were to investigate the potential application of a biological reactor (BioCAST) for the treatment of an ammonium-rich wastewater under the nitrite-limited feeding conditions ($NO_2$: 0 mg-N/L) through providing the favorable conditions for both nitrifying and anammox communities in a single bioreactor and to identify the anammox communities developed in the system. The bioreactor was initially seeded with a sludge containing nitrifiers and anammox bacteria and was fed with an ammonium-rich synthetic wastewater ($NH_3$-N: 350 mg/L) and operated at a hydraulic retention time of 2 days at 35°C. The results showed that under favorable conditions, nearly 80% and 98% of total nitrogen (TN) and $NH_3$-N were removed, respectively. Sequencing analyses revealed that *Candidatus brocadia* genus was the most abundant in the bioreactor. High ammonium removal and existence of anammox bacteria imply that the suitable conditions for the activity of nitrifying and anammox communities were provided in this system.

## 1 INTRODUCTION

The "Anammox" process is the oxidation of ammonium under anoxic conditions by ammonia oxidizers, at which nitrite ($NO_2$) is used as the electron acceptor and hydroxylamine ($NH_2OH$) and hydrazine ($N_2H_4$) and $N_2$ are produced as the intermediates and final products, respectively (Kuenen 2008). This process in combination with nitrification has been successfully used to treat ammonium-rich wastewater with a low carbon to nitrogen ratio (Dijkmall and Strous 2002; Duan et al. 2012; Mulder et al. 1995) since the autotrophic anammox bacteria do not need organic carbon sources (Li et al. 2008; van-Dongen et al. 2001). However, providing an environmental balance for the aerobic nitrifying and anoxic ammonia oxidizers in a single vessel system has proven to be extremely challenging.

Studies showed the formation of microbial granules in the anammox reactors (Lu et al. 2012; Tang et al. 2011). Microbial granules are formed when suspended biomass aggregates as a result of physicochemical and microbiological interactions. Those studies have showed that the anammox bacteria form granules (composed of the granule, subunit, microbial cell cluster and single cell) and excrete extracellular polymeric substrates (EPSs) (Lu et al. 2012; Tang et al. 2011) composed of mainly polysaccharides and proteins (Liu et al. 2009). Studies on the anammox granule characteristics such as size, settling property, density, morphology, color, and specific anammox activity revealed that the properties of anammox granules are influenced by several factors, including reactor configuration, hydrodynamic shear force, and nitrogen loading rate (Liu et al. 2009; Zhang et al. 2015). Furthermore, it was proposed that the granule characteristics can be used as an indicator of the anammox reactor operation (Lu et al. 2012, 2013; Zhang et al. 2015).

The BioCAST technology is an airlift biological wastewater treatment system, proven to effectively provide multi-environment conditions for the treatment of wastewater containing organic materials, nitrogen, and phosphorous (Alimahmoodi et al. 2012; Yerushalmi et al., 2013). In this study, the multi-environment BioCAST bioreactor was modified by placing geotextile material in the microaerophilic zone as the medium for biofilm growth and attachment to enhance the retention of microorganisms. The nonwoven geotextile material was selected because it has been shown effective in biofilm formation in a previous study (Alimahmoodi et al. 2012). The main objective of the present study was to investigate the feasibility of this technology for the removal of ammonium from a synthetic wastewater with partial nitrification/anammox processes, by providing environmental conditions (e.g. dissolved oxygen and temperature) that are suitable for both the nitrifiers in the aerobic zone (which produce the nitrite required for the anammox microorganisms) and for the anammox bacteria in the anoxic zone (which can convert the remaining ammonia and the produced nitrite to nitrogen gas). Moreover, the granule microbial and physicochemical characteristics and the presence of anammox bacteria developed in this system were monitored to better overcome operating problems and optimize the BioCAST technology. This technology is promising due to providing stable and reliable operation conditions for a long period (more than one year), a homogeneous distribution of ammonium and biomass in the reactor, and successful and relatively rapid anammox start-up.

## 2 MATERIALS AND METHODS

### 2.1 BioCAST Reactor Design

The BioCAST reactor (Figure 1) is an airlift treatment system comprised of three main compartments/zones, namely aerobic, microaerophilic, and anoxic for biological treatment, and a clarification zone for solid-liquid separation. The aerobic zone is connected to the microaerophilic zone through eight adjustable opening structures (e.g. 1/4 - 1 inch) to the interconnected microaerophilic and anoxic zones. A cylindrical microbial support made of stainless steel and covered with a non-woven geotextile (estimated porosity of 0.88 (Rowe, 2001), density of 950 kg/m$^3$, specific surface area of 1500 m$^2$/m$^3$, Texel Inc., Quebec (Alimahmoodi et al. 2012)) was placed in the aerobic zone and strips of geotextile were placed in the microaerophilic zone as support materials for the bacterial attachment.

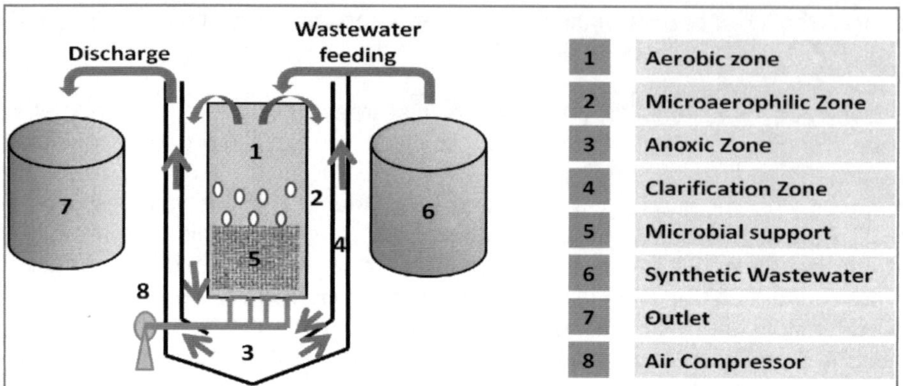


| Zone | Height (cm) | Diameter (cm) | Volume (L) |
|---|---|---|---|
| Aerobic | - | 16.7 | 17 |
| Microaerophilic | 91 | 35.7 | 61 |
| Anoxic | 100 | - | 22 |
| Clarification | - | 49.5 | 65 |

Figure 1: Schematic diagram of the BioCAST reactor

## 2.2 BioCAST Start-Up and Operation

The system start-up was carried out by using a sludge seed containing nitrifying and anammox bacteria obtained from Hampton Roads Sanitation District (USA). The bioreactor was operated continuously for over 500 days and it was fed with a synthetic wastewater prepared with the addition of $(NH_3)_2CO_3$ (different concentrations), and trace elements to a feeding tank (100 L) containing tap water, following the Van de Graaf et al. (1996) and Strous et al. (1997) procedures with modifications. The process was started with an $NH_4$-N concentration of 10 mg/L which was kept constant for a few days until ammonium removal efficiency reached above 80%. Subsequently, the concentration of $NH_4$-N was increased stepwise to 250 mg/L at a hydraulic retention time (HRT) of 4 days. When the nitrogen removal efficiency reached above 80% the HRT was reduced to 2 days. At this stage, the $NH_4$-N concentration increased stepwise to 400 ± 25 mg/L and operation was continued at this HRT using a peristaltic pump to maintain a flow rate of 2.1 L/h for over a year. Furthermore, the aeration was provided through air diffusers placed at the bottom of the aerobic zone and air at 13 kPa entered the reactor to supply oxygen for the aerobic microorganisms (final concentration of <2 mg/L) and to support the circulation of liquid between the aerobic, microaerophilic and anoxic zones. The microbial granules/sludge was wasted from the bottom of the anoxic zone under gravity flow using clear vinyl tubing. The temperature was maintained at 35°C using a 1000 W titanium heating system with a digital controller (JBJ TRUE TEMP) and pH was monitored by a portable pH meter (OAKTON PD-650). The bioreactor was wrapped with a black plastic sheet/aluminum foil to prevent the bacteria from exposure to light. The operating period and conditions of the BioCAST technology are shown in Figure 2.

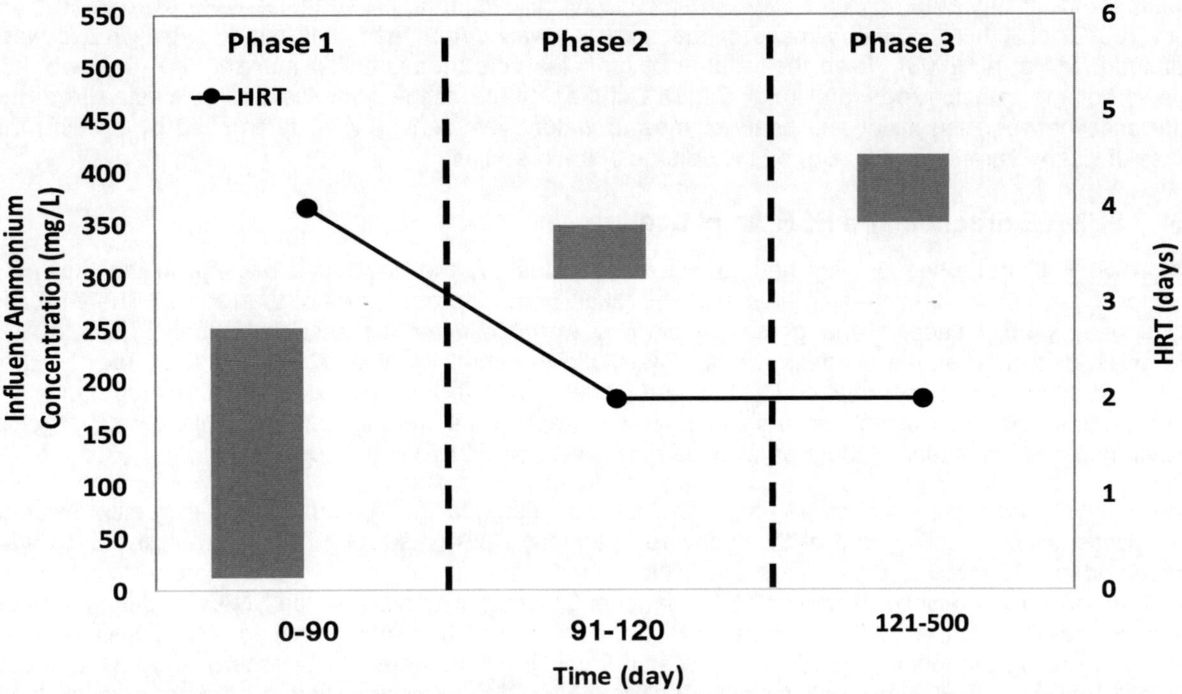


Figure 2: Influent ammonium concentration and bioreactor hydraulic retention time (HRT; right y-axis) from days 0 to 500. The operating period is divided into three phases defined by the influent ammonium concentration and HRT

## 2.3 Sample Collection and Analysis

Three sets of samples including liquid, granules, and biomass were collected for chemical, microbial, and molecular characterization and all measurements and tests were carried out at room temperature. The liquid sampling was conducted at least three times a week and samples were taken from the influent as well as the aerobic, microaerophilic, and anoxic zones to determine the total alkalinity, $NH_4$-N, $NO_2$-N, and $NO_3$-N concentrations (colorimetric methods TNT 870, 832, 840, 836; HACH Canada).

The granule sampling was conducted biweekly or monthly to determine granule morphological and settleability properties. Settling tests were performed by transferring one liter of granules from the anoxic zone of the bioreactor to an Imhoff cone filled up to the 1-L mark and the granule settling was recorded (as mL/L) every 5 min for 30 min. Total suspended solids (TSS) were determined by filtration (WhatmanTM, No.1, 42.5 mm) of 1 ml of wet granules from the anoxic zone using a vacuum filtration. Then the filter was placed in an Al dish and dried at 105°C (Blue M oven, USA) for 30 minutes and cooled in a desiccator and the filter was weighed (Denver Instruments Balance, 0.1 mg). The TSS was calculated from the difference in weight of empty filter and dried filter at 105°C (mg/L of the sample). Then the filter was placed at 550°C (Blue M oven, USA) for 20 minutes to remove the organic matter in the suspended solids and was weighed again. The VSS was calculated from the difference in weight between the dried (VSS) and ignited (550°C) matter (mg/L of the sample) (Standard Methods 2540 D/E, APHA, 2005).

Granule size distribution and morphological properties were determined following Lu et al. (2013) with modifications. Known amounts of granules were placed in a series of clean Petri dishes (e.g. 1 mL, n = 3) and on clean glass microscope slides (e.g. 0.1 mL, n = 3) using a disposable pipet. The Petri dishes and slides were air-dried for 10 min and an optical microscope (OMAX), a USB digital camera and ScopeImage 9.0 professional imaging software were used to capture and analyze the images (Lu et al. 2013).

Volume and density of the granules were measured as follows: first, the granules were dried in an oven with 103°C for 1 h and the dry mass of the granules was determined until nearly 2 g of mass were obtained (Mu et al. 2008). Then the volume of granules was measured by merging the granules in a known volume of deionized water (e.g. 5 mL) and the volume of the granules was calculated from the difference between the initial and final volumes of water. The density was determined by dividing the mass of washed granules (e.g. 2 g) by the volume of the granules.

## 2.4 DNA Extraction and PCR Amplification

The 16S rRNA gene sequencing method was used to identify the developed bacteria in the BioCAST reactor. First, the liquid containing biomass was taken from different zones of the reactor. The biomass from the microbial support and geotextile carriers were collected by scratching the surface of the geotextile biofilm. Then the biomass was transferred to the sterile centrifuge tubes (2 mL, autoclaved at 121°C, 20 min) and was centrifuged (12000 rpm, Fisher Scientific™ High-Speed Mini Centrifuge) until a 1 g of biomass was recovered and the biomass was kept in the freezer until the DNA extraction step (Environmental Engineering laboratory, Concordia University, BCEE).

The PowerLyzed Power Soil DNA Isolation kit was used for the genomic DNA extraction without modifications. The total volume of the recovered genomic DNA was around 100 µL and the DNA was preserved in a 10 mM Tris solution. The possible contaminants in the genomic DNA were removed using an OneStep™ PCR Inhibitor Removal Kit (Cedarlane, Canada). The purity of the DNA was determined by running the DNA on a 0.8% agarose gel. The extracted DNA was quantified and qualified (ratio (260/280)) using the Qubit R ds DNA BR assay (Qubit@ 2 Fluorometer, Invitrogen) and Infinite® 200 PRO NanoQuant (TECAN), respectively. The genomic DNA was amplified according to a protocol developed in our laboratory (Department of Biology, Concordia University, Canada). Two anammox-specific primer combinations including 519F/805R and 533/805R (Integrated DNA Technologies, Inc.) from published data (Sonthiphand and Neufeld 2013) and one universal primer set were used in this study. The specifications of the primers used in this study are summarized in Table 1. Nuclease-free water (Ambion) and the swamp water DNA were used as the negative and positive controls, respectively. The master mix per each reaction (total volume of 25 µl) contained a 10X buffer (2.5 µl), dNTP (0.5 µl), forward primers (1.25 µl), reverse primer (1.25 µl), taq polymerase (0.5 µl), genomic DNA template (either 0.5 or 1 µl, dilutions of $10^0$-$10^{-3}$), and nuclease-free water (18 or 18.5 µL). The amplification was started with one cycle of a hot start at 94ºC for 3 min and followed the conditions in Table 1 and terminated by one cycle at 72ºC for 10 min. The PCR products were run on an agarose gel (0.8%, Agarose A (BIOTECH), run time of 100 min, 60 V) and the gel was visualized by a UV (UltraCam digital imaging) to check the quality of PCR products. The genomic DNA was submitted to the MR DNA molecular research laboratory (Shallowater, TX, USA) for sequencing.

Table 1: Primers specifications and PCR conditions used in this study

| Primers | Sequences 5'-3' | PCR conditions | | | | | | |
|---|---|---|---|---|---|---|---|---|
| | | Denaturation | | Annealing | | Extension | | |
| | | Temp (ºC) | Time (sec) | Temp (ºC) | Time (sec) | Temp (ºC) | Time (sec) | Max cycle # |
| Universal bacteria | | | | | | | | |
| 27F | AGAGTTTGATCMTGGCTCAG | 94 | 45 | 55 | 30 | 72 | 90 | 30 |
| 1492R | TACGGYTACCTTGTTACGACTT | | | | | | | |

| | | | | | | | | |
|---|---|---|---|---|---|---|---|---|
| Anammox bacteria | | | | | | | | |
| 513F | CAGCMGCCGCGGTAA | 94 | 45 | 50 | 30 | 72 | 90 | 30 |
| Bakt_805R | GACTACHVGGGTATCTAATCC | | | | | | | |
| 533F | TGCCAGCAGCCGCGGTAA | | | | | | | |
| Bakt 805R | GACTACHVGGGTATCTAATCC | | | | | | | |

## 3 RESULTS AND DISCUSSION

### 3.1 BioCAST Performance

The overall performance of the BioCAST technology at HRT of 2 days and at 35°C is summarized in Table 2. Measurements of $NH_4^+$, $NO_2^-$ and $NO_3^-$ in the influent and different zones of the bioreactor showed that a maximum removal of 98% and 80% for ammonium and total nitrogen (Figure 3 A) were achieved in this system under the selected operating conditions (Figures 3 B and C). The 16S rRNA sequence data indicated the existence of Planctomycetes phyla in this system. High ammonium removal and the dominancy of *Candidatus brocadia* species (Figure 3 D) in the system imply the success of the bioreactor in providing favorable environmental conditions for the effective coexistence of nitrifying and anammox bacteria in this system.

Table 2: Ammonium removal from synthetic wastewater with BioCAST technology in Phase-3

| Parameter | Value |
|---|---|
| Influent $NH_4^+$ concentration (mg $NH_4^+$-N/L) | 350.71 ± 6.61 |
| Effluent $NH_4^+$ concentration (mg $NH_4^+$-N/L) | 8.21 ± 7.9 |
| Influent $NO_2^-$ concentration (mg $NO_2^-$-N/L) | 0.0 |
| Ammonium removal (%) | 97.8 ± 2.29 |
| Total nitrogen removal (%) | 80.21 ± 4.23 |

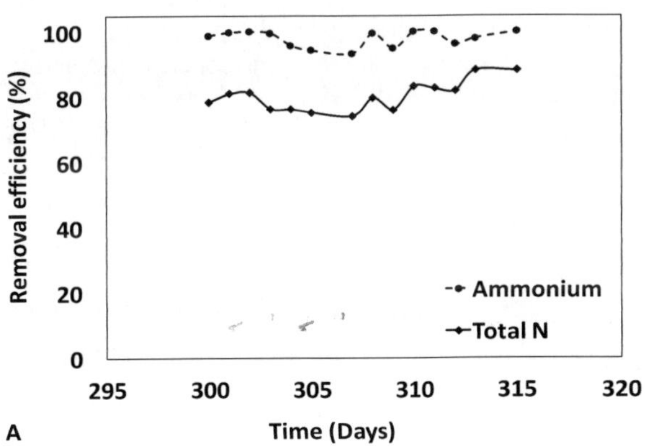


A

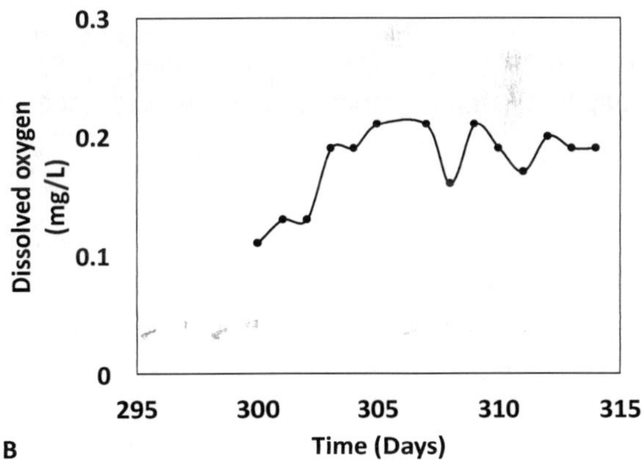


B

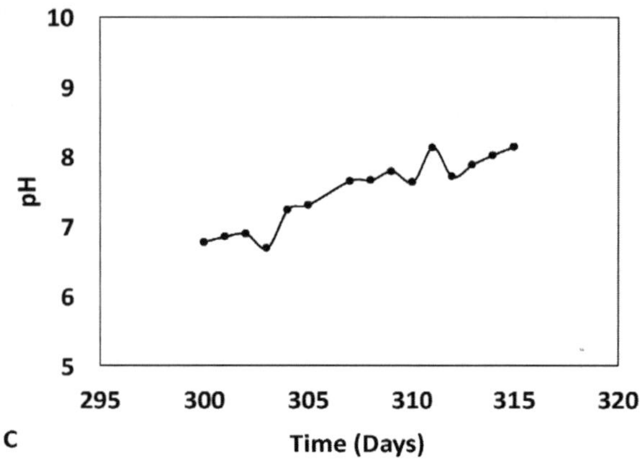


C

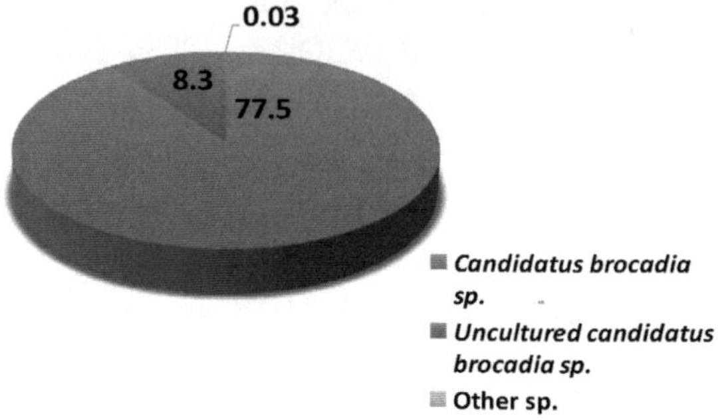


D

Figure 3: Performance of the BioCAST reactor in Phase-3 ($NH_4$-N: 350.7 ± 6.6 mg/L, $NO_2$: 0 mg/L, HRT: 2 days, temperature 35ºC): (A) $NH_4$-N and TN removal efficiency; (B) level of dissolved oxygen (DO) in the anoxic zone; (C) variations of pH in the bioreactor; (D) percentage of anammox species developed in the BioCAST as determined by 16S rRNA gene sequencing

## 3.2 Biofilm Development

Microbial biofilm development can be used as an indicator of overall biological reactor performance. Microbial biofilms can be comprised of a single heterotrophic and/or autotrophic species or mixed-species and developed through several steps including cell-surface and cell-cell interactions, and mature biofilm development as an adaptive strategy in response to changes in environmental conditions such as the availability of nutrients to microbial species, temperature, osmolarity, pH, iron, and oxygen (O`Toole et al. 2000).

Figures 4 and 5 show the gradual development of biofilm on the microbial support in the aerobic zone and the geotextile carriers in the microaerophilic zone following the start-up of the BioCAST system. The biofilm development suggests the positive effect of the microbial support and geotextile structure on the attachment and growth of the biomass and the appropriate environmental conditions including dissolved oxygen, pH, temperature, and nutrients for their growth (Figure 4).

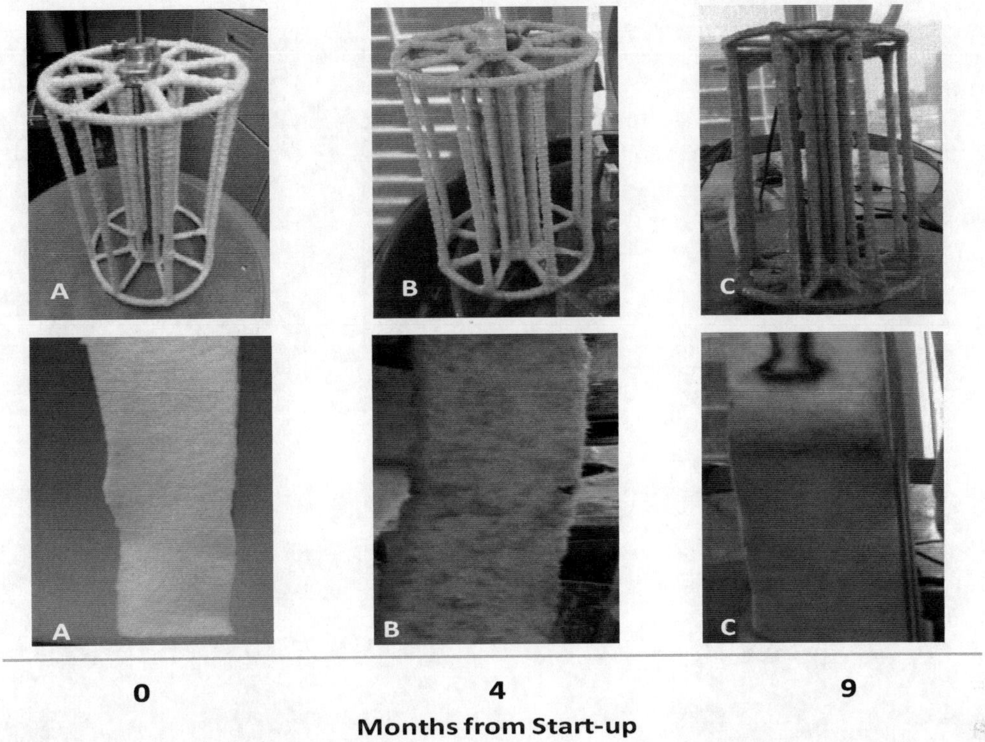


Figure 4: Growth of microbial communities on the microbial support installed in aerobic zone (top) and geotextile strips in the microaerophilic zone (bottom); (A) day 0, (B) day 120, (C) day 270

### 3.3 Granule Formation and Morphology

In the present study, various tests were performed to determine the physicochemical characteristics of anammox granules developed in the bioreactor. The granules were monitored because it can show the overall performance of bioreactors, the sludge-liquid separation, and consequently the effluent quality (Pereboom 1997). The microbial granule properties usually depend on the granule size and concentration (Liu et al. 2009).

The VSS measurements (Figure 5 A) showed that under stable operating conditions, when no excess sludge was withdrawn, the cell concentration increased from an initial 487 mg/L to over 1500 mg/L (e.g. 780 mg/L at day-30 after the start-up period) and the anammox granules exhibited an excellent settling capacity, so that following a 5-min settling time, the granules were settled and a clear supernatant was observed.

Visual observation (Figure 5 B) and light microscopic (Figure 5 C and D) of the granules confirmed the presence of single and clusters of brown and red granules with irregular shapes and sizes. Analysis showed that the single granules had an average surface area of 26.5 $\mu m^2$ (n = 24) with a range of 3.82 - 156.5 $\mu m^2$ and the clusters of granules had an average surface area of 2366.2 $\mu m^2$ (n =3). The granule density was 24.6 ± 4.3 $kg/m^3$ (n = 6).

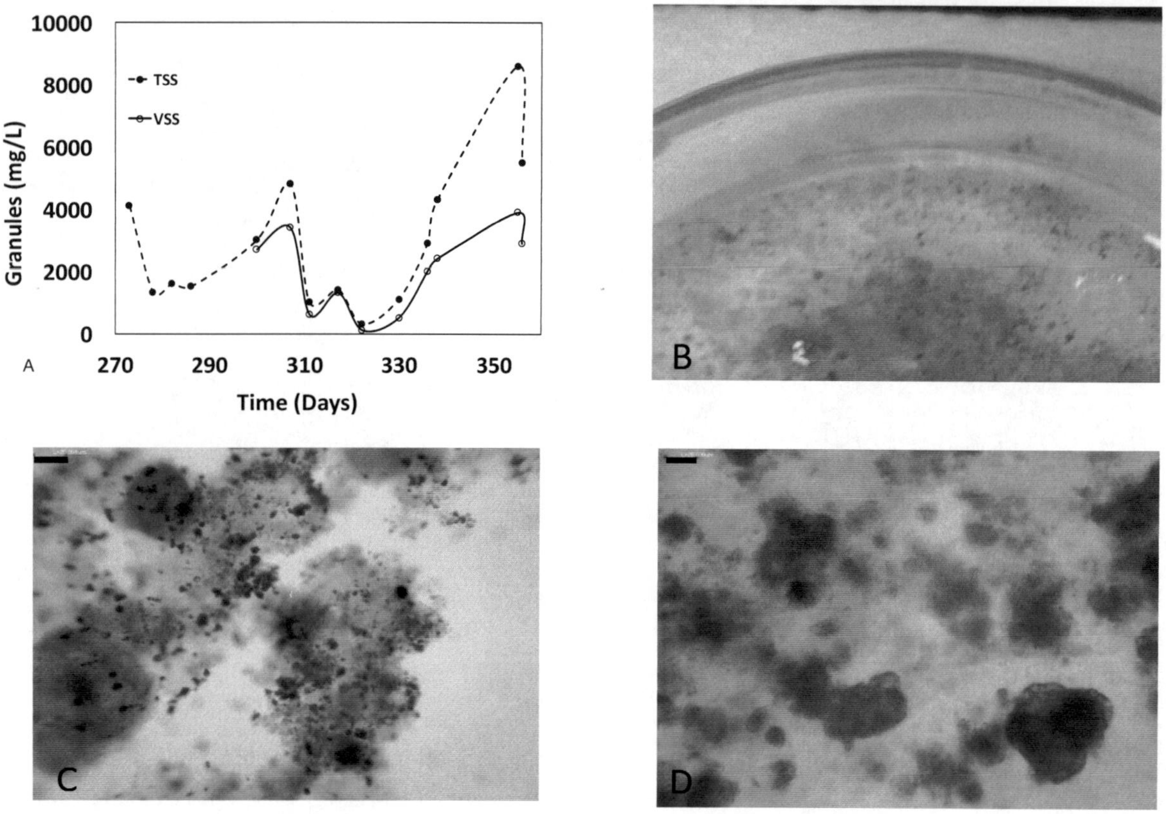


Figure 5: (A) TSS and VSS in the bioreactor (B) photographs of the granular sludge (C and D) light microscopic of granular sludge (magnification :4X and 40X, respectively, scale bar: 20 μm)

## 4 CONCLUSIONS

The potential application of a biological reactor (BioCAST) for the treatment of ammonium-rich synthetic wastewater by nitrification and anammox processes was investigated. The removal of ammonium coupled with the development of anammox granular sludge with high settleability and gradual growth of distinctly red biomass on microbial support covered with the geotextile in the aerobic zone and geotextile carriers in the microaerophilic zone highlighted the following points. First, the appropriate environmental conditions for partial nitrification and anammox processes were effectively established in this bioreactor. Second, the geotextile facilitated rapid growth of biofilm due to the creation of favorable conditions in the interior spaces of the geotextile for anaerobic bacteria and anoxic ammonia oxidizers and aerobic condition in the exterior part of the biofilm for aerobic nitrifying bacteria. In the next steps of this study, the environmental conditions that trigger the formation of anammox biofilm and distribution and activity of anammox and other coexisting bacteria in the biofilm and suspended medium to better understand the overall BioCAST performance and to further optimize this technology will be investigated.

**Acknowledgements**

This work was supported by the Natural Sciences and Engineering Research Council (NSERC), BIOCAST and Concordia University.

**References**

Alimahmoodi, M. Yerushalmi, L. and Mulligan, C. N. 2012. Development of biofilm on geotextile in a new multi-zone wastewater treatment system for simultaneous removal of COD, nitrogen and phosphorus. *Bioresource Technology*, **107**, 78–86.

Dijkmall, H. and Strous, M. 2002. Process for the Treatment of Waste Water Containing Ammonia, United States Patent Patent N0.: US 6,485,646 B1.

Duan, X. Zhou, J. Qiao, S. Yin, X. Tian, T. and Xu, F. 2012. Start-up of the anammox process from the conventional activated sludge in a hybrid bioreactor. *Journal of Environmental Sciences (China)*, , **24**(6), 1083–1090.

Kuenen, J. G. 2008. Anammox bacteria: from discovery to application. *Nature Reviews Microbiology*, **6**(4), 320–326.

Li, A. Sun, G. and Xu, M. 2008. Recent Patents on Anammox Process. *Recent Patents on Engineering*, **2**(3), 189–194.

Liu, X. W. Sheng, G. P. and Yu, H. Q. 2009. Physicochemical characteristics of microbial granules. *Biotechnology Advances*, **27**(6), 1061–1070.

Lu, H.-F. Ji, Q.-X. Ding, S. and Zheng, P. 2013. The morphological and settling properties of ANAMMOX granular sludge in high-rate reactors. *Bioresource Technology*, **143**, 592–597.

Lu, H.-F. Zheng, P. Ji, Q.-X. Zhang, H.-T. Ji, J.-Y. Wang, L. Ding, S. Chen, T. Zhang, J. Tang, C. and Chen, J.-W. 2012. The structure, density and settlability of anammox granular sludge in high-rate reactors. *Bioresource Technology*, **123**, 312–317.

Mu, Y. Ren, T. T. and Yu, H. Q. 2008. Drag coefficient of porous and permeable microbial granules. *Environmental Science and Technology*, **42**(5), 1718–1723.

Mulder, A. van de Graaf, A. A. Robertson, L. A. and Kuenen, J. G. 1995. Anaerobic ammonium oxidation discovered in a denitrifying fluidized bed reactor. *FEMS Microbiology Ecology*, **16**(3), 177–183.

O`Toole, G. Kaplan, H. B. and Kolter, R. 2000. Biofilm Formation As Microbial Development. *Annu. Rev. Microbiol*, **54**, 49-79.

Pereboom, J. H. F. 1997. Strength characterisation of microbial granulesJ.H.F. Pereboom. *Water Science and Technology*, **36**(6–7), 141-148.

Sonthiphand, P. and Neufeld, J. D. 2013. Evaluating Primers for Profiling Anaerobic Ammonia Oxidizing Bacteria within Freshwater Environments. *PLOS ONE*, **8**(3).

Tang, C. J. Zheng, P. Wang, C. H. Mahmood, Q. Zhang, J. Q. Chen, X. G. Zhang, L. and Chen, J. W. 2011. Performance of high-loaded ANAMMOX UASB reactors containing granular sludge. *Water Research*, **45**(1), 135-144.

van Dongen, U. Jetten, M. S. M. and van Loosdrecht, M. C. M. 2001. The SHARON-Anammox process for treatment of ammonium rich wastewater. *Water Science & Technology*, **44**(1), 153-160.

Van de Graaf, A.A. Mulder, A. de Bruijn, P. Jetten, M.S. Robertson, L.A. and Kuenen, J.G., 1995. Anaerobic oxidation of ammonium is a biologically mediated process. *Applied and Environmental Microbiology*, **61**(4), pp.1246-1251.

Yerushalmi, L. Alimahmoodi, M. Afroze, N. Godbout, S. and Mulligan, C. N. 2013. Removal of carbon, nitrogen and phosphorus from the separated liquid phase of hog manure by the multi-zone BioCAST technology, *Journal of Hazardous Material,* **254-255,** 364-371.

Zhang, Y. Zhang, X. and Zhang, W. 2015. Research advances in Anammox granular sludge. *AASRI International Conference on Industrial Electronics and Applications*. Atlantis Press, 460-463.

Leadership in Sustainable Infrastructure
*Leadership en Infrastructures Durables*

**Case Study**

Vancouver, Canada
May 31 – June 3, 2017/ *Mai 31 – Juin 3, 2017*

# ANAEROBIC SULFATE REDUCTION AND METAL PRECIPITATION IN UPFLOW HYBRID REACTOR: COMPARISON OF SINGLE AND TWO STAGE PROCESSES

Shahsavari, Shahrokh[1], Seth, Rajesh[1,2] and Biswas, Nihar[1]
[1] University of Windsor, Ontario, Canada
[2] rseth@uwindsor.ca

## 1 Introduction

Biological anaerobic sulfate reduction to sulfide by sulfate reducing bacteria (SRB) is an effective and economically attractive option in removing heavy metals from acid mine drainage (AMD), an acidic and sulfate-rich waste stream produced by the mining industry. The metal removal process can be accomplished either in a single-stage reactor in which the biological sulfate reduction to sulfide and metal precipitation as metal sulfide occur simultaneously, or in a two-stage reactor where it happens sequentially. The single stage process may be expected to be more cost-effective and simpler to operate. However, studies conducted with suspended growth cultures by this group and others suggest that above a certain level of metal loading, the process of sulfate reduction and the corresponding metal precipitation is adversely affected (Shahsavari et al., 2015, Kieu et al., 2011 & Utgikar et al., 2002). Therefore, to make the single stage process more efficient and viable, strategies would need to be developed and tested so that the accumulated precipitates could not affect the process of sulfate reduction by SRB. In this study, a modified up-flow anaerobic hybrid reactor (UAHR) was studied as a single stage process and the results were then compared to those of a two stage process to examine the effectiveness of the UAHR configuration.

## 2 Materials and Methods

Two identical UAHRs (configuration shown in Figure 1) were inoculated with anenriched culture of SRB developed from a mixture of anaerobic sludges from industrial (ethanol) and municipal wastewater treatment plants in Chatham, Ontario. In the start-up phase, the reactors were operated semi-continuously on a synthetic wastewater (Sodium Sulfate ($Na_2SO_4$): 4500 mg/L, Dextrose ($C_6H_{12}O_6$):2850 mg/L, Ammonium chloride ($NH_4Cl$): 1000 mg/l, Potassium phosphate monobasic ($KH_2PO_4$): 100 mg/L and Potassium phosphate dibasic ($K_2HPO_4$): 400 mg/L) at an hydraulic retention time (HRT) of 40 days. After establishing a steady state condition, one UAHR was used to represent the single-stage process for which the synthetic wastewater feed was modified to include metal (copper). The other UAHR represented the first stage of the two-stage process and the feed. Batch precipitation experiments were conducted with a synthetic copper solution with the effluent from this reactor to represent the second-stage of the two-stage process. The performance of the two processes was compared over different hydraulic, organic, and sulfate loading rates by varying the HRT between 40 and 2.5 days at a temperature of 33±3°C. The parameters such as pH, Oxidation Reduction Potential (ORP), total organic carbon (TOC), sulfate and sulfide concentration were measured according to standard methods (APHA, 2005). The volatile fatty acids (VFA) and alkalinity were determined following the methods suggested by

Dilallo et al.(1961). The solids inventory in both reactors were also carried out at the end of the reactor operation.

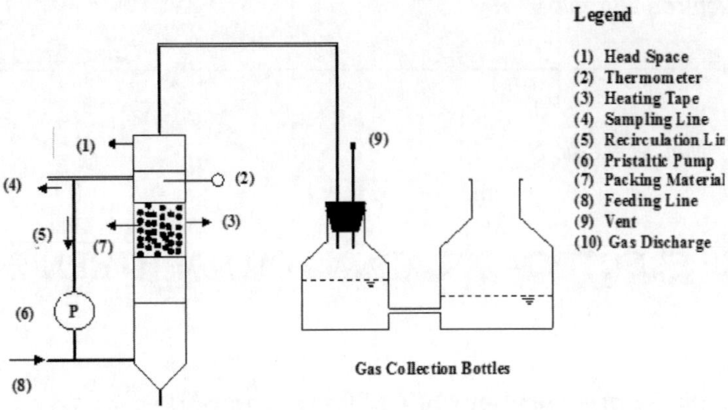


Figure 1: Schematic of UAHR Configuration

## 3 Results

Close to complete sulfate removal was achieved in both reactors at HRT level as low as 10 days. The sulfate removal efficiencies decreased with decrease in HRT to 5 and 2.5 days but were stil similar at around 79% in both reactors However, the rate of copper removal were different. The rate of copper removal in the single stage process was found to reach upto 625 mg/L/d, which is approximately two times of that of the two stage process. In previous studies with single-stage suspended growth systems, sulfate reduction was shown to be inihibited by metal precipitates formed (Shahsavari et al., 2015, Kieu et al., 2011 & Utgikar et al., 2002). The results from the current study show that the proposed UAHR configuration was successful in overcoming the inhibition of SRB by the metal precipitates. The solids inventory inside single stage reactor showed > 99% of precipitated copper in the bottom third of the reactor. It is postulated that the improved performance of the single-stage process in the current study is due to the SRB present in the upper section of the UAHR reactor that did not suffer from the inhibitory effects of metal precipitates.

## 4 Conclusion

The results show that both sulfate reduction and copper precipitation in the single stage process were similar to or better than the two-stage process over the entire duration of the study.

## References

APHA. (2005). Standard methods for the examination of water and wastewater. American Public Health Association (APHA): Washington, DC, USA.

DiLallo, R., & Albertson, O. E. (1961). Volatile acids by direct titration. Journal (Water Pollution Control Federation), 356-365.

Kieu, Hoa TQ, Elizabeth Müller, and Harald Horn. "Heavy metal removal in anaerobic semi-continuous stirred tank reactors by a consortium of sulfate-reducing bacteria." Water research 45.13 (2011): 3863-3870

Shahsavari, S., Hamadani, Y., Seth, R. and Biswas, N. (2015) " Inhibitory Effect of Metal Precipitation on Anaerobic Biological Sulfate Reduction Process." WEFTEC 2015, Chicago, Illinois, United States

Utgikar, Vivek P., et al. "Inhibition of sulfate-reducing bacteria by metal sulfide formation in bioremediation of acid mine drainage." Environmental toxicology 17.1 (2002): 40-48.

Leadership in Sustainable Infrastructure
*Leadership en Infrastructures Durables*

**Case Study**

Vancouver, Canada
May 31 – June 3, 2017/ *Mai 31 – Juin 3, 2017*

# APPLICATION OF ELECTRO-OXIDATION FOR NAPHTHENIC ACIDS (NAS) DEGRADATION

Abdalrhman, Abdallatif Satti [1] and Gamal El-Din, Mohamed [2, 3]
[1,2] Department of Civil & Environmental Engineering, University of Alberta, Edmonton, AB, Canada T6G 1H9
[3] mgamalel-din@ualberta.ca

## 1 Research Overview

As a result of the currently used processes for bitumen extraction from oil sands and subsequent treatments, large volumes of Oil Sands Process-Affected Water (OSPW) are generated. It has been confirmed by different studies that OSPW is acutely, sub-chronically and chronically toxic to a variety of aquatic organisms (Wiseman et al. 2013). Among the different constituents in OSPW, a group of aliphatic and alicyclic carboxylic acids known as naphthenic acids (NAs) are believed to be responsible for the toxicity of OSPW (Wang et al. 2013). NAs are highly toxic, recalcitrant and persist in the environment for many years. The treatment of OSPW is considered a great challenge facing the oil sands industry today. Various chemical, physical and biological processes have been investigated for the treatment of OSPW and degradation of NAs. However, so far, effective and cost-efficient treatment approaches applicable to full scale have not been found.

Electro-oxidation has emerged as promising process for wastewater treatment. It involves the application of direct current (DC) voltage by using suitable type of electrodes and an external electricity source. In electro-oxidation, pollutants can be oxidized either directly by exchanging electrons with the anode surface, or by indirectly through the mediation of some electroactive species generated at the electrodes. It has been applied successfully for the degradation of a variety of recalcitrant organic pollutants (Panizza and Cerisola 2009). The objective of this study was to investigate the effectiveness of applying electro-oxidation for the degradation of NAs. The study focused on evaluating the performance of electro-oxidation under low current (energy) operating conditions for degrading NAs and reducing their toxicity, in addition to understanding the involved mechanisms and reaction pathways.

## 2 Innovation

This study proposed the use of electro-oxidation for the treatment of OSPW and the degradation of NAs. Considering the nature and structure of NAs and the characteristics of OSPW, with its high electrical conductivity, electro-oxidation seemed to be a potential effective and cost efficient option for OSPW treatment and NAs degradation. Applying anodic oxidation by using cheap electrodes materials, such as graphite, and under low voltage conditions should preferentially degrade NAs of higher cyclicity and carbon number and decreased the number of rings and molecular weight without resulting in complete mineralisation. Therefore, the application of electro-oxidation should enhance OSPW biodegradability and reduce the toxicity.

Low-voltage electro-oxidation can be considered as a promising treatment option for OSPW given that it can lead to improved biodegradability and reduced toxicity of OSPW while maintaining the economic feasibility. The lower voltages required for the treatment will result in a sustainable and environmental friendly process that can be operated by solar energy, and the exclusion of the need for chemicals addition will prevent the production of any additional hazardous sludges.

## 3 Lessons Learned

The preliminary results from this study have shown positive NA degradation results by using graphite as the anode material for the treatment of a model NAs, cyclohexanecarboxylic acid (CHA), with initial concentration of 45 mg/L. Different voltages in the range from 2.9 to 10.5 V were applied for CHA degradation in short periods of time (30 to 180 minutes). It was found that the degradation rate increased with increasing the applied voltage, achieving degradation rates of 29.5%, 53%, 72.5% and 81.1% at voltages of 3.3, 6, 7.5 and 10.5 V, respectively (Figure 1). Changing the electrolyte medium from NaCl to $Na_2SO_4$ has shown no effect in terms of CHA removal, which can be considered as an indication that the dominant degradation mechanism, in the case of graphite electrodes, is the direct oxidation at the anode surface, which is consistent with what was previously reported in the literature.

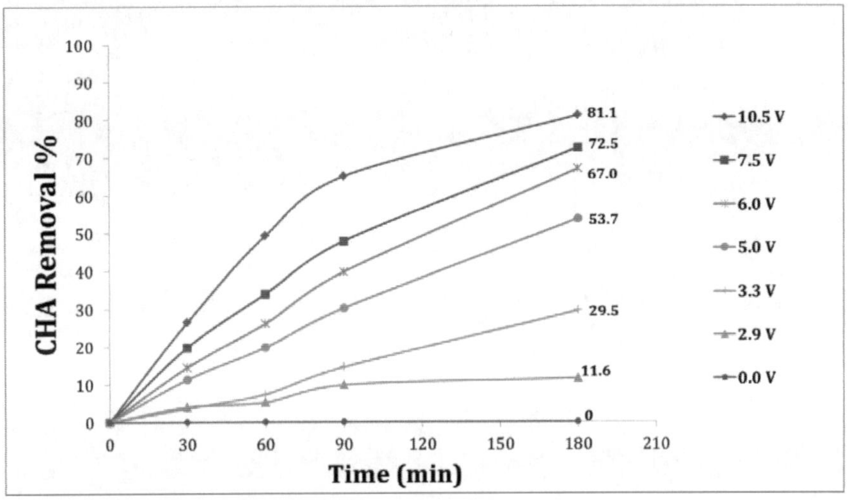


Figure 1: Cyclohexanecarboxylic acid (CHA) degradation at different applied voltages.


## Acknowledgements

The financial supports provided by a Natural Sciences and Engineering Research Council of Canada (NSERC) Senior Industrial Research Chair (IRC) in Oil Sands Tailings Water Treatment, the Helmholtz-Alberta Initiative through Alberta Environment and Parks' ecoTrust Program, Shell Enhanced Learning Fund (SELF), and the NSERC CREATE Research and Training via the Institute for Water, Energy and Sustainable Systems are greatly appreciated. Dr. Pamela Chelme-Ayala helped in reviewing the abstract. Dr. Mingyu Li and Dr. Selamawit Messele helped in performing the HPLC analysis.


## References


Wiseman, S.B., He, Y., Gamal-El Din, M., Martin, J.W., Jones, P.D., Hecker, M. and Giesy, J.P. 2013. Transcriptional responses of male fathead minnows exposed to oil sands process-affected water. *Comparative Biochemistry and Physiology Part C: Toxicology & Pharmacology,* **157**(2): 227-235.

Wang, N., Chelme-Ayala, P., Perez-Estrada, L., Garcia-Garcia, E., Pun, J., Martin, J.W., Belosevic, M. and Gamal El-Din, M. (2013) Impact of ozonation on naphthenic acids speciation and toxicity of oil

sands process-affected water to Vibrio fischeri and mammalian immune system. *Environ Sci Technol* **47**(12), 6518-6526.
Panizza, M. and Cerisola, G. 2009. Direct and mediated anodic oxidation of organic pollutants. *Chemical reviews* **109**(12): 6541-6569.

Leadership in Sustainable Infrastructure
*Leadership en Infrastructures Durables*

Vancouver, Canada
May 31 – June 3, 2017/ *Mai 31 – Juin 3, 2017*

# ASSESSMENT OF MICROBIAL DENITRIFYING CAPACITY IN VEGETATED BIOREACTORS TREATING GREENHOUSE EFFLUENT

Fatehi Pouladi, Soheil[1,5], Anderson, Bruce C.[1], Wootton, Brent[2], Weber, Kela P.[3], Rozema, Lloyd[4] and Bissegger, Sonja[3]
[1] Department of Civil Engineering, Queen's University, Kingston, ON, Canada
[2] Centre for Alternative Wastewater Treatment, Fleming College, Lindsay, ON, Canada
[3] Department of Chemistry and Chemical Engineering, Royal Military College of Canada, Kingston, ON, Canada
[4] Aqua Treatment Technologies, Campden, ON, Canada
[5] 12sfp@queensu.ca

**Abstract:** The discharge from food production greenhouses (greenhouse effluent) contains high nutrient concentrations, which if directly released, can have adverse effects on the environment. Wood-chip bioreactors are increasingly popular passive water treatment systems favoured for their economical denitrification in treating agricultural field tile drainage. Microbial communities are central to denitrification, however little is known about the maturation of microbial communities in wood-chip bioreactors treating greenhouse effluents. Multiple subsurface flow wood-chip bioreactors, each vegetated with a different plant species, together with an unplanted unit, received synthetic greenhouse effluent with elevated nitrate concentrations. The hybrid bioreactors were operated for over 2 years, during which water samples were collected from the inlet, outlet and within the reactors. The denitrification rate in the bioreactor planted with Typha angustifolia (narrowleaf cattail) was shown to be higher with increasing microbial activity and metabolic richness, measured by the carbon utilization patterns in Biolog® EcoPlates. The increased denitrifying gene (*nirS*) copies (determined by quantitative polymerase chain reaction, qPCR), and near-complete nitrate removal were observed in the *T. angustifolia* and unplanted reactors after 16 and 23 months of operation respectively. The findings suggested that an acclimation period of at least one year can be expected in unseeded bioreactors planted with *T. angustifolia*, while bioreactors without vegetation may require a longer time to maximize their denitrification capacity. These results are important for the designers and operators of wood-chip bioreactors, which are expected to be more commonly applied in future.

Keywords: Denitrification; wood-chip bioreactor; greenhouse; microbial community

## 1 INTRODUCTION

Denitrifying bioreactors are passive treatment systems supplied with an external carbon source such as wood chips that are used for nitrate removal from wastewaters. The early studies showed that the available carbon source in the riparian zones near surface waters had an important effect on the denitrification of the ground water that occurred in the soil (e.g. Schipper et al., 1991). These observations led to the innovation of in-situ systems with externally supplied solid carbon for enhanced nitrate removal (e.g. Robertson and Cherry, 1995). In addition to providing an electron donor for heterotrophic denitrification, these carbon substrates create a favourable anoxic environment by serving as an energy source for the aerobic microorganisms (Schipper et al., 2010). As reviewed by Addy et al. (2016), there have been many recent studies focused on using denitrifying bioreactors to reduce the adverse impact of the nitrates released from agricultural tile drainage. However, the nitrate concentrations found in the discharges from food production greenhouses (greenhouse effluent) are notably greater than those in the typical agricultural runoff. These high nutrient levels in the greenhouse effluent are typically due to the unused portion of the dissolved nutrients that are added to the irrigation water. High nitrate concentrations can lead to hypoxia in surface waters and cause potential health implications if consumed in drinking water. Therefore, detrimental environmental and health impacts can occur from the continuous release of the untreated greenhouse effluent.

Most of the studies on denitrifying bioreactors are comprised of unplanted systems, where employing cover vegetation has received little attention. This is potentially due to the common understanding that plants in passive treatment systems, such as constructed wetlands (CWs) can facilitate oxygen transfer (Kadlec and Wallace, 2009), which in turn might compromise the anoxic conditions necessary for denitrification. However, up to 99% nitrate removal from the greenhouse effluent was recently achieved in a series of pilot-scale vegetated bioreactors (Fatehi Pouladi et al., 2016a).

Despite the high performance of denitrifying bioreactors in reducing nitrate loads, the evidence to confirm the presence or function of microbial communities, presumed as the key driver of the reported denitrification, is limited. The few published studies investigating the microbial communities in these bioreactors (e.g. Warneke et al., 2011; Healy et al., 2015) were primarily conducted on unplanted reactors designed for treatment of discharges other than the greenhouse effluent. Among the microbial assessment methodologies employed for investigating treatment wetlands, the tools developed for functional assessment provide an understanding of what the existing microbial community is able to accomplish without directly identifying the species. This is a useful approach, as the presence of microbial species with a certain gene does not necessarily indicate that the environmental conditions are favourable for that gene to be expressed, and result in the expected function (Weber, 2016). The quantitative polymerase chain reaction (qPCR) and Community-Level Physiological Profiling (CLPP) are two examples of these methods than can assess the functional capacity of microbial communities.

The qPCR technique is a standard and powerful molecular biological method in which a targeted gene is amplified and measured in real time. This method provides information about the activity of the targeted enzymes and their potential function, such as those involved in the denitrification process. In a CLPP method using Biolog® microplates, the environmental sample is inoculated with various types of carbon sources and the colour change developed through the production of NADH over the incubation period provides information regarding metabolic function of the mixed microbial communities (Weber and Legge, 2010). In many recent studies, the parameters derived from the CLPP method have helped researchers to understand the microbial function of treatment wetlands of different scales and designs (e.g. Weber and Legge, 2011; Bissegger et al., 2014; Button et al., 2015).

The aim of the initial stage of this study was to investigate the potential development of microbial communities in vegetated wood-chip bioreactors with no prior seeding by measuring the mixed metabolic function of microbial communities via the CLPP method. Moreover, the study aimed to evaluate the bioreactors for their utilisation patterns of specific organic substrates in the CLPP Biolog® EcoPlates. The other important goal was the characterization of the denitrifying capacity of the bacterial community by investigating the denitrifying gene copies present in the bioreactors. These research objectives were defined to describe whether the nitrate removal exhibited from the water quality measurements could be explained by the microbial activity assessments. As each of the vegetated bioreactors was planted with a different plant species including halophytes and non-halophytes, the role of the plants in the microbial development of denitrifying bioreactors was also of interest. Overall, the study investigated the microbial denitrification capacity of vegetated wood-chip bioreactors as a potentially effective mechanism in treating the effluents from greenhouse operations.

## 2 MATERIALS AND METHODS

### 2.1 Timeline and Reactor Design

The pilot-scale project was built and operated at the laboratory of the Department of Civil Engineering at Queen's University (Kingston, ON, Canada). Each of the 5 bioreactors was constructed from open-top 220-L barrels (diameter: 56 cm, height: 90 cm) and filled with Maple hard wood chips, acquired from Quebec, Canada, in a single 80-cm layer. A vertical PVC tube (inner diameter: 2.54 cm) was positioned in the centre of each reactor through the wood-chip media as a platform to support 3 interstitial water sampling tubes (inner diameter: 1.6 mm), which were installed at 20, 40 and 60 cm below the top surface of the wood chips.

The synthetic greenhouse effluent was prepared in the laboratory per the common greenhouse effluent characteristics reported in the literature and by greenhouse operations in Ontario. The bioreactors were operated in vertical subsurface-flow (VSSF) mode. The influent was continuously distributed over the surface of each reactor by means of a peristaltic pump (hydraulic loading rate: 12 cm $d^{-1}$) and a square distribution grid, while the outflow was discharged from the bottom. The 4 vegetated bioreactors together with the unplanted (control) unit were in continuous operation indoors (ambient air temperature: 24° C) under a 1,000 W Metal Halide grow light set to a 16/8 on/off cycle to simulate the sunlight. More details about the experiment, including its design, schematic of the reactors, plant selection criteria and influent characteristics have previously been published (Fatehi Pouladi et al., 2016a, 2016b).

The project timeline, various test dates and operational phases are shown in Figure 1. The total operation was divided in the High Loading (average influent $NO_3$-N: 307 mg $L^{-1}$) and Low Loading phase (average influent $NO_3$-N: 202 mg $L^{-1}$). The active plant species were grouped as A, B and C. The A selection included softstem bulrush (*Schoenoplectus tabernaemontani* C.C. Gmel. Palla), switchgrass (*Panicum virgatum* L.), narrowleaf cattail (*Typha angustifolia* L.) and Canada wildrye (*Elymus canadensis* L.). The species present in the B group were *S. tabernaemontani* and *T. angustifolia* as the other two plants could not survive the experimental conditions. The C group including *S. tabernaemontani*, *T. angustifolia*, prairie cordgrass (*Spartina pectinata* Bosc ex Link) and saltgrass (*Distichlis spicata* L. Greene) resumed the operation starting in June 2015.

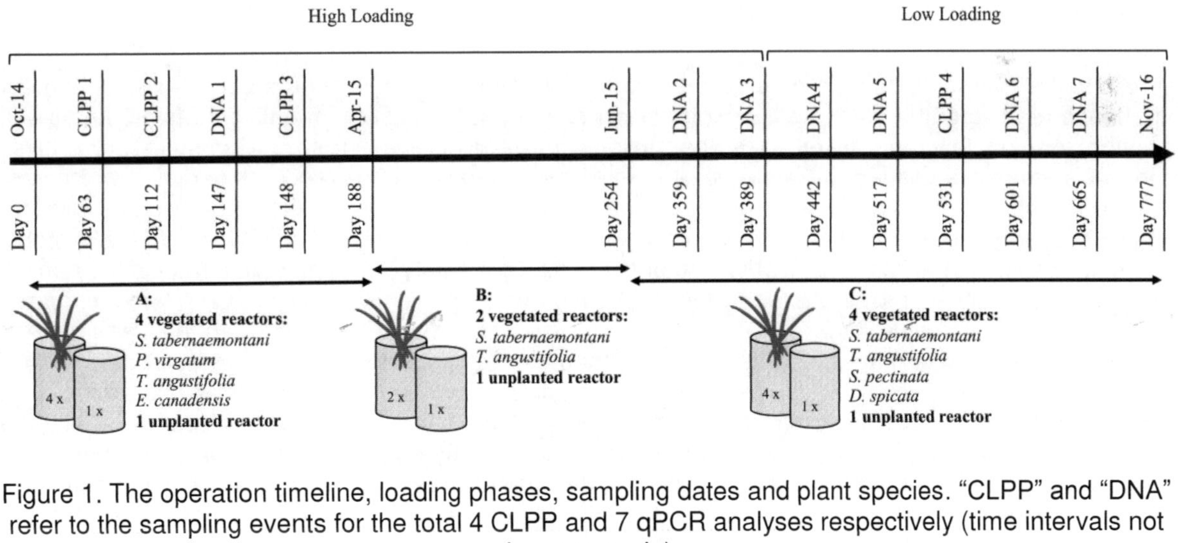


Figure 1. The operation timeline, loading phases, sampling dates and plant species. "CLPP" and "DNA" refer to the sampling events for the total 4 CLPP and 7 qPCR analyses respectively (time intervals not drawn to scale).

### 2.2 Microbial Assessment

The interstitial water samples were drawn from the spatial vertical profile using a set of dedicated 50 mL plastic syringes and were then transferred to air-tight sterile polypropylene tubes. Samples were consequently transported to the laboratory of the Environmental Sciences Group (ESG) at the Royal Military College of Canada (Kingston, ON, Canada) for further preparation.

### 2.2.1 CLPP

Biolog® EcoPlates (BIOLOG Inc., Hayward CA, USA) with 31 different carbon sources and one blank well in triplicates (96 wells per plate) were inoculated with each water sample at room temperature on the same day the samples were collected. Sample preparation was performed in an aseptic environment using sterile laboratory consumables and 70% ethanol spray for sterilizing the tools and bench tops. 100 μL of the unfiltered samples was pipetted to each well, after which the plates were stored in the dark during the incubation period. The optical densities of the plates were read using a microplate absorbance reader (TECAN Infinite®, Tecan US Inc., Morrisville, NC, USA) at an absorbance of 590 nm at several intervals after inoculation.

The first two CLPP tests (CLPP 1 and 2), were performed per the commonly practiced aerobic protocol where sample collection, preparation, inoculation and incubation procedure were exposed to natural oxygen in air. CLPP 3 and 4 were conducted anaerobically to better preserve the anaerobic microbial communities and inhibit aerobic respiration. Nitrogen gas was used to purge oxygen from the sample tubes by de-airing them prior to sample collection. The tubes were then sealed using multi layers of adhesive tape. The plate inoculation was done in an oxygen-free environment inside a glove box (or a glove bag) purged with a high flow of nitrogen gas. All the laboratory equipment including the samples and plates were stored inside the de-aired chamber while a small continuous flow of nitrogen ensured a low-oxygen environment. The plates, petri dishes and other tools were similarly purged prior to use, and the inoculated plates were sealed using adhesive tape around each plate ensuring the top lids and bottom surfaces remained intact. The plates were stored in the chamber covered with opaque black plastic bags during the incubation period, and were only removed temporarily to record their absorbance values. The lids remained taped to the plates during the entire process.

The CLPP data analysis was performed according to the previously established methods (Weber et al., 2007; Weber and Legge, 2010). The absorbance readings at 48 to 50 h were selected as the suitable time point for further analysis. The absorbance reading of each plate, associated with one sample from a specific reactor and depth resulted in 96 carbon source utilisation patterns (CSUPs). The overall average well colour development (AWCD) of one sample was calculated as:

[1] $AWCD = \frac{1}{31}\sum_{i=1}^{31}(A_i - A_0)$

where $A_i$ is the average of the triplicate absorbance readings of the carbon source i, and $A_0$ is the average absorbance reading of the blank well with no carbon source. In the case of negative numbers for ($A_i$-$A_0$), indicating very low recorded absorbance, the values were recorded as zeros. For the statistical analysis where replicates were needed, 3 AWCD values were associated with each reactor's depth, in which case $A_i$ and $A_0$ represented the absorbance values of one of the three replicates for the carbon source i and the blank well respectively, rather than the average value of the three replicates. The richness of the microbial community was described as the metabolic potential of the community, calculated as a ratio of the number of the carbon sources that were utilized by the inoculated microorganisms, over the total number of carbon sources, presented in percentage (%). A carbon source was counted as 'utilized' when the well's blanked value ($A_i$-$A_0$) was greater than 0.25. The carbon substrates in each Biolog® plate have been grouped in 5 major guilds, namely carbohydrates, carboxylic/acetic acids, polymers, amines/amides and amino acids (Weber and Legge, 2009, adapted from Zak et al., 1994). Moreover, 8 of the total 31 carbon substrates have recently been identified for their common occurrence in various plant root exudates (Button et al., 2016). The plant root exudate richness of a sample (%) was also reported as the microbial community's richness by calculating the percentage ratio of the utilized number of the carbon sources associated with plant root exudate over the total number of substrates (8).

### 2.2.2 qPCR

The expression level of the nitrite reductase genes *nirK* and *nirS* was determined by real-time RT-PCR using the SYBR Green detection system. The primers and programs used for the amplification of *nirK* and *nirS* have been previously designed and optimized (Ligi et al., 2014). The influent and interstitial water samples were first filtered using glass-fiber filter papers (0.2 μm pore size, diameter: 47mm, Millipore,

Etobicoke, ON, Canada). Filters with the captured residues were then placed in sterile tubes and stored at -20 ºC until the DNA extraction was conducted. The total community DNA extraction was performed by rolling the thawed filters inside the extraction vials provided in the FastDNA SPIN Kit for Soil (MP Biomedical, Solon, OH, USA), following the manufacturer's protocol. The nucleic acid concentrations were measured by pipetting 1 µL of the extracted samples using the NanoDrop 2000 spectrophotometer (Fisher Scientific, Toronto, ON, Canada). The samples were analyzed undiluted in duplicates on Eco™ Real-Time PCR System (Illumina, San Diego, CA, USA) using Maxima SYBR Green Master Mix (Fisher Scientific, Toronto, ON, Canada) and 0.4 µM of forward and reverse primers. The reaction efficiencies calculated by EcoStudy Software (Illumina, San Diego, CA, USA) were 100 ± 10 % with an $R^2 > 0.990$. On each plate, a standard curve (0.025 to 2.5 ng) and no template control were run with the samples. The qPCR standard curves were prepared using 6 serial dilutions (25.00, 6.25, 1.56, 0.40, 0.10 and 0.025 ng) of the DNA extracted from thermophilic sludge from Ravensview wastewater treatment plant (Kingston, ON, Canada).

The gene copy concentrations were derived relative to the standard curves. Each value was normalized for both the nucleic acid concentration of the extracted DNA sample, as well as the volume of the filtered water sample. The normalized values for the interstitial or influent water samples were calculated as:

[2] $nirS\ or\ nirK = nirS_1\ or\ nirK_1/(V \times N_C)$

where *nirS* (or *nirK*) represents the sample's normalized amplified target gene copy quantity (no unit), $nirS_1$ (or $nirK_1$) is the recorded amplified target gene copy quantity (ng) of the sample, V is the initial water sample volume (µL) and $N_C$ is the nucleic acid concentration (ng $\mu L^{-1}$) of the respective sample after DNA extraction. The normalized gene copies of all the samples were standardized against that of the influent sample, which was selected as the calibrator to report the gene copy fold-change as shown in Eq. 3:

[3] $nirS\ or\ nirK\ fold\ change = nirS\ or\ nirK\ /(nirS_{inf}\ or\ nirK_{inf})$

where '*nirS* or *nirK* fold change' (no unit) describes the times-increase of the amplified target gene quantity to that of the calibrator (influent) and $nirS_{inf}$ or $nirK_{inf}$ is the normalized amplified target gene copy quantity (no unit) of the influent sample per Eq. 2. It is important to note that the reported *nirS* and *nirK* quantities are not absolute and are relative numbers. Nitrate concentrations in the influent and outflow grab samples were analyzed at an external laboratory according to Standard Methods (APHA, 2005). All the statistical analyses including one-way ANOVA, the Tukey's and Dunnett's tests were performed using XLSTAT software (©Addinsoft, New York, NY, USA) and a difference was reported significant when *p-value*<0.05.

## 3 RESULTS AND DISCUSSION

The AWCD values of the CLPP analyses and the richness of the mixed microbial communities were generally highest in the bioreactor planted with *S. tabernaemontani* in the first CLPP test conducted in the early stages of the experiment, and in the *T. angustifolia* reactor in the rest of the measurements (Table 1 and Figure 2). Each of these reactors were identified as significantly different from the other units for their metabolic activity within each CLPP analysis. A statistical analysis of the AWCD values with depth (top, middle and bottom) as an independent factor suggested some paired significant differences in CLPP 1 and CLPP 3. However, these differences were small, showing no apparent depth trend for the activity values. Moreover, the depth variations as shown in Figure 2 made it difficult to identify a consistent pattern in the AWCD values along the vertical profile. The AWCD numbers in the *T. angustifolia* reactor increased from 0.32 in CLPP 1 to 1.07 in CLPP 4 (over 3-times increase), while the denitrification rate of this reactor reached its maximum of 99% (32 g N $day^{-1}$ $m^{-3}$) after about 16 months of continuous operation (Day 491: Figure 3.a and 4.a).

The water quality measurements indicated the denitrification rate in the *T. angustifolia* reactor remained above 30 g N $day^{-1}$ $m^{-3}$ (99% $NO_3$-N removal) during a period of at least 9 months after the last CLPP test was performed on Day 531 (Figure 4.a), suggesting that suitable conditions for successful denitrification continued to be maintained. Similarly, the greatest microbial activity measurement of 0.72 in the *S. tabernaemontani* bioreactor in CLPP 1 was concurrent with the highest nitrate reduction of 99% (49 g N $day^{-1}$ $m^{-3}$) observed in this reactor in the beginning of the experiment. However, the nitrate removal rate in this reactor decreased rapidly as the operation progressed, and the metabolic activity dropped to 0.35 in CLPP 4 as well. The decreased activity and denitrification of this reactor were likely due to the loss of organic carbon from wood chips after the initial leaching, which deprived the heterotrophic community of the energy required for their metabolism (outflow $BOD_5 < 2$ mg $L^{-1}$ on Day 559).

Table 1. Mean AWCD values ± standard deviation and mean total richness (%) in each bioreactor (n: 3 depths) and influent. Reactors grouped with similar lower-case letters did not show significant statistical difference in their AWCD values within each CLPP test. Asterisk (*) indicates a significant difference from the unplanted reactor.

| | CLPP 1 | | CLPP 2 | | CLPP 3 | | CLPP 4 | |
|---|---|---|---|---|---|---|---|---|
| | AWCD | Rich. (%) | AWCD | Rich. (%) | AWCD | Rich. (%) | AWCD | Rich. (%) |
| Influent | 0.36 | 45 | 0.36 | 39 | 0.20 ± 0.25 | 29 | 0.24 ± 0.23 | 42 |
| Control | 0.54 ± 0.18 b | 48 | 0.24 ± 0.09 c | 35 | 0.32 ± 0.01 c | 32 | 0.51 ± 0.16 c | 45 |
| *E. canadensis* | 0.54 ± 0.04 b | 58 | 0.10 ± 0.02 d* | 13 | 0.20 ± 0.08 d* | 26 | - | - |
| *P. virgatum* | 0.33 ± 0.02 c* | 39 | 0.30 ± 0.08 b, c | 42 | 0.30 ± 0.01 c | 45 | - | - |
| *S. tabernaem.* | 0.72 ± 0.02 a* | 64 | 0.35 ± 0.08 b* | 42 | 0.42 ± 0.13 b* | 45 | 0.35 ± 0.06 d* | 35 |
| *T. angustifolia* | 0.32 ± 0.01 c* | 39 | 0.54 ± 0.05 a* | 55 | 0.55 ± 0.11 a* | 55 | 1.07 ± 0.13 a* | 74 |
| *S. pectinata* | - | - | - | - | - | - | 0.64 ± 0.02 b* | 48 |
| *D. spicata* | - | - | - | - | - | - | 0.32 ± 0.03 d* | 32 |

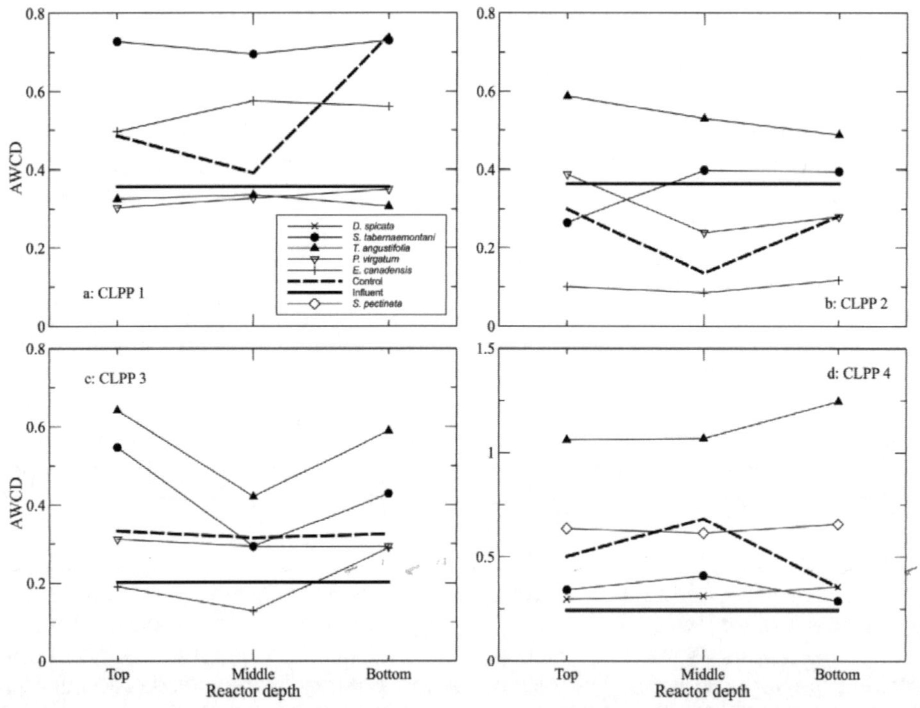


Figure 2. AWCD values for each bioreactor and influent measured by 4 CLPP tests on water samples from the reactors' top, middle and bottom depths.

Of the four CLPP tests, the results of CLPP 4 can be considered the most representative of the matured state of the system, as it followed an anaerobic protocol, and was conducted approximately 17 months after the project start-up. Despite the statistical difference that was reported between the control reactor and all the other units (Table 1), the *T. angustifolia* reactor was the only unit with considerably higher metabolic activity, while the other reactors showed lower, or in case of *S. pectinata*, only slightly higher numbers than control. In addition to the AWCD values, the carbon utilization richness associated with plant root exudates for the *T. angustifolia* reactor displayed a consistent increase from 37% in CLPP 1 to 81% in CLPP 4 (Figure

3.b), suggesting that the planted reactor could utilize a higher number of root exudate carbon substrates as the bioreactor matured. This can potentially indicate that the presence of *T. angustifolia* and its roots played a role in the maturation of the microbial community that developed within the reactor. These results should however be interpreted with caution, as the metabolic activity and root exudate richness values of the influent also ranged from 0.20 to 0.36 and 25% to 62% respectively over the same period, which can be an indication of a natural microbial community establishment inside the influent reservoirs. In addition, the observed activity in the influent samples might suggest that a background chemical reaction occurred between the substrates in the EcoPlates and the nutrient solution, which did not happen in the blank wells.

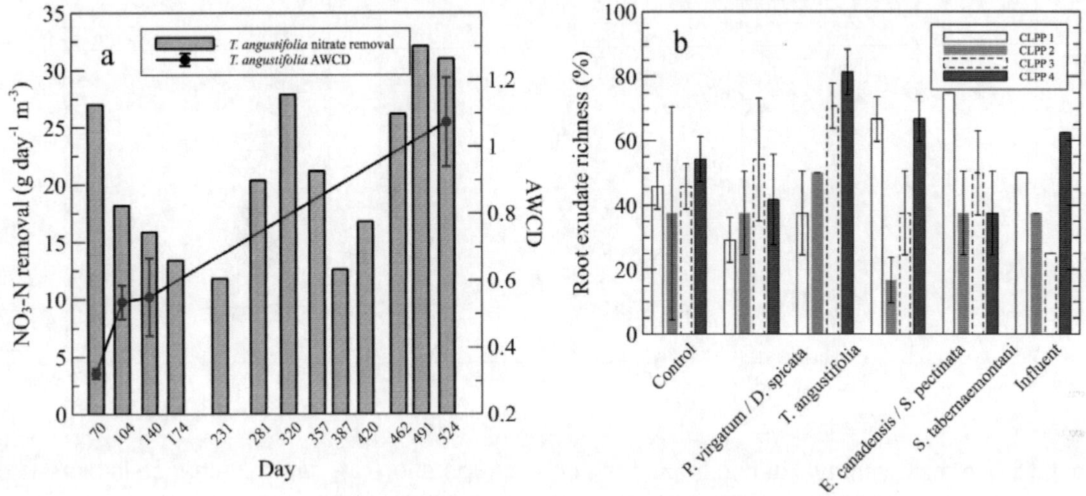


Figure 3. a: Temporal trend of $NO_3$-N removal rate of the bioreactor planted with *T. angustifolia* (left vertical axis) and the same reactor's AWCD (right axis). b: Average microbial community richness based on the utilized carbon sources associated with plant root exudates. Error bars represent the standard deviation of 3 depths in both figures.

The *nirS* gene copies were positively correlated with the denitrification rate (Figures 4.b and c). Similar to the mixed microbial activity results, the highest *nirS* copies were measured in the bioreactor planted with *T. angustifolia*. The average fold-change of *nirS* in this reactor revealed a sudden boost in the second half of the operation with an average of 1,732, 715 and 679-times increase on Days 517, 601 and 665 respectively (Figure 5.a). Similar to Healy et al. (2015), who investigated the denitrification genes along their experimental wood-chip columns, we did not observe a consistent trend in the denitrification gene copies with the depth.

The nitrate removal rate and the ln (*nirS* fold-change) were positively correlated in the *T. angustifolia* reactor (Figure 5.b, $R^2$=0.6). A correlation between the total copies of nitrite reductase genes (∑*nirS*, *nirK* and *nosZ*) and nitrate removal rate was previously shown in a study of denitrifying bioreactors with carbon-rich media (Warneke et al., 2011), where the reductase gene copies were found to be lower in hard and soft wood chips in comparison with the other carbon substrates such as maize cobs, wheat straws and sawdust. The *nirK* fold-changes in our study were found to be negligible and hence not reported. However, *nirK* populations in other denitrifying systems were typically found in abundance, with a *nirS*/*nirK* ratio ranging from 1 to 13 in different environmental conditions (Warneke et al., 2011; Ligi et al., 2014). The low *nirK* copies in our system can be attributed to the fact that our bioreactors were constructed and operated in a laboratory setting and received synthetic influent in reasonably controlled conditions. Thus, it was expected that the developed microbial communities were not as diverse as reported in other systems.

In the *T. angustifolia* reactor, the higher activities of the mixed microbial communities interpreted by the AWCD values, together with the higher nitrite reductase gene (*nirS*) populations are strong markers to indicate the elevated denitrification performance in this reactor was caused by microbial pathways involving denitrifying bacteria. However, future work to analyze the absolute quantities of *nirS* and *nirK* would be useful to determine the actual denitrification rates. It has been shown that CWs planted with *Typha latifolia*, another species of the same family, encouraged the growth of *nirK* containing bacteria, while the addition of the plant's litter could stimulate the growth of the bacteria expressing *nirS* genes (Chen et al., 2014). These findings reinforce the possibility that the dead and decaying biomass of *T. angustifolia* in our study might have contributed to the higher growth of the bacterial community. Since the results of the *T. angustifolia* unit stood out in all the different analyses, the presence of *T. angustifolia* can be regarded as an important factor in providing the favourable conditions for growth of heterotrophic bacteria.

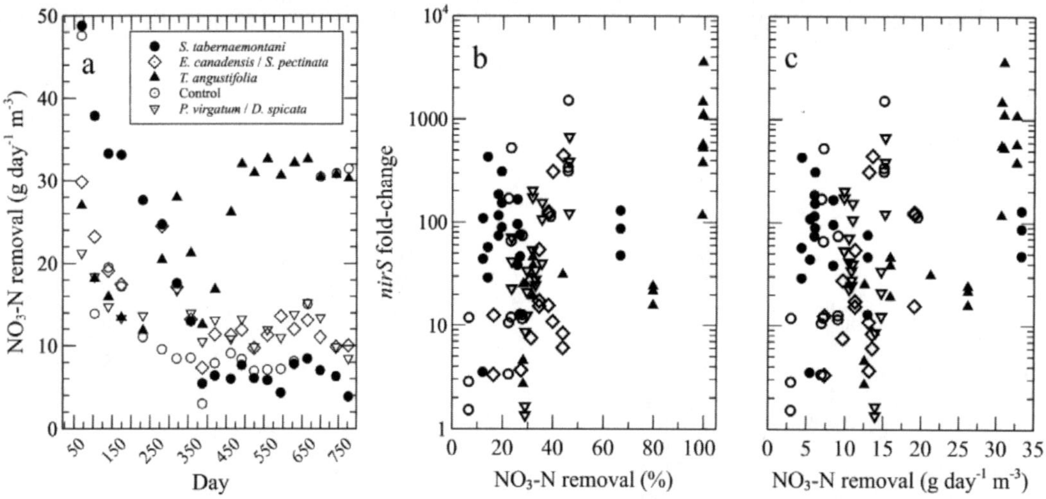


Figure 4. Nitrate mass removal during the entire period of operation (a), *nirS* fold-change in bioreactors vs. nitrate treatment rate (b), and nitrate mass removal rate (c).

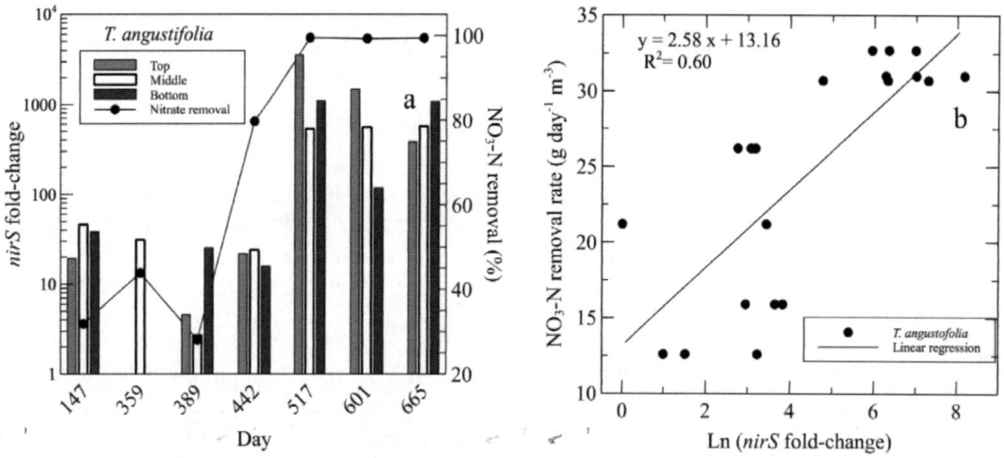


Figure 5. a: Temporal *nirS* fold-change in *T. angustifolia* bioreactor at 3 depths (left axis) and bioreactor's nitrate removal rate (right axis); b: *T. angustifolia* bioreactor's nitrate removal rate as a function of ln (*nirS* fold-change).

An important observation that ensued was a relatively high *nirS* copies in the unplanted reactor (Figures 4.b and c) despite nitrate maximum removal of 46% around the time when the final DNA sampling was performed (Day 665). It should however be noted that the nitrate reduction in this reactor reached 99% (30 g N day$^{-1}$ m$^{-3}$) on Day 699, about one month after the final DNA testing (Figure 4.a). This delayed high denitrification performance demonstrated the unplanted reactor required a longer acclimation time to attain its full denitrification capacity. Furthermore, the recognizable higher *nirS* fold-change observed in this reactor was likely captured during the exponential growth phase of its bacterial community, which proved an effective measure predicting the next phase, where denitrification rate was maximized and nitrate concentrations became the limiting growth factor.

It was earlier suggested that the substantially higher release of organic carbon (shown by outflow $BOD_5$ and COD measurements) in the *T. angustifolia* reactor was the main driver for the near-complete denitrification (Fatehi Pouladi et al., 2016a). The degradation of the wood chips can be a major factor in leaching organic matter that provides food and energy source for the heterotrophic denitrifying bacteria. The appearance of white substances on the surface of the wood chips in our bioreactors could hint at the presence and growth of fungi species as well. Fungi has been shown to contribute to co-denitrification and production of nitrogen gas in soils (Long et al., 2013). In addition to a potentially direct role, the fungi species may have an indirect role by aiding the denitrifying bacteria by degrading the cellulose and lignin in wood chips and making the organic carbon bio-available for the bacteria (Appleford et al., 2008; Porter, 2011). Thus, it is possible that fungi growth assisted the denitrification performance of the *T. angustifolia* and unplanted reactor by decomposing the wood chips. Therefore, further investigations are needed and recommended as future steps to gain deeper insights into the function of the fungi and denitrifying bacteria in enhancing denitrification in direct or indirect ways. Furthermore, although the interstitial water sampling method has previously been used in microbial studies of CWs to avoid destructive core sampling techniques (e.g. Button et al., 2015), the microbial characteristics of the biofilm attached to the wood chips are still of great interest. This component will be carried out in the next phase of the project, which will provide a basis for comparison with the current investigation of the interstitial water samples.

The results of this study provide useful information for the greenhouse owners and managers who are interested in employing low-cost treatment systems for nutrient management of their food production operations. The long-term nature of the pilot-scale experiments presented in this paper can inform designers about the dynamic nature of wood-chip bioreactors in long-term operations. It was also shown that bioreactors vegetated with capable species such as *T. angustifolia*, can host and develop a more active microbial community than other species such as *D. spicata* and result in a better denitrification performance with a shorter start-up period.

## 4 CONCLUSIONS

The results of this study demonstrated that high denitrification performance of wood-chip bioreactors in treating greenhouse effluent could be linked to the mixed microbial community and bacterial population encoding the denitrifying gene of *nirS*, thus confirming the biologically mediated denitrification in these bioreactors. The presence of *T. angustifolia* resulted in a shorter acclimation period, whereas the unplanted reactor followed the same trend with an approximately 7-month delay. Based on these findings, the presence of *T. angustifolia*, its rhizosphere and decaying biomass, as well the growth of fungi on wood chips were discussed as the potential factors that enhanced microbial denitrification. A closer look at the fungi population and the microbial characteristics of the biofilm attached to the media substrate were recommended as the future steps.

## ACKNOWLEDGEMENTS

The authors would like to acknowledge the financial support provided by the College-University Idea to Innovation (CU-I2I) Grant from the Natural Sciences and Engineering Research Council of Canada (NSERC) that made this research possible. NSERC Discovery Grants to Kela Weber are gratefully acknowledged.

## REFERENCES

Addy, K., Gold, A.J., Christianson, L.E., David, M.B., Schipper, L.A. and Ratigan, N.A. 2016. Denitrifying bioreactors for nitrate removal: A meta-analysis. *Journal of Environmental Quality*, **45**: 873–881.

APHA, AWWA, WEF, 2005. *Standard Methods for the Examination of Water & Wastewater*, 21st ed. American Public Health Association, American Water Works Association, and Water Environment Federation, Washington, DC, USA.

Bissegger, S., Rodriguez, M., Brisson, J. and Weber, K.P. 2014. Catabolic profiles of microbial communities in relation to plant identity and diversity in free-floating plant treatment wetland mesocosms. *Ecological Engineering*, **67**: 190–197.

Button, M., Nivala, J., Weber, K.P., Aubron, T. and Müller, R.A. 2015. Microbial community metabolic function in subsurface flow constructed wetlands of different designs. *Ecological Engineering*, **80**: 162–171.

Button, M., Rodriguez, M., Brisson, J. and Weber, K.P. 2016. Use of two spatially separated plant species alters microbial community function in horizontal subsurface flow constructed wetlands. *Ecological Engineering*, **92**: 18–27.

Chen, Y., Wen, Y., Zhou, Q. and Vymazal, J. 2014. Effects of plant biomass on denitrifying genes in subsurface-flow constructed wetlands. *Bioresource Technology*, **157**: 341–345.

Fatehi Pouladi, S., Anderson, B.C., Wootton, B. and Rozema, L. 2016a. Evaluation of passive reduction of salts and nitrate from greenhouse effluent by planted bioreactors. *The 13th IWA Specialized Conference on Small Water and Wastewater Systems*, The International Water Association, Athens, Greece. http://uest.ntua.gr/swws/proceedings/pdf/IWA_2016_Manuscript_Fatehi_Pouladi.pdf.

Fatehi Pouladi, S., Anderson, B.C., Wootton, B. and Rozema, L. 2016b. Evaluation of phytodesalination potential of vegetated bioreactors treating greenhouse effluent. *Water*, **8**.

Healy, M.G., Barrett, M., Lanigan, G.J., João Serrenho, A., Ibrahim, T.G., Thornton, S.F., Rolfe, S.A., Huang, W.E. and Fenton, O. 2015. Optimizing nitrate removal and evaluating pollution swapping trade-offs from laboratory denitrification bioreactors. *Ecological Engineering*, **74**: 290–301.

Appleford, J.M., Rodriguez, L.F., Cooke, R., Zhang, Y., Kent, A.D. and Zilles, J. 2008. Characterization of microorganisms contributing to denitrification in tile drain biofilters in Illinois. 2008. *American Society of Agricultural and Biological Engineers (ASABE)*. Providence, RI, USA.

Kadlec, R.H. and Wallace, S. 2009. *Treatment Wetlands*, 2nd ed. CRC Press, Taylor & Francis Group, Boca Raton, FL, USA.

Ligi, T., Truu, M., Truu, J., Nõlvak, H., Kaasik, A., Mitsch, W.J. and Mander, Ü. 2014. Effects of soil chemical characteristics and water regime on denitrification genes (*nirS*, *nirK*, and *nosZ*) abundances in a created riverine wetland complex. *Ecological Engineering,* **72**: 47–55.

Long, A., Heitman, J., Tobias, C., Philips, R. and Song, B. 2013. Co-occurring anammox, denitrification, and codenitrification in agricultural soils. *Applied and Environmental Microbiology*, **79**: 168–176.

Porter, M.D. 2011. Microbial community dynamics in denitrifying biofilters receiving agricultural drainage (Master of Science Dissertation). Department of Civil and Environmental Engineering, University of Illinois at Urbana-Champaign, Champaign, IL, USA.

Robertson, W.D. and Cherry, J.A. 1995. In situ denitrification of septic-system nitrate using reactive porous media barriers: Field trials. *Groundwater*, **33**: 99–111.

Schipper, L.A., Cooper, A.B. and Dyck, W.J. 1991. Mitigating nonpoint-source nitrate pollution by riparian-zone denitrification, In: Bogárdi, I., Kuzelka, R.D., Ennenga, W.G. (Eds.), *Nitrate Contamination: Exposure, Consequence, and Control*, 401–413, Springer Berlin Heidelberg, Heidelberg, Germany.

Schipper, L.A., Robertson, W.D., Gold, A.J., Jaynes, D.B. and Cameron, S.C. 2010. Denitrifying bioreactors—An approach for reducing nitrate loads to receiving waters. *Ecological Engineering*, **36**: 1532–1543.

Warneke, S., Schipper, L.A., Matiasek, M.G., Scow, K.M., Cameron, S., Bruesewitz, D.A. and McDonald, I.R. 2011. Nitrate removal, communities of denitrifiers and adverse effects in different carbon substrates for use in denitrification beds. *Water Research*, **45**: 5463–5475.

Weber, K.P., Grove, J.A., Gehder, M., Anderson, W.A. and Legge, R.L. 2007. Data transformations in the analysis of community-level substrate utilization data from microplates. *Journal of Microbiological Methods*, **69**: 461–469.

Weber, K.P. and Legge, R.L. 2011. Dynamics in the bacterial community-level physiological profiles and hydrological characteristics of constructed wetland mesocosms during start-up. *Ecological Engineering*, **37**: 666–677.

Weber, K.P. and Legge, R.L. 2010. Community-level physiological profiling, In: Cummings, S.P. (Ed.), *Bioremediation: Methods and Protocols*, 263–281, Humana Press, Totowa, NJ, USA.

Weber, K.P. and Legge, R.L. 2009. One-dimensional metric for tracking bacterial community divergence using sole carbon source utilization patterns. *Journal of Microbiological Methods*, **79**: 55–61.

Weber, K.P. 2016. Microbial community assessment in wetlands for water pollution control: Past, present, and future outlook. *Water*, **8**.

Zak, J.C., Willig, M.R., Moorhead, D.L. and Wildman, H.G. 1994. Functional diversity of microbial communities: A quantitative approach. *Soil Biology and Biochemistry*, **26**: 1101–1108.

Leadership in Sustainable Infrastructure

*Leadership en Infrastructures Durables*

Vancouver, Canada

May 31 – June 3, 2017/ *Mai 31 – Juin 3, 2017*

# ASSESSMENT THE PRESENCE OF HEAVY METAL (HM) CONCENTRATION IN RIVER WATER AND FISH SPECIES AND ASSOCIATED HUMAN HEALTH HAZARD DUE TO THE FISH CONSUMPTION OF TURAG RIVER IN BANGLADESH

Rashid, Abdur[1, 4], Siddika, Afia[2] and Rahman, Md Tauhid Ur [3]

[1,2,3]Department of Civil Engineering, Military Institute of Science and Technology (MIST), Mirpur-1216, Dhaka, Bangladesh

[4] arashid7002@gmail.com

**Abstract:** Dhaka city, the capital of Bangladesh is surrounded by mainly four rivers. One of the main rivers around Dhaka city is Turag River which is highly polluted. Wastes and effluents generated from industries and houses are directly discharged into this river without any treatment. Fishes eat wastes and deposit in their tissues as a form of Heavy Metal (HM) and through food chain HM transforms to human body and create health hazard. This study focuses on assessment of concentration of HM and potential human health hazards due to consumption of two fish species namely Banded Snakehead (*Channa striata)* and Stinging Catfish (*Heteropneustes fossilis*). Concentration of four HM i.e. chromium (Cr), cadmium (Cd), lead (Pb) and arsenic (As) in river water and two fish species was measured by atomic absorption spectrophotometer and presence were further confirmed by Scanning Electron Microscopy. Average concentration of HM in river water varied in descending order: Cr(16.67-20.07)>As(8.84-116.2)>Pb(6.4-8.56)>Cd(0.42-0.84) μg/L, respectively. Both non-carcinogenic risk (NCR) and carcinogenic risk (CR) on human health due to HM were calculated. It is found that both total target hazard quotients (TTHQ) and total hazard index (THI) exceeds the limiting value for both female and children consumers reflecting higher potential of developing NCR. Though for male, both THI and TTHQ are below limiting value but is not considered safe as their hazard index (i.e. THI and TTHQ) are approaching to limiting values. Again, considering CR, arsenic indicated potential risk. Therefore, to consume these fish species from this river is not safe to consume.

## 1. Introduction

At present the aquatic environments contamination by heavy metals is a great alarm in the world as because they are toxic and long-term persistence creates subsequent accumulation in aquatic habitats. The contaminated heavy metals accumulate in microorganisms, flora and fauna, and in turn, enter into the human food chain and result in health hazards. The important causes that creates pollution of aquatic environments are manifold: untreated industrial effluentsdischarge, unplanned and rapid urbanization,use of pesticides and fertilizers, atmospheric deposition of dust/aerosols, and waste water practice in irrigation. Uncontrolled application of sewage sludge and vehicular emissions also creates aquatic environment pollution. In the developing world, urban rivers are generally used as endpoints of industrial effluents and municipal sewage discharges. Likewise, random industrialization and unplanned urbanization in Dhaka city have evidently influenced the water quality in Turag river (Islam et al., 2015).

Dhaka city which is the capital of Bangladesh is surrounded by mainly four rivers; namely the Buriganga on the south, the Tongi Khal (Drainage channels are locally known as 'Khal') on the north, the Turag on the west and the river Balu on the east (Fig. 1a). All the rivers around Dhaka city are highly polluted. Now-a-days, Dhaka city generates about 3500 to 4000 metric tons of solid wastes per day. Moreover, a huge number of industries have set up in and around the City during the last few years and the number of new industries are alarmingly increasing which produces huge waste and directly discharge into these rivers without any treatment (Ahmad, M. K et al. 2010). Many dyes and textile industries are established beside these rivers which are the main sources of heavy metal (UNEP-2001). Again, due to discharging untreated domestic wastes into the river leads to increase of heavy metals in river water (Khadse et al., 2008). Fishes eat wastes from the water of these rivers and deposit the pollutants in their tissues as a form of heavy metals. Lead (Pb), barium (Ba), cadmium (Cd), mercury (Hg), chromium (Cr), and arsenic (As) are mainly well-known as toxic heavy metals. The heavy metals are accumulated in tissues of fishes and bones through absorption, and transmission of these metals occurs in the human body via the food web. Heavy metals, unlike organic pollutants, cannot be chemically degraded or biodegraded by microorganisms.

Now-a-days pollution of heavy metal in fish is a great concern worldwide, as because not only a great threat to fish, but also have a huge negative effect on human health due to consumption. Among the heavy metals, it was found that Cd, Cr, Pb and As is the most toxic and create health hazards to human health. For example, Pb causes renal failure and liver damage (Lee et al., 2011). Moreover, prolonged exposure to Pb may cause mental retardation to human body and even death (Al-Busaidi et al., 2011). Cadmium directly attacks on kidney and creates the symptoms of chronic toxicity that includes impaired kidney function, poor reproductive capacity, tumors, hepatic dysfunction and hypertension (Al-Busaidi et al., 2011; Rahman and Islam, 2010). Cr and Cu cause nephritis, anuria and extensive lesions in kidney (Rahman and Islam, 2010). The type of effects can be toxic (acute, chronic or sub-chronic), carcinogenic, neurotoxic, teratogenic or mutagenic (European Union, 2002). Cr and Pb are toxic in nature and may cause carcinogenic impact on humans (Malik et al., 2010). Therefore, trace metal contamination in fishes is increasing global attention.

Bangladesh is one of the most important inland fishing nations of the world. In 2014, Bangladesh produced 1,591,190 tons of fish which were fifteenth highest in the world (FAO, 2016). At present 3.8 percent of GDP is being contributed by fisheries sector, which is 55% of animal protein intake in Bangladesh. Therefore, the heavy metal accumulation in fish and their associated health hazard is a crying need to identify. Although a number of studies have investigated heavy metal concentrations in fishes (Dhanakumar et al., 2015, Islam et al., 2015), but so far, no research has been conducted on profile of heavy metals, particularly in flesh and bones of the fish species, Stinging Catfish (*Heteropneustesfossilis*) (locally known as 'Shing') and Banded Snakehead (*Channa striata)* (Locally known as 'Taki')and their evaluation of associated health risks of the Turag river. Therefore, the present study focuses on (1) determination of concentration levels of Cd, Pb, Cr and As in water and in the edible muscle tissues (flesh) and bones of Stinging Catfish (*Heteropneustesfossilis*) & Banded Snakehead (*Channa striata)* and bio-concentration of heavy metals and (2) examination of associated human health risk potential (non-carcinogenic and carcinogenic) due to consumption.

## 2. Materials & methods

### 2.1 Study site

The study site is confined within Turag River, which surround the Dhaka City, the capital of Bangladesh. A huge number of fishermen catch fishes from these rivers which is consumed by people of Dhaka city. Since Turag River runs beside Dhaka, therefore this river was selected for study purpose. Stinging catfish and banded snakehead are the main fish species found from this river. Total 9 water samples and fish species stinging catfish and banded snakehead were collected from (N 23°53'09", E 90°23'32.6") and (N 23°52'51", E 90°23'44.3") from Turagriver. The study site is shown in Fig. (1b).

### 2.2 Sample collection and preparation

Fish species, including Stinging catfish and Banded Snakehead were collected directly from Turag at the end of March 2016 by catching fishes using casting fishing nets. The month March is the dry season in Bangladesh. In 2016, there was only 48 mm precipitation in the month of March in Dhaka (AccuWeather, 2016). As both the rivers are highly polluted and there was very little rain in Dhaka in march 2016, the level of Dissolved Oxygen (DO) was extremely low (0.45-0.60 mg/L). Therefore, it is too difficult to survive for fish in the lower level of DO except fishes like Stinging Catfish and Banded Snakehead.

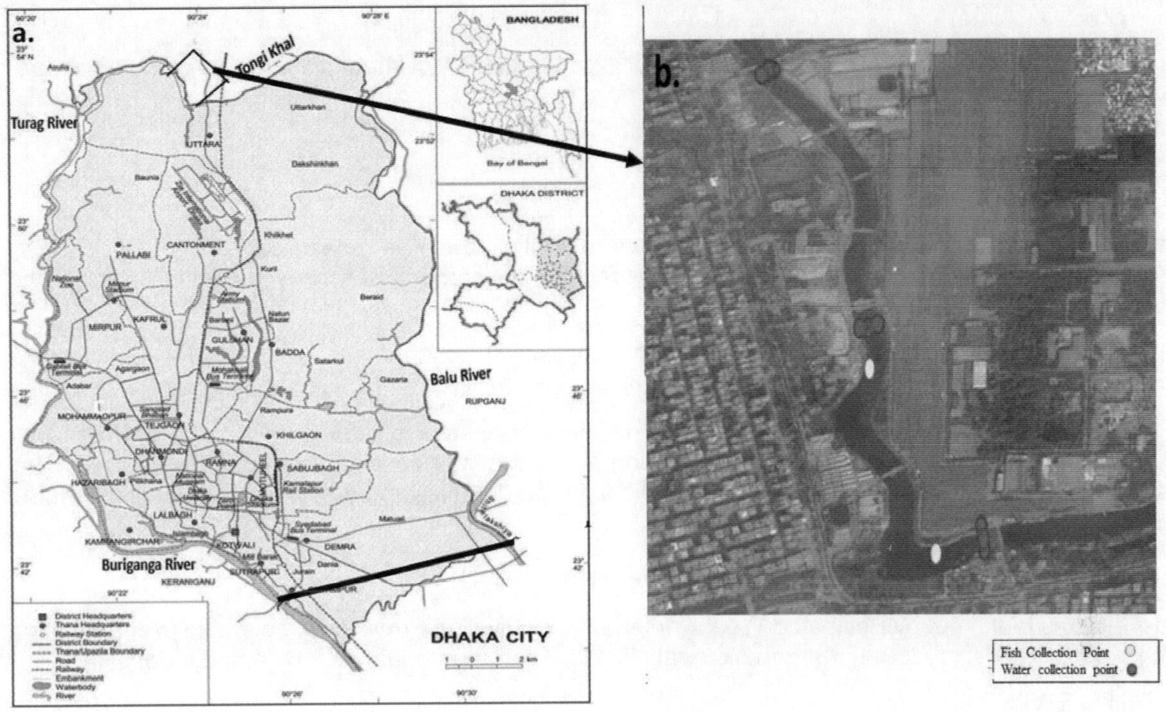


Fig. 1. Location of rivers around Dhaka City (a), Sampling points of Turag River (b)

Therefore, only two types of fishes were found from Turagriver. After collecting, the samples were kept in the refrigerator at -18$^0$C for 3 days after the arrival of environment laboratory, MIST, Dhaka, Bangladesh. The samples were divided into two major parts as bones and edible muscle tissue (flesh) using a steam cleaner stainless steel knife. Samples were washed with distilled water and made oven dry for twenty four hours at temperature 105 degree Celsius. Then it was ground and sieved using a USS #10 sieve to ensure more grinding.

### 2.3 Metal Extraction

1 g dry weight of the sample was taken for digestion. 10 ml of 1:1 nitric acid ($HNO_3$) was added and heated to 105$^o$C and refluxed for 15 minutes. The sample was cooled and 5 ml concentrated $HNO_3$ was added and refluxed for more 30 minutes. The sample was heated at 105$^o$C for two hours and cooled. Then 2 ml of water and 3 ml of 30% $H_2O_2$ was added and heated for two more hours. The sample was cooled and diluted to 100 ml with water and then filtered by the membrane filter (pore size =0.45µm). Finally, solutions were made for testing and preserved in glasses for analysis by AAS. Some dried ground samples were preserved for analysis by SEM-EDS.

### 2.4 Testing methods

The test was carried out according to the method 3050 B (USEPA, 2007) for fish species. Only one season data (in the month of March 2016) were measured. The amount of toxic metal analysis was carried out by atomic absorption spectrophotometer (AAS) (Model AA-6800, made in Japan) and the result was further confirmed by Scanning Electron Microscopy with Energy-dispersive X-ray spectroscopy (SEM -EDS) (JEOL JSM-7600F Field Emission Scanning Electron Microscope (made in USA)).

### 2.5. Bio-concentration factor (BCF)

Bio-concentration factor (BCF) which is the ratio between bioaccumulation of metal in fish to water provides the ability of fish to accumulate chemicals from river water. If the value of BCF >1 indicates the fish has potential to accumulate the metal. It is not considered as significant unless the value is >100. (USEPA, 1991). The formula for BCF:

[1] $BCF = C_{fish}/C_{water}$

Where, $C_{fish}$ = concentration of metal in fish; $C_{water}$ = concentration of metal in river water (Islam et al., 2015)

## 2.6. Health Risk Assessment

### 2.6.1 Estimated Weekly Intake (EWI)

The weekly consumption values for each species were multiplied by the measured value of Cd, Cr, Pb, and As and were divided by the average of an adult female, male and child's body weight in Bangladesh. The equation for EWI (Ozkan and Nuray, 2016):

[2] EWI = weekly fish consumption (g) × concentration of contaminant (µg/g) /body weight (b.w.) (kg).

Fish consumption in Bangladesh per capita per day in urban area is 59.9 g (Bangladesh Bureau of Statistics, 2014). Therefore, weekly fish consumption per capita was considered 419.3 g/week (wk). The average adult female, male (age 14-65 years) and child (age 0-13 years) body weight in Bangladesh are 54 kg, 46 kg and 22 kg, respectively. (Nasreen et al., 2009).

### 2.6.2 Hazard Index (HI) and Total Hazard Index (THI)

Hazard Index is found out by Estimated Weekly Intake (EWI) dividing Provisional Tolerable Weekly Intake (PTWI) (Ozkan and Nuray, 2016; Antonijevic et al., 2011). Therefore, equation for Hazard Index (HI):

[3] HI = EWI/PTWI

Calculated weekly intake for a specific contaminant/provisional tolerable weekly intake (PTWI): Provisional tolerable weekly intake was reported 7 µg/kg body weight (bw)/week (wk) for Cd, 25 µg/kg bw/wk for Pb (Ozkan and Nuray, 2016; JECFA., 2011), 3.5 µg/kg bw/wk for As (Joint FAO WHO Food Standards Programme, 2011) and 304 µg/kg bw/wk for Cr (Islam et al., 2015).

Total Hazard Index (THI) is calculated with the following equations:

[4] THI: $HI_{Cd} + HI_{Cr} + HI_{Pb} + HI_{As}$; HI<1, acceptable risk (AMEC E&I, INC, 2010).

### 2.6.3 Non-Carcinogenic and Carcinogenic risk

Non-Carcinogenic risk is assessed by total hazard Quotient. A hazard quotient provides a ratio of the potential exposure of any substance and the level at which no adverse effects are expected. If the value of Hazard Quotient is less than 1, then it is expected to have no adverse health effects as a result of exposure. If the Hazard Quotient is greater than 1, then it is likely to have adverse health effects. Hazard Quotient exceeding 1 does not mean that adverse effects will occur. The equation used to determine Target Hazard Quotient (THQ):

[5] THQ = (EF ×ED × FIR× C× 0.001)/ (RfD × WAB × TA)

EF is exposure frequency (365 days/year); (USEPA, 1989). ED is the exposure duration (years), over lifetime 70 years for non-carcinogens. FIR is the fish ingestion rate (g/day) which is approximately 59.9 g/day (fish consumption in Bangladesh per capita per day in Urban area 59.9 g) (Bangladesh Bureau of Statistics, 2014) and this consumption rate was used in health-risk assessment. *C* is the metal concentration in fish (µg/g); RfD is the oral reference dose [Cr = $3 \times 10^{-3}$ µg/g/day, Cd = $1 \times 10^{-3}$ µg/g/day, As = $3 \times 10^{-4}$ µg/g/day (USEPA 2016), Pb = $4 \times 10^{-3}$ µg/g/day (Ozkan and Nuray, 2016; Storelli and Barone, 2013)]. WAB is the average body weight. The average adult female, male (age 14-65 years) and body weight of the child (age 0-13 years) in Bangladesh are 54 kg, 46 kg and 22 kg, respectively. (Nasreen et al. 2009). TA is the averaging exposure time for non-carcinogens (365 days/year × ED). Total THQ (TTHQ) of heavy metals in individual fish is the sum of the following equation: [6] TTHQ (individual fish) =THQ (toxicant 1) +THQ (toxicant 2) +THQ (toxicant n)

Carcinogenic risk is estimated as probability of cancer developing over the lifetime exposure to that potential carcinogens. Acceptable risk levels for carcinogens $6\times10^{-6}$ (risk of developing cancer over a human lifetime is 6 in 1,000,000)for adults and $2\times10^{-6}$ (risk of developing cancer over a human lifetime is 2 in 1,000,000)for adolescents (AMEC E&I, INC., 2010). The equation used to determine the carcinogenic risk (CR) (Islam et al., 2015):

[7] $CR = (EF \times ED \times FIR \times C \times CSF_o \times 0.001)/ (WAB \times TA)$

CSFo is the oral carcinogenic slope factor of the integrated risk information system (USEPA, 2016) database, which were 1.5 and 8.5 × 10−3 $(mg/kg\text{-}day)^{-1}$ for As and Pb, respectively.

## 3. Results and discussion

Research on heavy metal concentration was carried for various tissues of fish such as muscles, gills, kidneys and liver (Dhanakumar et al., 2015) but this research was carried out by determining heavy metal concentration of water and fish species of bones and edible tissues as because Bangladeshi people consume only edible tissues and bones. In the following sections, results are mentioned.

### 3.1 Determination of trace metal of River water by AAS

Heavy metal concentration in water of Turagriver is presented in Table 1. In 2016, the average concentration of heavy metals in descending order is: Cr (16.67-20.07 µg/L) >As (8.84-16.2 µg/L) >Pb (6.4-8.56 µg/L) > Cd (0.42-0.84 µg/L). But in 2012, the concentration of heavy metals was, in descending order: As>Cr>Pb>Cd and Cr>As>Pb>Cd respectively. Considering the toxicity Reference Value (TRV) for fresh water (Islam et al., 2015; USEPA, 1999), it is observed in the Turag River that Cr, Cd and As <TRV and Pb>TRV in 2012 whereas in 2016 Cd and As<TRV and Cr and Pb>TRV. From Table 1 it is also clear that four heavy metals increased in 2016 from 2012 in Turag River because of alarmingly growing industries near river Turag. Again, Cr and Pb higher than the TRV value indicate water from this river is not safe for human use (Islam et al., 2015).

Table 1: Heavy metal concentration in river water of Turag (mean±SD)

| Duration | Cd µg/L | Cr µg/L | Pb µg/L | As µg/L | Reference |
|---|---|---|---|---|---|
| March, 2016 | 0.63±0.21 | 18.37±1.70 | 7.48±1.08 | 12.52±3.68 | This study |
| | 0.42-0.84 | 16.67-20.07 | 6.40-8.56 | 8.84-16.2 | |
| Feb-Mar, 2012 | 0.30 ± 0.007 | 11 ± 0.37 | 3.7 ± 0.05 | 12 ± 0.24 | (Islam et al., 2015) |
| | 0.29-0.31 | 10.63-11.37 | 3.65-3.75 | 11.76-12.24 | |
| TRV* | 2.2 | 11 | 2.5 | 150 | |

TRV*= Toxicity Reference Value for fresh water (USEPA, 1999)

### 3.2 Determination of trace metal of fish species by AAS

Four heavy metals of fish species were measured by AAS and the results are presented in Table 2. Among the three samples, highest value was noticed for Cr (2.71 µg/g) which was found in the bone of Banded Snakehead and lowest value was found in the flesh of Banded Snakehead. Ranking order of mean concentration of heavy metals are Cr (2.51 µg/g) >Pb (2.01 µg/g) >As (0.06 µg/g) > Cd (0.045 µg/g), respectively. Stinging catfish was found in the 0.04, 2.66, 2.11 & 0.05 µg/g, respectively for Cd, Cr, Pb and As. The highest value was noticed for Cr (2.66 µg/g) and lowest value was noticed for As (0.05 µg/g). Similarly, trace elements in bone and flesh of banded snakehead in the Turagriver (TTB & TTF) separately were found 0.07, 2.71, 1.59 and 0.06 µg/g and 0.03, 02.65, 1.98 and 0.06µg/g, respectively. In the bone and flesh of banded snakehead maximum value was found in Cr (2.71 and 2.65 µg/g) and minimum value was in As (0.06 µg/g).

Table 2: Concentration of heavy metal in fish species of Turag river

| Sample* | Cd µg/g | Cr µg/g | Pb µg/g | As µg/g |
|---|---|---|---|---|
| TS(B+F) | 0.04 | 2.66 | 2.11 | 0.05 |
| TTB | 0.07 | 2.71 | 1.59 | 0.06 |
| TTF | 0.03 | 2.66 | 1.98 | 0.06 |
| Mean | 0.05 | 2.67 | 1.89 | 0.06 |
| Std Deviation | 0.01 | 0.02 | 0.22 | 0 |
| | **0.03-0.06** | **2.65-2.7** | **1.67-2.11** | **0.05-0.06** |

*TS(B+F)=Turag-Shing-Bone+Flesh, TTB=Turag-Taki-Bone, TTF=Turag-Taki-Flesh

### 3.3 Determination of BCF

BCF is the indicator to evaluate the ability of aquatic organism to accumulate chemicals from river water. The values of BCF of Cd, Cr, Pb and As from Turag river were found to be 74.55, 145.44, 252.89 and 4.57, respectively. In all cases, the values of BCF>1 which indicate from Turag River fish species have the potential to accumulate metals. But for Cr, Pb>100 which indicate a significant value of BCF. Pb was found to be the most potential bio-concentrated element in the fish species.

### 3.4 Determination of HI and THI

It is calculated that consumption of heavy metal weekly is more of the children than adult female and male. Again estimated weekly intake of heavy metals consumed by an adult male, female and child was measured in mean following descending order: Cr (19.5-47.9 µg/g) >Pb (15.6-38.3 µg/g) > As (0.42-1.05 µg/g) >Cd (0.35-0.86 µg/g). (EWI) were compared with Provisional Tolerable Weekly Intake (PTWI) for Cd, Cr Pb and As which was the hazard index to determine risk to human health.World Health Organization established a 'Safe' tolerable intake for the heavy metals which is known as PTWI that is 7, 304, 25 and 3.5 µg/Kg b.w./week for Cd, Cr, Pb and As. HI for heavy metals of Cd, Cr, Pb and As is presented in Table 3. Considering HI for male, female and children is provided in following ranking order: Pb (0.62-1.53) >As (0.12-0.30) > Cr (0.064-0.158) > Cd (0.05-0.124). Since HI of individual heavy metal of Cr, Cd and As<1 except Pb for children (1.53), therefore Pb can be considered as most toxic heavy metal. Pb is well-known as cumulative poison. In human health many effects are observed, including Hematological effects, neurological and behavioral effects, cardiovascular effects, renal effects, and effects on the reproductive system due to accumulation of Pb. Results show that children are more vulnerable to the effects of Pb than adults. Impaired neurobehavioral development was considered to be the most critical effect (IPCS, 1995).Various systemic manifestations such as chronic obstructive and/or restrictive pulmonary disease, chronic lung disease characterized by chronic bronchitis, bronchiectasis; liver diseases such as non-cirrhotic portal fibrosis; polyneuropathy; peripheral vascular disease; hypertension; non-pitting edema of feet/hands; conjunctival congestion; weakness; and anemia are produced due to effect of As.High concentrations of arsenic, greater than or equal to 200 µg/L were found to be associated with a six fold increased risk for stillbirth during pregnancy. Skin, lung, and urinary bladder cancer are the important cancers associated with this toxicity. Specific skin diseases like pigmentation and keratosis occur due to chronic As toxicity. (Guha et al., 2011). Cr release of leather products poses problems to Cr-allergic individuals since such exposure may cause severe dermatitis. (Yolanda et al., 2014). Consumption of Cd contaminated food attacks on kidney which leads to renal dysfunction and formation of kidney stone, the skeletal system and the respiratory system in human health. (IPCS, 1992).In all cases, Children and adult women are most vulnerable (THI>1). Though THI<1 for adult men, but they are not out of danger.

Table 3: Hazard Index (HI = EWI/PTWI) and Total Hazard Index (THI) considering mean body weight of female, male and children

| Sample Number | Sample | Cd PTWI µg/Kg BW/wk | Cd HI | Cr PTWI µg/Kg BW/wk | Cr HI | Pb PTWI µg/Kg BW/wk | Pb HI | As PTWI µg/Kg BW/wk | As HI | THI |
|---|---|---|---|---|---|---|---|---|---|---|
| | | | | Considering Mean Body weight of female | | | | | | |
| 1 | TS(B+F) | 7 | 0.06 | 304 | 0.08 | 25 | 0.77 | 3.5 | 0.13 | 1.04 |
| 2 | TTB | 7 | 0.09 | 304 | 0.08 | 25 | 0.58 | 3.5 | 0.19 | 0.94 |
| 3 | TTF | 7 | 0.04 | 304 | 0.08 | 25 | 0.72 | 3.5 | 0.17 | 1.02 |
| | Mean | | 0.06 | | 0.08 | | 0.73 | | 0.16 | 1.03 |
| | | | | Considering Mean Body weight of male | | | | | | |
| 1 | TS(B+F) | 7 | 0.048 | 304 | 0.07 | 25 | 0.66 | 3.5 | 0.11 | 0.88 |
| 2 | TTB | 7 | 0.048 | 304 | 0.07 | 25 | 0.50 | 3.5 | 0.14 | 0.78 |
| 3 | TTF | 7 | 0.048 | 304 | 0.07 | 25 | 0.61 | 3.5 | 0.13 | 0.85 |
| | Mean | | 0.051 | | 0.07 | | 0.63 | | 0.12 | 0.86 |
| | | | | Considering Mean Body weight of Children | | | | | | |
| 1 | TS(B+F) | 7 | 0.117 | 304 | 0.17 | 25 | 1.61 | 3.5 | 0.28 | 2.17 |
| 2 | TTB | 7 | 0.182 | 304 | 0.17 | 25 | 1.22 | 3.5 | 0.34 | 1.91 |
| 3 | TTF | 7 | 0.085 | 304 | 0.17 | 25 | 1.51 | 3.5 | 0.32 | 2.08 |
| | Mean | | 0.128 | | 0.17 | | 1.44 | | 0.31 | 2.05 |

*TS(B+F)=Turag-Shing-Bone+Flesh, TTB=Turag-Taki-Bone, TTF=Turag-Taki-Flesh

### 3.5 Determination of THQ and TTHQ

Table 4: Target Hazard Quotient (THQ) and Total Target Hazard Quotient (TTHQ) considering mean body weight of female male and children.

| Ser. No | Sample | Cd THQ | Cr THQ | Pb THQ | As THQ | TTHQ |
|---|---|---|---|---|---|---|
| | | | Considering Mean Body weight of female | | | |
| 1 | TS(B+F) | 0.11 | 0.0023 | 0.69 | 0.22 | 1.02 |
| 2 | TTB | 0.17 | 0.0023 | 0.52 | 0.27 | 0.97 |
| 3 | TTF | 0.08 | 0.0023 | 0.64 | 0.25 | 0.98 |
| | Mean | 0.12 | 0.0023 | 0.66 | 0.24 | 1.01 |
| | | | Considering Mean Body weight of male | | | |
| 1 | TS(B+F) | 0.10 | 0.0019 | 0.58 | 0.19 | 0.87 |
| 2 | TTB | 0.15 | 0.0020 | 0.44 | 0.23 | 0.82 |
| 3 | TTF | 0.07 | 0.0019 | 0.55 | 0.22 | 0.84 |
| | Mean | 0.10 | 0.0019 | 0.56 | 0.20 | 0.86 |
| | | | Considering Mean Body weight of Children | | | |
| 1 | TS(B+F) | 0.23 | 0.0048 | 1.44 | 0.46 | 2.13 |
| 2 | TTB | 0.36 | 0.0049 | 1.08 | 0.57 | 2.02 |
| 3 | TTF | 0.17 | 0.0048 | 1.35 | 0.53 | 2.05 |
| | Mean | 0.26 | 0.0048 | 1.29 | 0.52 | 2.07 |

*TS(B+F)=Turag-Shing-Bone+Flesh, TTB=Turag-Taki-Bone, TTF=Turag-Taki-Flesh,

HI and THI were further confirmed by THQ and TTHQ (Table 4). Considering THQ, it was observed that all individual values of Cd, Cr and As were below the reference dose except Pb for children. Considering TTHQ it was observed that TTHQ>1 for women and children. Therefore, reference doses were crossed which indicate likely to cause adverse health effect for children and adult women. Though TTHQ<1 for adult men, which indicate not likely to cause any adverse health effects during life time but it approaches to 1. Therefore, adult male is not out of danger.

### 3.6 Determination of CR

Carcinogenic Risk for As and Pb due to consumption of these fishes were calculated. Mean Values of Pb and As was $2.23\times10^{-5}$ and $1.08\times10^{-4}$ for female, $1.89\times10^{-5}$ and $9.16\times10^{-5}$ for male and $4.97\times10^{-5}$ and $2.43\times10^{-4}$ for children. Cancer risk above $10^{-4}$ is considered unacceptable and below $10^{-6}$ is considered to be negligible (USEPA, 1989, 2010) and risk lying between $10^{-6}$-$10^{-4}$ are generally considered acceptable range (Islam and Ahmed 2014)(Islam et al., 2015). For Pb, carcinogenic risk was within acceptable range ($10^{-6}$-$10^{-4}$) whereas for As the values were found higher than acceptable value. Therefore, potential carcinogenic risk due to consumption of these fish species from these rivers cannot be ignored.

### 3.7 Determination of trace metals by SEM-EDS

The presence of heavy metals is further confirmed by SEM-EDS (Fig. 2).Total 13 heavy metals were selected to confirm the presence of heavy metals. With Turag River,mass% in the fish flesh and fish bone combined (TS-B+F) of *Heteropneustesfossilis* were found in the 4.85, 0, 30.84 and 7.66%, respectively for Cd, Cr, Pb and As. The highest value was noticed for Pb (30.84%) and no value was noticed for Cr.

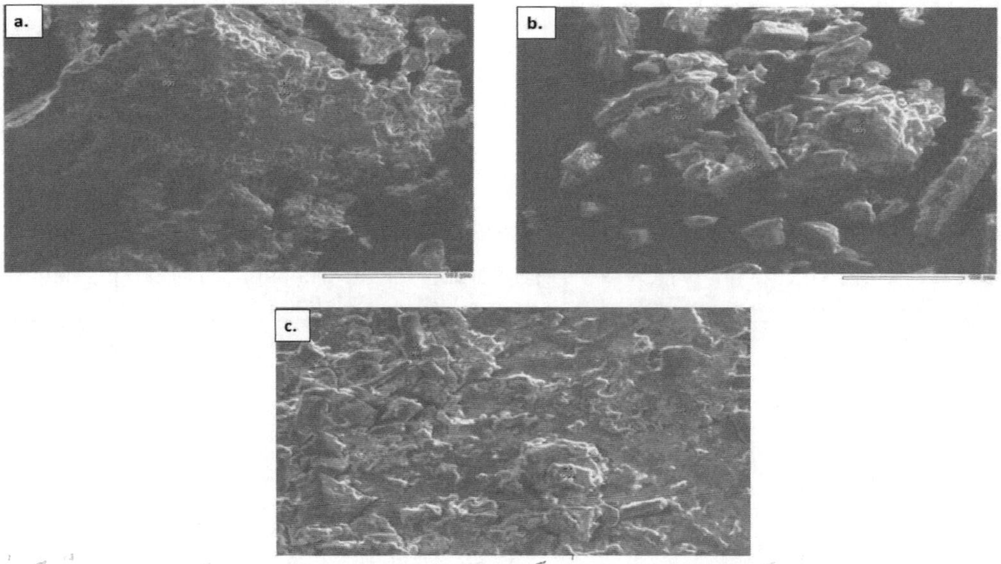


Figure 2. SEM images of surface characterization of sample (magnification 10000 μm).

TS(B+F)=Turag-Shing-Bone+Flesh (a), TTB=Turag-Taki-Bone (b), TTF=Turag-Taki-Flesh (c)

Similarly, trace elements in bone and flesh of *Channa striata* (TTB & TTF) separately were found 0, .48, 27.77 and 2.26% and 9.43, 0, 14.06 and 4.36%, respectively. In the bone and flesh of *Channa striata* maximum value was found in Pb (27.77%) and no value for Cd in bone and Cr in the flesh.

### 4. Conclusions

This study has shown the presence of four heavy toxic metals such as Cd, Cr, Pb and As both in water and fish of Turag River which were measured by AAS. The presence of Cd, Cr, As and Pb were further confirmed by SEM. Among the four heavy metals, Cr was found maximum in both the rivers as because of the presence of tannery and dying industries available near both the rivers and these industries directly discharges effluents into the water without treatment. Bio-concentrated Factor (BCF) shows that Pb is the most bio-concentrated element in both rivers. NCR is mainly associated with the development of hematological, neurological, behavioral, renal and cardiovascular effects. Whereas CR relates to the possibility of cancer development in different organs of the human body. Considering health point of view to human, for NCR, THI>1 for children and female which indicates the potential health risk, whereas THI<1 for male, which indicates less health risk for male in Bangladesh due to consumption of fish species. The health risk was further confirmed by TTHQ which shows identical results for children, adult female and male. Taking CR into account it was observed As is to be hazardous to human health. No step is taken to save Turagriver. More necessary steps such as discharge of effluents into the river after proper treatment of effluent treatment plant is to be ensured from industries. Further studies should investigate the ways to reduce heavy metals up to a tolerable limit acceptable to human health.

## Acknowledgements

This work was supported by Higher Education Quality Enhancement Project (CP-3143) funded by the Ministry of Education, Government of the People's Republic of Bangladesh and World Bank and implemented by University of Grand Commission of Bangladesh. The authors are grateful to the authorities of Military Institute of Science and Technology (MIST), Bangladesh to complete this study. Furthermore, authors are thankful to 7 Riverine Engineer Battalion. Sincerely acknowledged to the technician and engineers of Environment Laboratory, MIST and Department of Glass and Ceramic Engineering, Bangladesh University of Engineering and Technology (BUET) for their support to complete this research.

## References

AccuWeather, Inc. and AccuWeather.com, 2016.

Ahmad, M.K., Islam, S., Rahman, S., Haque, M.R., Islam, M.M., 2010. Heavy metals in water, sediment and some fishes of buriganga river, Bangladesh. *Int. J. Environ. Res. 4*, 321–332.

AMEC E&I, INC. (2010). Updated human health risk assessment report, Olin Mcintosh operable unit 2, 14 November 2011, MciNTOSH, Alabama: Author. Available online at: https://www.epa.gov/sites/production/files/201403/documents/nov_14_2011_revised_part3_updated_human_health.pdf.

Antonijevic, B., Jankovic, S., Curcic, M., Durgo, K., Stokic, E., Srdic, B., Tomic, N.D., 2011. Risk characterization for mercury, dichlorodiphenyltrichloroethane and polychlorinated biphenyls associated with fish consumption in Serbia. *Food Chem. Toxicol.* **49**, 2586–2593.

Al-Busaidi, M., Yesudhason, P., Al-Mughairi, S., Al-Rahbi, W.A.K., Al-Harthy, K.S., Al-Mazrooei, N.A., Al-Habsi, S.H., 2011. Toxicmetals in commercial marine fish in Oman with reference to national and international standards. Chemosphere 85, 67–73.

Bangladesh Bureau of Statistics, 2014. Statistical year book Bangladesh (34$^{th}$ed) (13 January 2016).

Bhuiya, A., Mostafa, G., 1993. Levels and differentials in weight, height and body mass index among mothers in a rural area of Bangladesh. *J Biosoc Sci.* 1993 Jan; 25(1):31-8.

Dhanakumar, S., Solaraj, G., Mohanraj, R., 2015.Heavy metal partitioning in sediments and bioaccumulation in commercial fish species of three major reservoirs of river Cauvery delta region, India. *Ecotoxicol. Environ. Saf.***113**, 145-151.

European Union, 2002. Heavy Metals in Wastes, European Commission on Environment, Available online at: http://ec.europa.eu/environment/waste/studies/pdf/heavymetalsreport.pdf.

FAO (Food and Agriculture Organization of the United Nations), 2016. Fisheries and aquaculture Department, National Aquaculture Sector over view Bangladesh.

FAO (Food and Agriculture Organization of the United Nations), 2016. Fishery and Aquaculture Statistics-2014. Rome.

Guha, M.D., Dasgupta, U.B., 2011. Chronic arsenic toxicity: Studies in West Bengal, India (Review), *Kaohsiung Journal of Medical Sciences*, Volume 27, Issue 9: 360-370.

Islam, M. S., Ahmed, M. K., Al-Mamun, M. H., Shigeki M., 2015. Assessment of trace metals in fish species of urban rivers in Bangladesh and health implications. *Environmental toxicology and pharmacology* 39, 347-357.

Programme on Chemical Safety, 1995. 2000 with reference to Environmental health criteria for inorganic lead, Food Additives Series 44.

JECFA (Joint FAO WHO Expert Committee on Food Additives), 2011. Safety evaluation of certain food additives and contaminants. WHO Food AdditSer 64.

Joint FAO WHO Food Standards Programme, 2011. Codex Committee on Contaminants in foods. Fifth Sessions.

Khadse, G.K., Patni, P.M., Kelkar, P.S., Devotta, S., 2008. Qualitative evaluation of Kanhanriver and its tributaries flowing over central Indian plateau. Environ. Monit. Assess. 147, 83–92.

Lee, K., Kweon, H., Yeo, J., Woo, S., Han, S., Kim, J.H., 2011. Characterization of tyrosine-rich Antheraeapernyi silk fibroinhydrolysate. Int. J. Biol. Macromol. 48, 223–226.

Malik, N., Biswas, A.K., Qureshi, T.A., Borana, K., Virha, R., 2010. Bioaccumulation of heavy metals in fish tissues of a fresh water lake of Bhopal. *Environ. Monit. Assess.* **160**, 267–276.

N. Mansouri, N. Khorasani, A.R. Karbassi , B. Riazi, M. Panahandeh , 2013, Assesing Human Risk of Contaminants in Anzali Wetland Fishes, International Journal of Application or Innovation in Engineering & Management (IJAIEM), Volume 2, Issue 11.

Nasreen I. K., David B., Ravi N., Gary O., 2009. Implementation of food frequency questionnaire for the assessment of total dietary arsenic intake in Bangladesh: Part B, preliminary findings. Environ. Geochem. Health. 31, 221–238.

Ozkan, O., Nuray, E., 2016. Evaluation of risk characterization for mercury, cadmium, lead and arsenic associated with seafood consumption in Turkey. Expo Health 8, 43–52.

Rahman, M.S., Islam, M.R., 2010. Adsorption of Cd (II) ions from synthetic waste water using maple sawdust. Energy Sources Part A: Recov. Util. Environ. Effects 32, 222–231.

Storelli, M.M., Barone, G., 2013. Toxic metals (Hg, Pb, and Cd) in commercially important demersal fish from Mediterranean Sea: contamination levels and dietary exposure assessment. *J Food Sci* **78**(2), 362–366.

UNEP (United Nations Environment Programme), 2001. Bangladesh State of the Environment Report p-48.

USEPA (US Environmental Protection Agency), 1989. Risk Assessment Guidance for Superfund, vol. I: Human Health Evaluation Manual. EPA/540/1–89/002, Office ofEmergency and Remedial Response, Washington, DC.

USEPA (US Environmental Protection Agency), 1999. Screening Level Ecological Risk Assessment Protocol for Hazardous Waste Combustion Facilities, vol. 3, Appendix E: Toxicity Reference Values. EPA 530-D99-001C,http://www.epa.gov/epaoswer/hazwaste/combust/eco-risk/voume3/appx-e.pdf.

USEPA (US Environmental Protection Agency), 2010. Risk-Based Concentration Table, Available at: <http://www.epa.gov/reg3hwmd/risk/human/index.htm>.

USEPA (US Environmental Protection Agency), 2016. Regional Screening Levels (RSLs) - Generic Tables (May 2016) Summery Table (TR=1E-6, THQ=1).

Yolanda S. H., Carola L., Inger O. W., 2014. Correlation between bulk- and surface chemistry of Cr tanned leather and the release of Cr (III) and Cr (VI).Journal of Hazardous Materials 280: 654–661.

Zarith S. B., Mohd Y. I., 2015. Determination of heavy metal accumulation in fish species in Galas River, Kelantan and Beranang mining pool, Selangor. Procedia Environ. Sci. 30, 320 – 32.

Leadership in Sustainable Infrastructure
*Leadership en Infrastructures Durables*

**Case Study**

Vancouver, Canada
May 31 – June 3, 2017/ *Mai 31 – Juin 3, 2017*

# BIOMASS WASTE-DERIVED ADSORBENTS FOR THE REMOVAL OF ORGANIC POLLUTANTS FROM WATER

An, Chunjiang[1,2], Huang, Gordon[1,2], Chen, Xiujuan[1], Shen, Jian[1], and Shi, Yarong[1]
[1] Institute for Energy, Environment and Sustainable Communities, University of Regina, Regina, S4S 0A2, Canada
[2] Faculty of Engineering and Applied Science, University of Regina, Regina, S4S 0A2, Canada
[3] chunjiang.an@uregina.ca

## 1. BACKGROUND

Many physical, chemical and biological approaches, such as coagulation/flocculation, advanced oxidation and membrane filtration have been used in the removal of pollutants from wastewater. Adsorption is an effective alternative for wastewater treatment and it has been widely employed to remove hazardous inorganic and organic pollutants present in the effluent. It is featured with some advantages such as simple design and toxic tolerance. A number of adsorbents including activated carbon, diatomite, clay, and adsorptive resin have been used for the treatment of wastewater. Biomass adsorbents derived from biomass wastes such as orange peel, rice husks, peanut husk, corn cobs and wheat straw are also applied in pollution control.

Flax biomass is often discarded in the land or incinerated outdoors, which not only result in a waste of bioresources, but also lead to environmental problems with inappropriate disposal. Flax biomass contain a certain amount of lignin and cellulose, showing the potential as adsorbent material. However, the adsorption of organic pollutants by flax biomass is not studied in previous works. Moreover, biomass wastes can be used to produce biochar. Biochar is carbon-rich material which can be obtained through pyrolyzing biomass at a given temperature. Biochar is featured by unique morphology and structure. It has the capacity to retain molecules of various sizes and chemical characteristics. Compared with other carbon-related adsorbents, biochar is considered as a cost-effective alternative for immobilizing organic contaminants such as polycyclic aromatic hydrocarbons, polychlorinated biphenyls, pentachlorophenol, imidacloprid and atrazine. This study investigated the removal of organic pollutants from water using biomass-derived adsorbents. Both equilibrium and kinetic adsorption studies were conducted and the effects of aqueous chemistry on adsorption performance were also explored. The results of this study have important implications for the future application of biomass adsorbent in pollution control and prevention.

## 2. RESULTS

When flax biomass is placed in water, it usually possesses a net negative charge as a result of ionization of some of its surface functional groups. Cationic gemini surfactants may interact with limited and specific anionic sites at the surface. Due to the high charge density of gemini surfactant with short spacer, they can bind strongly with the anionic groups on the flax biomass surface. Moreover, the action of short-range noncoulombic attractive van der Waals forces and hydrophobic interactions can also contribute to the binding process. The hydrophobic tails in gemini molecules could present strong hydrophobic forces to some function groups on flax surface. The surface of modified flax biomass was characterized by synchrotron infrared and SEM analysis. The absorbed amount of some pollutants such as azo dyes varied with the change of adsorbent dosage, pH and ionic strength. The adsorption isotherm data well fit to the Langmuir model. The adsorption process followed the pseudo-second-order kinetics and the liquid film diffusion models. Thermodynamic studies indicated that the adsorption process was spontaneous.

The adsorption of Tetrabromobisphenol A (TBBPA) on pinecone-derived biochars was also investigated. The surface structures and functional groups of biochars produced at different temperatures were characterized through synchrotron-assisted FTIR analysis. The adsorption isotherms were well described by the Langmuir model. The adsorption capacity of TBBPA varied as pH and TBBPA initial concentration changed. The influences of some inorganic ions such as ammonium, phosphate and nitrate on the adsorption of TBBPA by biochars were revealed through fractional-factorial assisted analysis. The results indicated the main effects include negative effects of phosphate, positive effects of ammonium and insignificant effects of nitrate ions and there are interactions among these influencing factors. These results are useful for developing economically viable and environmental friendly approaches for the removal of organic pollutants from water.

**Acknowledgements**
This research was supported by the Natural Science and Engineering Research Council of Canada and the Canada Foundation for Innovation (CFI). The authors are also thankful to the Canadian Light Source for providing support in synchrotron infrared measurements and analysis.

**References**

Ahmad, A., Mohd-Setapar, S.H., Chuong, C.S., Khatoon, A., Wani, W.A., Kumar, R., Rafatullah, M. 2015. Recent advances in new generation dye removal technologies: novel search for approaches to reprocess wastewater. *RSC Advances*, **5**: 30801-30818.

An, C., Yang, S., Huang, G., Zhao, S., Zhang, P., Yao, Y. 2016. Removal of sulfonated humic acid from aqueous phase by modified coal fly ash waste: Equilibrium and kinetic adsorption studies. *Fuel*, **165**: 264-271.

An, C., Huang, G. 2012. Stepwise adsorption of phenanthrene at the fly ash-water interface as affected by solution chemistry: Experimental and modeling studies. *Environmental Science & Technology*, **46**: 12742-12750.

An, C., Huang, G., Wei, J., Yu, H. 2011. Effect of short-chain organic acids on the enhanced desorption of phenanthrene by rhamnolipid biosurfactant in soil-water environment. *Water Research*, **45**: 5501-5510.

An, C., Huang, G., Yu, H., Wei, J., Chen, W., Li, G. 2010. Effect of short-chain organic acids and pH on the behaviors of pyrene in soil-water system. *Chemosphere*, **81**: 1423-1429.

Baskaralingam, P., Pulikesi, M., Elango, D., Ramamurthi, V., Sivanesan, S. 2006. Adsorption of acid dye onto organobentonite. *Journal of Hazardous Materials*, **128**: 138-144.

Li, H., Huang, G., An, C., Zhang, W. 2012. Kinetic and equilibrium studies on the adsorption of calcium lignosulfonate from aqueous solution by coal fly ash, *Chemical Engineering Journal*, **200-202**: 275-282.

Lin, L., Zhai, S., Xiao, Z., Song, Y., An, Q., Song, X., 2013. Dye adsorption of mesoporous activated carbons produced from NaOH-pretreated rice husks, *Bioresource Technology*, **136**: 437-443.

Kousha, M., Daneshvar, E., Sohrabi, M.S., Jokar, M., Bhatnagar, A. 2012. Adsorption of acid orange II dye by raw and chemically modified brown macroalga Stoechospermum marginatum. *Chemical Engineering Journal*, **192**: 67-76.

Mahmoud, D.K., Salleh, M.A.M., Karim, W.A.W.A., Idris, A., Abidin, Z.Z. 2012. Batch adsorption of basic dye using acid treated kenaf fibre char: Equilibrium, kinetic and thermodynamic studies, Chemical Engineering Journal, **181-182**: 449-457.

Sain, M., Fortier, D., Lampron, E. 2002. Chemi-refiner mechanical pulping of flax shives: refining energy and fiber properties, Bioresource Technology, **81**: 193-200.

Yu, J.X., Li, B.H., Sun, X.M., Yuan, J., Chi, R.A. 2009. Polymer modified biomass of baker's yeast for enhancement adsorption of methylene blue, rhodamine B and basic magenta. *Journal of Hazardous Materials*, **168**: 1147-1154.

# BIOTRANSFORMATION OF AROMATIC COMPOUNDS IN OIL SANDS PROCESS-AFFECTED WATER BY PSEUDOMONADS

Yanyan Zhang [1, 4], Jinkai Xue [2], and Mohamed Gamal El-Din [3,]
[1,2,3] Department of Civil and Environmental Engineering, University of Alberta, Edmonton, Alberta T6G 1H9, Canada
[4] yanyan2@ualberta.ca

The potential environmental impact of oil sands process-affected water (OSPW), contained in storage ponds until eventual release after treatment, has received extensive attention in recent years. Biological treatments, which exploit degradation by microorganisms, have been used successfully for wastewater reclamation in various industries. However, previous work has demonstrated the recalcitrance of OSPW to biological treatment. Aromatic naphthenic acids (NAs) in OSPW have been proved to cause more toxicity of OSPW than classic NAs. No study has yet been conducted for the biodegradation of aromatic compounds in OSPW, although synthetic aromatic alkanoic NAs have been used to investigate the mechanisms involved in aromatic NA degradation. *Pseudomonas fluorescens* and *Pseudomonas putida* have been demonstrated to have the capacity to degrade commercial NAs; however, no classical NAs in OSPW could be degraded. As Pseudomonads have showed the capability to biodegrade aromatic compounds, those two Pseudomonas species were used to treat aromatic compounds in OSPW in this study. Furthermore, biotransformation by Pseudomonads was also employed to remove the residual toxicity after ozonation. The objective of the current study was to determine the potential biodegradation of aromatic compounds in OSPW by *P. fluorescens* and *P. putida*. This degradation was monitored by assessing the synchronous fluorescence spectroscopy (SFS) spectrum before and after treatment to determine the biodegradation of aromatic compounds as described previously. Due to the recalcitrance of organic compounds in OSPW, an external carbon source was added to provide adequate bacterial nutrition. In addition, various concentrations of this external carbon source, and an external iron source, on biodegradation were evaluated. Furthermore, the role of the Pseudomonads in removing the residual toxicity of ozonated OSPW was also investigated.

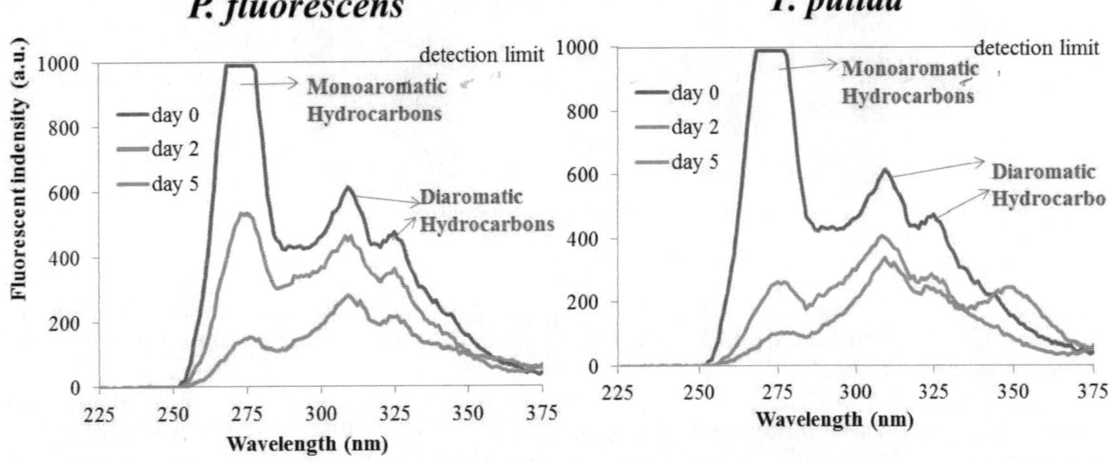


Fig. 1. The degradation of aromatic compounds after exposure to OSPW as indicated by the changes of SFS spectrum without the addition of $FeCl_3$ [1]

After exposure to raw OSPW, significant bacterial growth was observed for both *P. putida* and *P. fluorescens*. In the presence of $FeCl_3$, more biomass was produced as compared to those without $FeCl_3$ treatments. The OSPW SFS spectrum (Figure 1) revealed a maximum peak at 273 nm (one aromatic ring) and other peaks at 310 nm and 324 nm (two aromatic rings), which is consistent with previous OSPW work. Rowland et al. [2] indicated that these peaks around 282 nm and 320 nm coincided with dehydroabioatic acid (a specific monoaromatic compound) and 1,2,3,4-tetrahydrophenanthrene 1,2-dicarboxylic acid, dimethyl ester (a specific diaromatic acid), suggesting that those peaks could be attributed to mono and diaromatic compounds, respectively. The SFS spectra changed over the 5-day treatment period indicating the biodegradation of aromatic compounds (Figure 1). The intensity of the monoaromatics peak at 273 nm declined more significantly than the multiple-rings aromatics peaks at 310 nm and 324 nm. This indicates that the monoaromatics are more rapidly biodegraded which is in agreement with a previous study. For the killed controls, the SFS spectra did not change, indicating that aromatic compounds were not being removed due to biomass adsorption. To further confirm the degradation of aromatic compounds, GC-MS was applied to determine the BTEX, PAHs and PAH and petroleum hydrocarbon (PHC) fractions after treatment. After biodegradation, PHC fraction F2 (C10-C16) became undetectable after biodegradation that indicates the effectiveness of Pseudomonads degradation of this fraction.

Results of the Microtox® *V. fischeri* assay (Figure 2) showed that all Pseudomonads treatments have significantly reduced toxicity of raw OSPW (P value <0.05). This was unexpected as there was no significant reduction in the NAs after biodegradation of raw OSPW without $FeCl_3$ addition, which suggests that the toxicity reduction must be due to the degradation of other compounds. As indicated previously, biodegradation reduced the overall aromatics as indicated by SFS and FT-IR results. In addition, this reduced toxicity may be the result of the biotransformation, versus complete biodegradation, of the NAs to less toxic compounds in these treatments. The addition of $FeCl_3$ improved the toxicity reduction for both Pseudomonads for both raw and ozonated OSPW treatments. Although an ozonation dose of 80 mg/L reduced the classical NAs concentrations in OSPW by more than 90%, the toxicity of ozonated OSPW was still significant (16.7% effect to the luminescence of *V. fischeri*). This result further indicates that classical NAs may contribute a smaller portion of the OSPW toxicity than previously considered, making the toxicity reduction performance of ozonation more limited. However, the biodegradation of the ozonated OSPW showed promising results with decreases in the toxicity by both Pseudomonads with and without $FeCl_3$ addition (P value < 0.02). Of note is the elimination of toxicity after biodegradation for both species in the presence of $FeCl_3$. This indicates that biodegradation is an excellent candidate for removing the residual toxicity after ozonation.

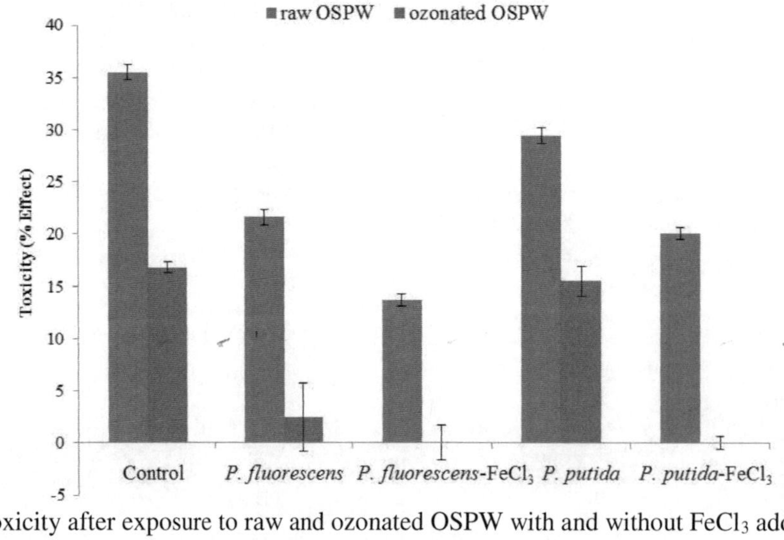


Fig. 2. Microtox® toxicity after exposure to raw and ozonated OSPW with and without $FeCl_3$ addition on Day 2. [1]

A limitation of recalcitrant compounds' biodegradation may be attributed to the lack of available carbon for nutrition. To investigate this limitation, the SFS peaks including 273, 310 and 324 nm (discussed previously) of various sodium acetate concentrations were considered for the Pseudomonads over a 4-day duration for treatments. When the concentration of sodium acetate was less than 0.5 g/L, only limited biodegradation was observed indicating that the external carbon source played an important role in mitigating ring cleavage. When the concentration of sodium acetate was higher than 1 g/L, biodegradation of aromatic compounds started to increase with increasing acetate concentration and the highest reduction of peak intensities were observed when the acetate concentration was 4 to 5 g/L. Interestingly, the results of toxicity tests showed that the addition of acetate greater than 1 g/L did not further reduce toxicity despite the continued reduction in peak intensities.

In summary, *P. fluorescens* and *P. putida* showed their capability in the biodegradation of OSPW aromatic compounds, especially for the petroleum hydrocarbons (C10-C16). Although no significant NAs were degraded during this process, the toxicity of OSPW was reduced significantly after biodegradation. The external carbon source played an important role in ring cleavage. The addition of iron could accelerate the growth of biomass and reduction of toxicity. Our study indicates biotransformation by Pseudomonads is a complementary approach to ozonation to eliminate the residual toxicity of ozonated OSPW.

**Reference**

1. Zhang, Y., K.N. McPhedran, and M.G. El-Din, *Pseudomonads biodegradation of aromatic compounds in oil sands process-affected water.* Science of the Total Environment, 2015. **521**: p. 59-67.
2. Rowland, S.J., et al., *Steroidal aromatic Naphthenic Acids in oil sands process-affected water: Structural comparisons with environmental estrogens.* Environmental Science and Technology, 2011. **45**(22): p. 9806-9815.

Leadership in Sustainable Infrastructure
*Leadership en Infrastructures Durables*

Vancouver, Canada
May 31 – June 3, 2017/ *Mai 31 – Juin 3, 2017*

# CIGARETTE BUTT WASTE MANAGEMENT AND SMOKING POLICY ON CAMPUS

Ng, Kelvin T.W.[1], and Richter, Amy[2]
[1,2] Faculty of Engineering and Applied Science, University of Regina, Canada
[2] kelvin.ng@uregina.ca

**Abstract:** Indoor smoking-free policies for educational institutions and universities have been widely adopted in Canada. However, illegal littering of cigarette butts is frequently observed at campus building entrances, benches, bus stops, and other open areas beside designated smoking areas. Discarded cigarette butts are largely non-biodegradable, and the environmental impacts associated with these toxic wastes are not fully understood by the research community and policy makers. Previous research focuses largely on people's attitudes towards littered cigarette butt waste, but studies on the actual amount of cigarette butts littered on campus are very limited. The objectives of this study are to develop methods to quantify littered cigarette butts at the University of Regina main campus, and to examine smokers' behaviors and their potential impacts on University smoking policies. A total of seventeen hotspots on campus were identified, and the number of cigarette butts were estimated using three square reference frames with different sizes (0.5 m×0.5 m, 0.75 m×0.75 m, and 1 m×1 m). Preliminary trials suggested the mid-sized frame (0.563 $m^2$) was more appropriate, and was adopted for subsequent data collection. Field data was collected from Regina main campus from May to August 2016. It is estimated that there were about 20,500 littered cigarette butts on campus at any given time. Using provincial statistics, the daily generation rates was estimated to be 4,200 CB/d. This study provides some of the first cigarette butt waste field data at a Canadian campus. Such data and analyses likely provide a solid framework for updating our current smoking policy.

## 1 Introduction

Cigarette butts (CBs) pose an interesting challenge for disposal, since there are millions of points of disposal; and because of the requirement to segregate, collect and dispose of the butts in a safe manner, and the toxic and hazardous nature of the waste (Barnes, 2011). In spite of many anti-litter events and campaigns, smokers continue to litter across the globe. According to a recent US survey with a sample size of 2,000 respondents, over half of American smokers (55.7%) reported littering their cigarettes butts (CBs) on the ground, in a sewer/gutter, or down a drain in the past month (Rath et al., 2012). Micevska et al. (2006) estimated that 24-32 billion CBs were littered annually in Australia. Similar results were obtained in a large-scale street-littering study in the Middle East. Arafat et al. (2007) interviewed 1,000 residents in West-bank-Palestinian Territories and found that CBs represent 30% of littered items that respondents regularly litter. Becherucci et al. (2014) reported that about 43% of the urban litter collected in their field study in Mar del Plata, Argentina was CBs. A field study in Ueda City, Japan revealed that CBs represented more than 70% of street waste from drivers and pedestrians (Moriwaki et al. 2009). In fact, waste audits done in North America showed that CB waste accounted for 10% of all waste (Marah and Novotny, 2011) and CB waste is the most common waste worldwide (Sawdey et al., 2011). Published studies on Canadian solid waste management are mostly focused on efficiency of management system and impacts of waste policy and program (Wang et al. 2016; Bruce et al. 2016; Richter et al 2017), studies on littering of CBs are very limited. However, CBs can be commonly observed in storm drains, sidewalks, parks and entrance of buildings in many Canadian towns and cities. Owing to their weight, CBs and other light-weight litter dropped to the ground will likely be carried away by runoff water into streams, river and lakes (Arafat et al. 2007).

Potential fire hazards associated with CB litter are identified and discussed by a number of studies (Arafat et al. 2007; Rath et al. 2012). In addition, CBs are toxic to the environment and harmful to the ecosystem if not disposed properly. Most CBs are made of cellulose acetate and are degradable only under a set of ideal conditions (Puls et al., 2011; Novotny et al., 2009). However, chemicals retained by the CB filters such as heavy metals and polycyclic aromatic hydrocarbons (Micerska et al., 2006; Mohajerani et al., 2016) may adversely impact local aquatic ecosystems and other sensitive environmental receptors (Novotny et al., 2009; Lee and Lee 2015). For example, experimental work by Moriwaki et al. (2009) suggested that arsenic would leach out from CBs into the environment through rainwater at a concentration level exceeding local water quality requirements.

In their study, Arafat et al. (2007) found that the level of education attained correlated well to the type of littering items, and CBs were the most common litter item among people with post-graduate education. They also found that CBs and glass bottles were commonly littered among respondents aged 21-30 years (Arafat et al., 2007). According to Statistics Canada (2010), in 2006 over 75% of college and university students were between 17 and 27 years of age and over 90% were under age 40. As such, college and university campuses remain an appropriate study field for Canadian CB littering studies on smoking policy. Levy et al. (2012) developed a numerical model based on SimSmoke to examine the effects of various smoking policies over time in Finland. They found that smoke-free policies such as smoking restrictions in the workplace yielded a 6% reduction in the prevalence rate of smoking in all ages and both genders. A study of littered cigarettes at a University in the Southeastern US found that in 30 days, 7861 CBs were littered, despite having a no-smoking area within 25 ft (7.62 m) of entrances. There was very low compliance with non-smoking policies, most likely because there was no enforcement of this type of policy (Seitz et al., 2012). In fact, only when there were 100% smoke free policies at colleges, was any reduction in CB waste noted. Lack of signage and placement of butt receptacles in non-smoking areas are likely a reason for decreased compliance, likely due to lack of awareness and conflicting messaging (Lee et al., 2013).

The long term impacts of smoking are well documented and reported in literature. However, it is found that the data on CB litter on Canadian campus were largely unknown. In this study, University of Regina main campus is selected as a study area. The objectives of the project are (i) to develop method to quantify littered cigarette butts at the University of Regina main campus, and (ii) to examine smokers' behaviors and their potential impacts on University smoking policies.

## 2 Methodology

### 2.1 Study area

University of Regina (U of R) is a publicly funded, mid-sized educational institution located in the capital city of Saskatchewan. In the 2015-2016 academic year, the U of R reported serving 14,360 students in degree programs and continuing education courses, with about 3,000 staff (Office of Resource Planning, 2016). The University and the Federated Colleges offered programs in 10 faculties. The gender ratio in 2015 Winter semester was 39 male students to 61 female students, which was assumed to be representative of the gender ratio during the 2016 summer semester. The majority (78%) of the students were full time students, and about 13.7% of the total students were international students (Office of Resource Planning, 2016). The study area selected was located at the main campus for a total area of about 23.3 ha (233,000 $m^2$), as shown in Figure 1. A total of 14 buildings were included in the study area and the buildings were mostly interconnected, owing to long and often harsh Saskatchewan winters. The summer was a logical choice for data collection, since quantification would have been difficult during other seasons when snow, leaves, and other obstructions would lead to errors in quantification.

Smoking policies for educational institutions and universities have been widely adopted in Canada. The four most commonly adopted policies are: (i) 100% smoke-free indoors and outdoors, (ii) perimeter policies prohibiting smoking, (iii) designated smoking area policies, and (iv) smoke-free indoors only. U of R currently employs 3 out of these 4 measures (Policy, 2016). Smoking and the use of tobacco products is prohibited in all U of R buildings, vehicles, outdoor University areas used for sports, meetings or other gatherings. Smoking is also prohibited within 6 m of all ventilation intakes, doors and windows. About 26 designated smoking areas have been established for smoking and tobacco use on campus, among them, 15 areas are located inside the study area (within the red box, Figure 1). Receptacles are provided in these designated areas to limit the drifting of second hand smoke indoors. Advertising of smoking and tobacco products within the U of R is also prohibited (Policy, 2016). However, approved tobacco products may be burned as part of religious and spiritual ceremonies, or in a controlled environment for scientific research purposes. The hotspots chosen in this study are denoted with a red dot, and their location ID written in black beside each.

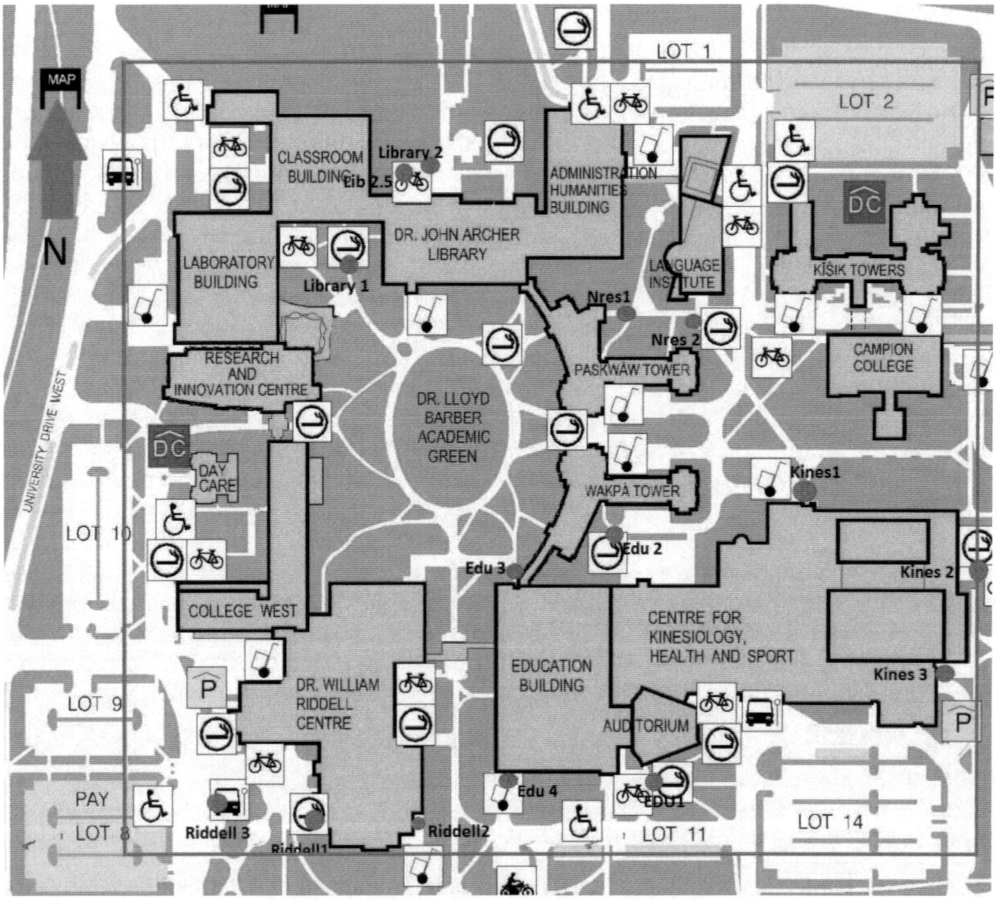

Figure 1: U of R Main campus and study area (modified from Campus map 2016)

## 2.2 Definitions and preliminary trials

In this study, we define CBs as any butt or disposable filter of a smoked tobacco products including cigarettes, cigars and pipes. Only CBs littered outdoor were considered in the present study. A total of three preliminary trials were conducted in May 2016 (i) to identify the number of "hotspots" within the study area, and (ii) to evaluate the performance of the reference frames. Hotspots are defined as a small area (typically about 0.7m×0.7m or smaller) with a relatively high concentration of CB litter using observations and visual comparisons. A total of 16 hotspots were identified on paved surfaces, open fields, and surface drainage devices in the 3 trials. CBs from or very near (<0.1 m) University smoking receptacles and designed smoking areas were excluded in this study to avoid counting of the unintentional CB litter.

In their study, Lee et al. (2013) manually collected and stored CBs using gloves and multiple layers of store bags. To minimize health and safety concerns of field staff, we have developed a CB quantification technique using a reference frame method to quantify the number of CBs at each hotspot. Lightweight and durable PVC pipes and nylon string were used to construct the weatherproof reference frames. The sizes of the small, medium and large reference frames were 0.5 m×0.5 m, 0.75 m×0.75 m, and 1 m×1 m, respectively. Figure 2 shows a picture of the medium frame with the CBs. Three locations (Edu1, Edu2 and Edu3) with different pedestrian traffic, distribution of litter and architectural features near the Education Building were used to evaluate the performance of the reference frames. The Education building was selected due to its mixed occupancy use (wet labs, computer labs, student lounge, library, staff office, meeting space, lecture halls, classrooms) and a decent distribution pattern of CBs. The research assistant feedback and CB estimates from the frames were evaluated and examined in the preliminary trials.

Figure 2: The use of the medium reference frame for CB quantification (Picture taken on August 3, 2016, location: Library_1).

## 2.3 Data collection and analysis

In addition to the preliminary trials, a total of 12 field measurements were conducted from May 25 to Aug 10, 2016 using standardized data collection forms and the proposed CB quantifying techniques. Detailed campus maps and a small handheld GPS device (Garmin eTrex 10) were provided to each field staff to ensure the accuracy of sampling locations. The reference frame was placed directly on top of the hotspot and oriented according to the CB spatial distribution (such that the numbers of CBs observed within each of the four quadrants were similar). Photographs were taken at each hotspot using a HTC Desire 601 camera with a minimum picture resolution of 1456×2592 pixels. The use of photography in data collection minimized human errors due to fatigue of the field staff and provided an easy reference for data verification. Pictures were brought back to office for data analysis, where the number of cigarettes in each picture was counted and recorded in the standardized data collection forms. MS Picture Manager was used to adjust the contrast and saturation of the pictures to (i) eliminate sun glaze or reflections of moist surfaces, and to (ii) facilitate CB counting on difficult backgrounds such as dirty and partially degraded CBs on grey gravel. The weather conditions (wind speed, precipitation, temperature, etc.) and solar intensity were also recorded for each field measurement as they may affect the number and distribution of CBs at hotspots.

## 3 Results and Discussion

### 3.1 Number of hotspots and reference reframes

A total of 16 hotspots were identified in preliminary trials. Observations from preliminary trials suggested that the areas of the hotspots ranged from tiny (about 0.3 m×0.3 m, commonly found near CB receptacles or around designated smoking areas) to huge (over 6.0 m×6.0 m, mostly from parking spaces, benches and bus stops), with a majority of the hotspots less than 1.0 m×1.0 m. As such, 3 different reference frames were designed and constructed to quantify the CB data. Table 1 compares the results from these frames. The average numbers of CBs for small, medium and large frames were 21.0, 33.7, and 38.0, respectively. To further explore the relationships between the number CBs and the frame sizes, the CB densities (defined as the number of CBs per unit frame area) were plotted against the frame sizes on Figure 3.

Table 1: Number of CBs collected at 3 different locations

| Location ID | Small frame (Area=0.25 $m^2$) | Medium frame (Area=0.563 $m^2$) | Large frame (Area=1.00 $m^2$) | Average |
|---|---|---|---|---|
| Edu1 | 13 | 18 | 23 | 18.0 |
| Edu2 | 27 | 33 | 37 | 32.3 |
| Edu3 | 23 | 50 | 54 | 42.3 |
| Average | 21.0 | 33.7 | 38.0 | |

The CB densities were found to be sensitive to hotspot locations, similar to the findings from Marah and Novotny (2011), who developed a weighted GIS overlay to determine geographic patterns of CB waste in urban areas. For example, the average CB density of Edu1 was 35.7 CB/$m^2$, less than half (about 45.6%) of the average Edu3 CB density. For frame areas ranging from 0.5 $m^2$ to 1.0 $m^2$, the relationships were consistently tending towards being linear at all locations. However, the trends were inconclusive for frame areas less than 0.5 $m^2$. The slope of the Edu3 curve was flat and horizontal when the area of the reference frame was less than 0.5 $m^2$, and the opposite trend was observed for Edu1 and Edu2. The finding was consistent with the field observation that a hotspot size of 0.7 m×0.7 m was most abundant within the study area. The use of the small reference frame (0.5 m×0.5 m) is not recommended due to the difficulties associated with selection of a representative frame location and orientation on the hotspot. The medium frame with an area of 0.563 $m^2$ was selected for further study due to its weight and ease of use for frame orientation and data measurement.

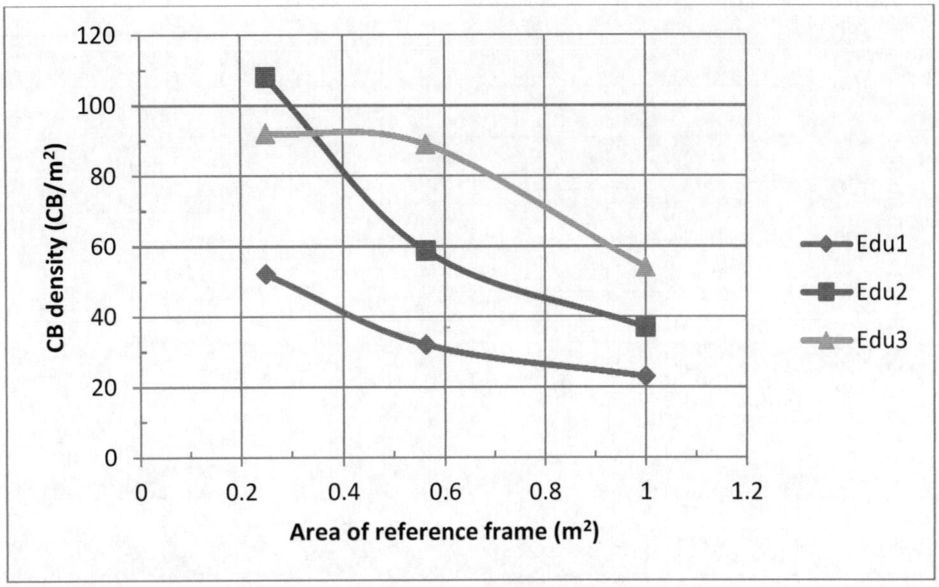


Figure 3: Relationships between CB density and reference frame area

## 3.2 Number of CBs and indoor smoking policy

Using the medium reference frame, CBs at the hotspots were measured. During the study period, one additional hotspot was added to better capture smokers' behaviours and CB littering practices. There were also times where no data was recorded as either it was being occupied by a group of smokers, or objects such as construction materials blocked the frame from being correctly placed. The mean (calculated by taking the average of the number of CBs at each location for each of the 12 days when data was recorded over the study period) and standard error (calculated by comparing the mean to the deviation of each number, and then squaring the result) of CB densities at the 17 hotspots are shown in Figure 4. For most locations, the numbers of CBs were quite consistent during the 3-month study period at about 38.3 CB/$m^2$. The CB densities at Library1, Library2, and Kines3, however, were noticeably higher, with values ranging from 100 to 140 CB/$m^2$. On the other hand, Figure 1 showed that the amount and distribution of nearby designated smoking areas and receptacles were similar to other buildings. The elevated CBs may be explained by the building usage and occupancy. Unlike other buildings on campus, Dr. John Archer Library (Library1 and Library2) and Centre for Kinesiology, Health & Sport building (Kines3) have long hours of operation including both weekdays and weekends, and are regularly visited by off campus guests and public for various educational and recreational programs. It is hypothesized that the CBs found in these areas may originate from off campus visitors.

It is not clear why considerably lower amount of CBs were observed in Riddell2 and N.Res1. The CB densities were about 18-20 CB/$m^2$, less than half of the average Campus amount. Architectural features such as stairs, glass walls, and no smoking signage may discourage littering behaviors and therefore may be partially responsible for the lower CB counts (Lee et al. 2013). However, more research is needed before definite conclusions can be made.

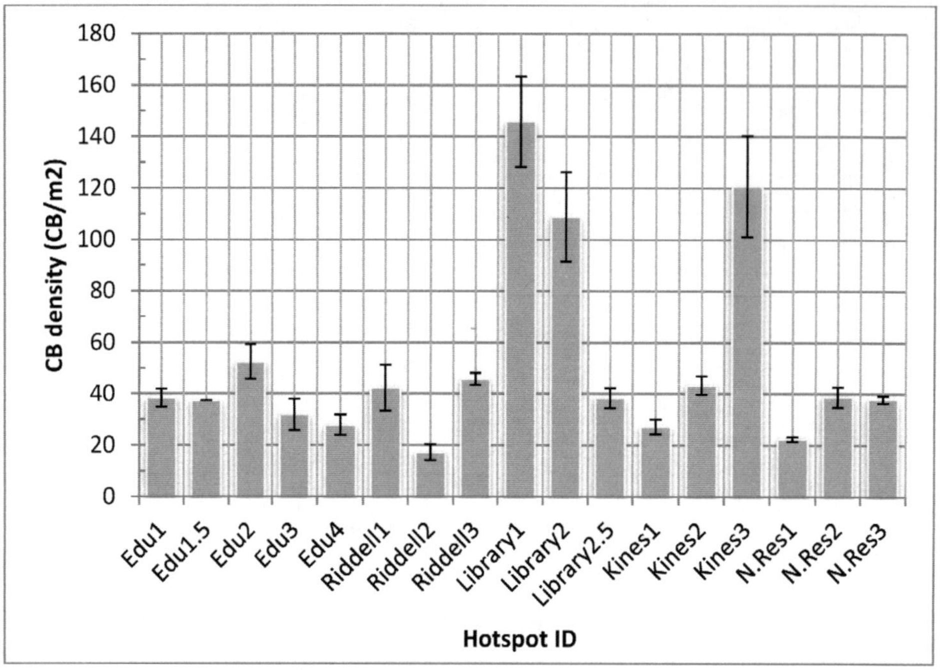


Figure 4: Average CB densities during study period with standard error from mean

The standard error of the mean values reported for Library1, Library2 and Kines3 ranged from 17.4 $CB/m^2$ to 19.6 $CB/m^2$, among the highest in the study (Figure 4). This follows from the discussion above, that these buildings have longer operation hours and host more events that likely bring more off-campus visitors. This being said, the distribution of these events (for example, sporting events) may not be constant, explaining a higher standard error. As well, since the library is most likely used more during final exams, and finals are much more variable during the summer semester because of 6 week versus 12 week courses, it is likely that visits to the library are much less predictable in the summer semester. Figure 5 examines the changes in CB densities to expose the behaviors of CB generation and cleanup rates at U of R campus. Overall, decreasing trends were observed at Library1, Library2 and Kines3, as well as other sites during the study period (curves not shown). Correlation results using climatic data suggest that wind speed, temperature and precipitation had little effect on the CB counts, at least for the hotspots considered in the present study. It is evident that facilities staff or volunteers had performed cleanup regularly during the study period. The maximum reduction in CB densities was attained at Kines3 where a reduction of 154 $CB/m^2$ to 32 $CB/m^2$ occurred between July 20 and July 27. A weekly cleanup rate of about 40 - 60 $CB/m^2 \cdot wk$ was generally observed at these 3 sites. Given the average CB density of 38.3 $CB/m^2$, results suggested existing cleanup protocol was effective in removing CBs on campus. Again, considering that these places are likely more visited than other buildings on campus, it is not surprising that clean-up would occur more frequently, in order to preserve the image of the University.

Since the 3 hotspots in Figure 5 experienced the highest fluctuations of CB counts, the maximum CB generation rates at a given hotspot may be indirectly estimated from the curves. By observing the up-and-down pattern of the Kines3 curve between May 30th and July 9th, a weekly generation rate of 60-80 $CB/m^2 \cdot wk$ was estimated. Caution must be taken to generalize this CB generation rates, as discussed in the subsequent section.

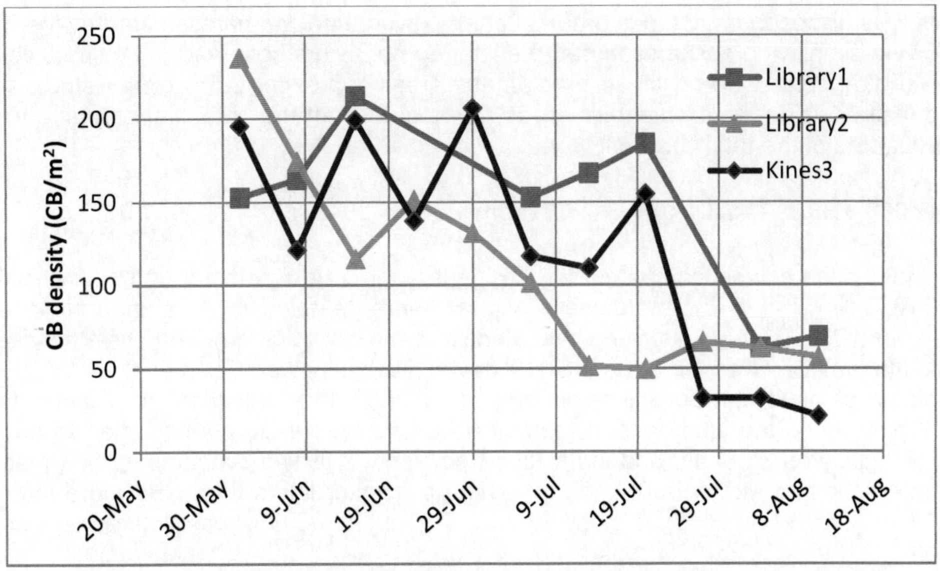


Figure 5: Change of CB densities with respect to time at 3 hotspots

Data from this study found a positive and significant correlation ($m=0.533$, $R^2=0.672$, $p<0.001$) between the number of hotspots and entrances at a given building (Figure 6). According to the smoke-free indoor policy, smoking is banned within 6 meters from any entrance or window of U of R buildings (Policy 2016). Field observations confirmed that the rules and regulations were properly enforced during the study period at U of R. The majority of hotspots were identified near entrances of key buildings and there were no CBs found indoors, suggesting partial success of the indoor smoke-free policy. On the other hand, considerable CB hotspots were found within 6 m of the buildings. Seitz et al. (2012) also found that there was very low compliance with policies for a 25 ft (7.62 m) no-smoking perimeter around entrances.

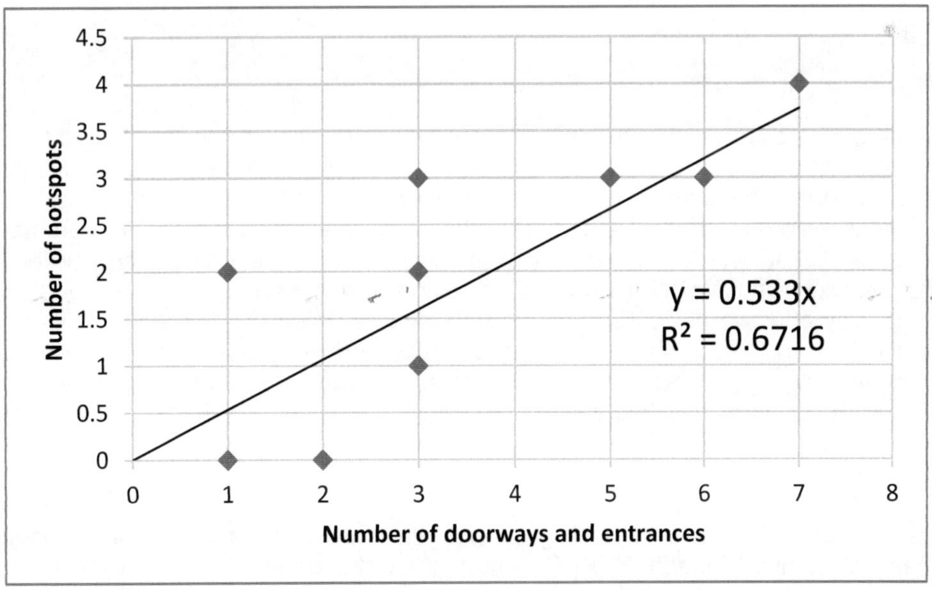


Figure 6: Relationship between numbers of hotspots and building entrance

The number of CBs disposed on campus grounds at any given time can be estimated by the hotspot data using the following simplifying assumptions: (i) the littering behaviors observed at hotspots centered near the entrances of key buildings and these problematic areas represent 0.5% of the study area, (ii) the average CB density outside the problematic areas are about 1% of the hotspots, and (iii) the study area represents about 65% of the total University area:

$$[1]\ \left(38.3\frac{CB}{m^2}\times 0.005+38.3\frac{CB}{m^2}\times 0.01\times 0.995\right)\times 23250m^2\times\frac{1}{0.65}=20481CB\approx 20{,}500\ CB$$

The above field estimate was compared to the generation rates derived from provincial statistics. According to Reid et al. (2015), there were slightly more female smokers than male smokers in Saskatchewan, and the average smoking prevalence rate was 17.6%. Given that the average daily cigarette consumption rates was 13.7 cigarette per day in Saskatchewan (Reid et al. 2015), the total daily CBs generated on campus can be estimated by assuming: (i) the total amount of students, staff and visitors on campus during the study period (summer) can be conservatively assumed to be 20% of the total number of registered students and staff, (ii) all smoking activities on campus took place outdoors because of the smoke-free indoor policy, (iii) an average smoker litters their CBs more than half of the time when outdoor (Seitz et al. 2012). It is estimated that about 4,200 CBs were littered on campus daily.

$$[2]\ 17360ppl\times 0.2\times 0.176\frac{smoker}{population}\times 13.7\frac{cigarette}{day}\times 0.5=4186\frac{CB}{d}\approx 4{,}200\frac{CB}{d}$$

Please note that the estimate from equation 1 was the total CBs on campus at any given time, and the estimate from equation 2 was the daily CB generation rate. Figure 5 suggests that the CB counts were decreasing during the study period. As such, the cleanup rate was expected to be higher than the estimated generation rate. Please note that cleanup performed by facilities management crews and volunteers were not explicitly considered in the present study.

The summer semester was selected as the study period due to difficulties with field work during long and harsh Saskatchewan winters. In addition, quantification of CBs in winter is more challenging due to snow coverage and the presence of melting salt and dirt. Due to the nature of the work, the estimates are not intended to be a precise representation of the actual amount of CB littered on campus. However, the results provided some important bases for future CB studies on smoking behaviors and smoking policy at Canadian campus.

## 4 Conclusions

In this study, a method was developed to quantify the amount of CBs on campus. Using the proposed method, it is estimated that there were about 20,500 CBs on the U of R campus.. The results are based on the field data collected in the summer semester at the University of Regina main campus, and care must be taken to generalize the results elsewhere. The daily generation rate was estimated using provincial data at about 4,200 CB/day. Fewer cigarettes smoked near entrances and windows may reduce second hand smoke exposure for the campus community. It is found that the U of R smoke-free indoor policy was partially effective as no CBs were found indoors. However, a considerable amount of hotspots were identified within 6 m of building entrances. A CB study in United States (Lee et al., 2013) concluded that campus smoking policies may have an important impact on smoking behaviour. U of R data reported in this study supports this observation.

## 5 Acknowledgement

This work is financially supported by a research grant (RGPIN-385815) from the Natural Sciences and Engineering Research Council of Canada (NSERC). The authors are grateful for their support. The views expressed herein are those of the writers and not necessarily those of our research and funding partners.

## 6 References

Arafat, H. A., Al-Khatib, I. A., Shwahneh, R. D. H. 2007. Influence of socio-economic factors on street litter generation in the middle east: effects of education level, age, and type of residence. *Waste Management & Research*, **25**(4), 363-70. DOI: 10.1177/0734242X07076942

Barnes, R.L. 2011. Regulating the disposal of cigarette butts as toxic hazardous waste. *Tobacco Control*, **20**, i45-i48.

Becherucci, M. E.,Pon, J. P. S. 2014. What is left behind when the lights go off? Comparing the abundance and composition of litter in urban areas with different intensity of nightlife use in Mar del Plata, Argentina. *Journal of Waste Management,* **34**(8), 1351–1355. DOI: 10.1016/j.wasman.2014.02.020

Bruce, N., Asha, A., and Ng, K. T. W. 2016. Analysis of Solid Waste Management Systems in Alberta and British Columbia using Provincial Comparison. *Canadian Journal of Civil Engineering*, **43**(4), 351-360. DOI:10.1139/cjce-2015-0414

Campus map, University of Regina. 2016. Available at https://www.uregina.ca/contact/maps-directions/

Lee, J., Ranney, L.M. and Goldstein, A.O. 2013. Cigarette butts near building entrances: What is the impact of smoke-free college campus policies? *Tobacco Control,* **22**, 107-112.

Lee, W. and Lee, C. C. 2015. Developmental toxicity of cigarette butts – An underdeveloped issue. *Ecotoxicology and Environmental Safety*, **113**, 362–368. DOI 10.1016/j.ecoenv.2014.12.018

Levy. D. T., Blackman, K., Currie, L. M., Levy, J., Clancy, L. 2012. SimSmokeFinn: How far can tobacco control policies move Finland toward tobacco-free 2040 goals? *Scandinavian Journal of Public Health*, 40(6), 544-552. DOI: 10.1177/1403494812456635

Marah, M., and Novotny, T. 2011. Geographic patterns of cigarette butt waste in the urban environment. *Tobacco Control,* **20**, i42-i44.

Mohajerani, A., Kadir, A.A., Larobina, L. 2016. A practical proposal for solving the world's cigarette butt problem: Recycling in fired clay bricks. *Waste Management*, **52**, 228–244. DOI:/10.1016/j.wasman.2016.03.012

Moriwaki, H., Kitajima, S., Katahira K. 2009. Waste on the roadside, 'poi-sute' waste: its distribution and elution potential of pollutants into environment. *Waste Management*, **29**(3),1192-1197. DOI: 10.1016/j.wasman.2008.08.017

Novotny, T.E., Lum, K., Smith,E., Wang V. and Barnes, R. 2009. Cigarettes Butts and the Case for an Environmental Policy on Hazardous Cigarette Waste. *International Journal of Environmental Research and Public Health*, **6**(5), 1691–1705.

Office of Resource Planning, University of Regina. 2016. Supplemental Reports for Fall 2016. Available at http://www.uregina.ca/orp/statistics/registration/index.html

Policy, University of Regina. 2016. Smoking on Campus. Available at https://www.uregina.ca/policy/browse-policy/policy-GOV-100-010.html

Puls, J., Wilson, S.A. & Hölter, D. 2011. Degradation of Cellulose Acetate-Based Materials: A Review. *Journal of Polymers and the Environment,* **19**, 152. DOI:10.1007/s10924-010-0258-0

Reid, J.L., Hammond, D. Rynard, V.L., and Burkhalter, R. 2015. Tobacco Use in Canada: Patterns and Trends, 2015 Editions. Waterloo, ON: Propel Centre for Population Health Impact, University of Waterloo. Available at: https://uwaterloo.ca/tobacco-use-canada/sites/ca.tobacco-use-canada/files/uploads/files/tobaccouseincanada_2015_accessible.pdf

Richter, A., Bruce, N., Ng, K. T. W., Chowdhury, A. and Vu, H. L. 2017. Comparison between Canadian and Nova Scotian waste management and diversion models – A Canadian Case Study. *Sustainable Cities and Society*, **30**, 139-149. DOI: 10.1016/j.scs.2017.01.013

Sawdey,M., Lindsay, R.P. Novotny, T.E. 2011. Smoke-free college campuses: no ifs, ands or toxic butts. *Tobacco Control,* **20**, i21-i24.

Seitz, C.M., Strack, R.W., Orsini, M.M., Rosario, C., Haugh, C., Rice, R., Wyrick, D.L. Wagner, L. 2012. Quantifying Littered Cigarette Butts to Measure Effectiveness of Smoking Bans to Building Perimeters. *Journal of American College Health* **60**(4), 331-334.

Statistics Canada. 2010. Trends in the Age Composition of College and University Students and Graduates. 81-004-X, Vol. 7, No. 5

Wang, Y., Ng, K. T. W., and Asha, A. 2016 Non-hazardous Waste Generation Characteristics and Recycling Practices in Saskatchewan and Manitoba, Canada. *Journal of Material Cycles and Waste Management*, **18**(4): 715-724. DOI: 10.1007/s10163-015-0373-z

Leadership in Sustainable Infrastructure
*Leadership en Infrastructures Durables*

Vancouver, Canada
May 31 – June 3, 2017/ *Mai 31 – Juin 3, 2017*

# CO-COMBUSTION OF MUNICIPAL SEWAGE SLUDGE WITH CASTER SEED SHELL IN BUBBLING FLUIDIZED BED COMBUSTOR UNDER OXYGEN-ENRICHED CONDITION: EXPERIMENTAL INVESTIGATION

Singh, Ravi Inder [1, 3], and Kumar, Rajesh [2]
[1] Assistant Professor, Birla Institute of Technology and Science, Pilani, India
[2] Research Scholar, Birla Institute of Technology and Science, Pilani, India
[1, 3] raviinder@pilani.bits-pilani.ac.in

**Abstract:** In India, municipal sewage sludge (MSS) disposal is one of the most complex environmental problems. Incineration is one of the promising technique to undertake this problem which utilized the heating value of the MSS to produce energy and reduced the sludge volume to small-stabilized ash. Generally, MSS having low calorific value and high moisture contents. The co-incineration of MSS with coal or other fuels overcome the problem of burning low calorific MSS. Oxy-fuel fluidized bed combustion technology provides the most favorable environments for the combustion of solid fuels having the moisture content up to 60%, which results in maximum amount of carbon dioxide, which is easy to control. Co-combustion with oxy-fuel fluidized bed has potential for negative $CO_2$ emission level for power generation. Co-combustion of municipal sewage sludge (MSS) with caster seeds shell (CSS) has been investigated in oxygen-enriched bubbling fluidized bed (BFB) combustor. The tests are performed with two different ratios of MSS/CSS is 25%/75% and 50%/50%. This work extensively investigates the temperature profile, flue gas emission, and the performance of co-combustion MSS with CSS under air-fired and oxy-enriched condition.

## 1 INTRODUCTION

Due to industrial revolution and urbanization a hue amount of municipal sewage sludge (MSS) is generated in India. Handling of this huge amount of MSS is one of big challenge. To takle such problem and to provide a healthy environment to human being Indian government launch a scheme "Clean India Mission". To successfully accomplish this goal, India requires to implement or test new advance technologies for energy security, clean environment, and waste management. Four methods generally used for the disposal of MSS are land filling, dumping in sea, used as a fertilizer in agriculture field and incineration under controlled environment. What still these technique has some major issues like (Gupta, Yadav, and Kumar 2015): as land filling takes up a lots of land and contaminates, dumping into sea is harmful, some of MSS products are not suitable for the agriculture, and MSS has low calorific value and high moisture content which make it difficult to burn efficiently. Among these all methods incineration of MSS in a fluidized bed combustor is one of the promising method which gives the advantage of burning low calorific and high moistures MSS under controlled environment. Fuel flexibility is the one of major advantage of the FBC. Incineration of MSS in FBC utilized the heating value of the MSS to produce energy and reduced the sludge volume to small-stabilized ash. Werther et al. (Werther and Ogada 1999) discussed the three groups of thermal processing of sewage sludge like; mono combustion incineration, co-combustion and alternative process. The co-incineration of MSS with coal or other fuels overcome the problem of burning low calorific MSS. Three types of bed are used for combustion are bubbling fluidized bed (BFB), circulating fluidized bed (CFB), and pressurised fluidized bed (PFB). BFB & CFB are used extensively for MSS incineration worldwide where's PFB is under initial stage. Some other advanced combustion technologies like oxy-fired or oxygen-enriched fluidized bed combustor are under developing stage and it can take years to utilize the benefits of this technology(Singh and Kumar 2016). Co-combustion of municipal sewage sludge with biomass under oxygen-enriched or oxy-fired condition has potential to overcome the existing problem and gives some advantages over conventional technologies (Jang et al. 2016).

The main objective of present work is to investigate the combustion and emission characteristics of co-combustion of MSS with biomass in a bubbling fluidized bed combustor under oxygen-enriched (O-E) conditions. Castor seeds shell (CSS) an agriculture waste is used for co-combustion with MSS which is available in surplus quantity in Rajasthan, India ("Biomass Fuel Report Prepared by ABI Energy Consultancy Services Private Limited" 2015). The satisfactory combustion of selected blend will increase the share of renewable energy resource, reduce the problem of MSS disposal and provide some clean energy. A ratio of 25%MSS/75%CSS and 50%MSS/50%CSS is examined in BFB combustor and observations are made of the temperature profile, flue gas emission, and combustion efficiency.

## 2 METHODOLOGY

### 2.1 Fuel property:

In this study MSS and CSS (shown in Figure 1) are used for co-combustion. Two different ratio of MSS/CSS investigated are 25%/75% and 50%/50%. MSS used is collected from the sewage treatment plant Bits-pilani, pilani campus, India in dry form and CSS an agriculture waste is collected from the pilani village, India. The CSS is dried into direct sun light for at least 6 hour before use and then used directly. The particles of the CSS are less then 5mm and not require any further treatment. The proximate and ultimate analyses of MSS, CSS and blends are shown in the Table 1. The 25%MSS/75%CSS has the 16.98 MJ/kg gross calorific value (GCV) where's 50%MSS/50%CSS has 13.9 MJ/kg GCV. Sulphur concentration is not found in all blends.

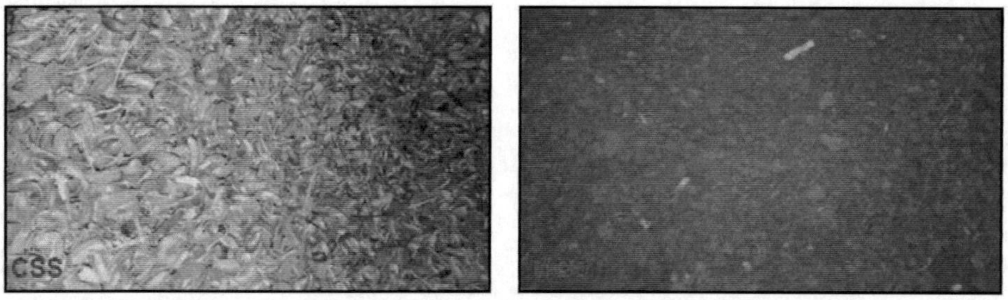


Figure 1: Fuel used for co-combustion (CSS and MSS)

Table 1: Properties of fuel used

| | MSS | CSS | 25%MSS/75%CSS | 50%MSS/50%CSS |
|---|---|---|---|---|
| Proximate Analysis (Wt.%) | | | | |
| Moisture | 4 | 8.1 | 7.075 | 6.05 |
| Volatile matters | 17 | 69.22 | 56.166 | 43.11 |
| Ash | 59 | 8 | 20.75 | 33.5 |
| Fixed carbon | 20 | 14.68 | 16.01 | 17.34 |

| | | | | |
|---|---|---|---|---|
| Ultimate Analysis (Wt. %) | | | | |
| Carbon | 13 | 40.135 | 33.35 | 26.5 |
| Hydrogen | 1.8 | 4.991 | 4.19 | 3.35 |
| Nitrogen | 1.3 | 1.104 | 1.1805 | 1.2 |
| Sulphur | -- | -- | -- | -- |
| Oxygen | 14 | 30.668 | 26.5 | 22.3 |
| GCV (MJ/kg) | 7.8 | 20 | 16.98 | 13.9 |

## 2.2 Experimental setup and procedure:

The work described in this paper is performed in a 1.8 m long square bubbling fluidized bed combustor made of stainless steel SS310 is shown in Figure 2 a schematic view & in Figure 3 real image of test rig. The test rig consists two cyclones, one heat exchanger, hopper, ID fan, two blowers, data logger, and chimney. Combustor is insulated with ceramic fiberboard with inner side and with glass wool from outer side to minimize the heat loss. Three doors are provided in each zone of combustor as shown in Figure 3. A premixed fuel blend of MSS/CSS at different specified ratio (mass %age) is entered into the splash zone of combustor through hopper and screw feeder arrangement. To control the supply of fuel blend the screw feeder is driven by variable speed stepper motor. Pre-produced Ash in bed is used as a bed material. Primary air and oxygen for fluidization is passed into the bed from bottom of combustor through distributer nozzle. To burn the fuel particle which were transported from the combustor bed or splash zone to freeboard a secondary air or oxygen is supplied into the combustor freeboard though side. Temperature is measured at different location along the height of the combustor by placing 9-thermocouple on combustor at suitable height. LPG is used to initiate the combustion process. The port "S" in the freeboard is used to take measurement of gas concentration by employing "testo 350" portable gas analyser having pump volumetric flow rate 1 l/min (controlled), standard litre ±0.1l/min . The net thermal capacity of this system is below 20 kW. The combustor is made of stainless steel (SS310) to resist with higher temperature.

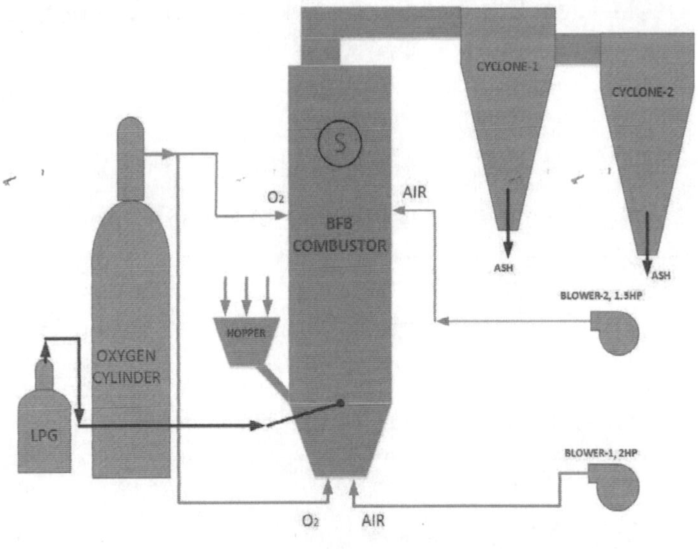


Figure 2: Experimental Setup

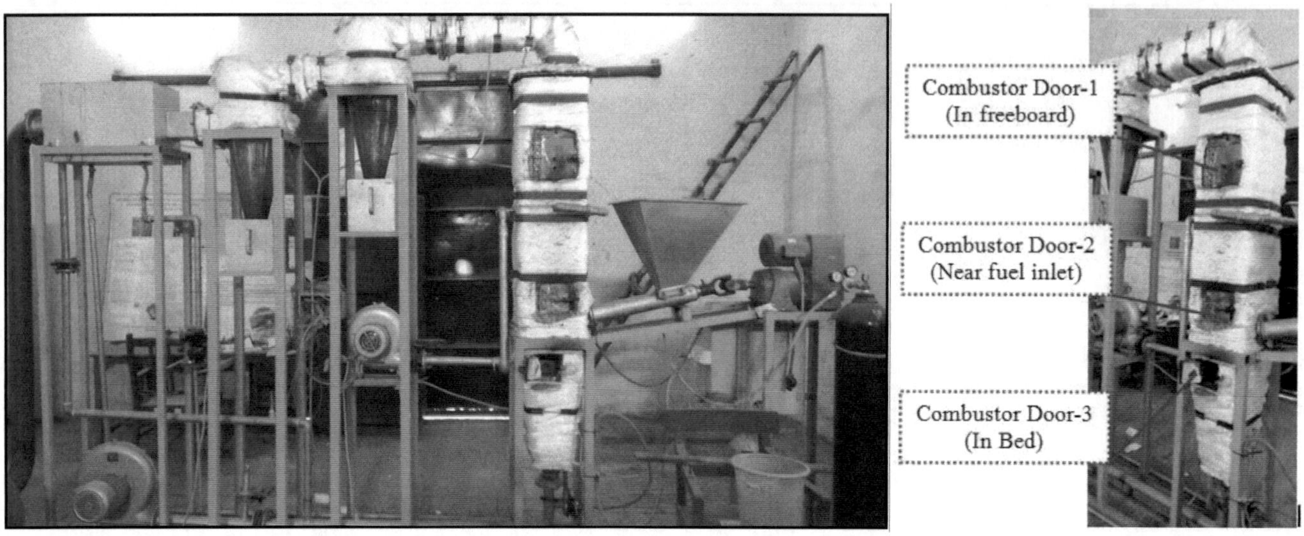


Figure 3: Actual picture of Bubbling Fluidized Bed Unit

Table 2: Experimental Setup Description

| | |
|---|---|
| Plant Type | 20 kW Bubbling fluidized bed combustor |
| Auxiliary systems | Gas supply system, solid feeding, gas analyzer, internal heat exchanger, ID fan. Cyclone |
| Test Conducting Option | Oxygen-Enriched (O-E) combustion mode, Air-Fired (A-F) combustion mode |
| Combustor Material | Stainless steel |
| Combustor Dimensions | 1.8 m long square Combustor |
| Fuel Used: | Municipal Sewage Sludge (MSS), CSS (Caster Seed Shell) |
| Bed material | Ash |
| Reactant Gas | Air, $O_2$ |
| Instrument | Thermocouple, Data Logger, “Testo 350” professional portable gas analyzer |

### 2.3 **Performance measure**

Combustion efficiency measure of how efficiently energy from the fuel is converted into useful energy. Here the combustion efficiency is calculated by the following equation

$$\textit{Combustion efficiency}, \quad \eta_C = 100 - S_L \qquad (1)$$

Flue gas passing out of the stack is calculated by "Siegert" formula as follows(Kumar and Singh 2016) ;

$$S_L = k_S \times \left[ \left( T_{fg} - T_a \right) / CO_2 \right] \qquad (2)$$

The Siegert constant value for 25%MSS/75%CSS is 0.46 and for 50%MSS/50%CSS is 0.49.

($S_L$- Stack loss, $k_s$-"Siegert" Constant, $T_{fg}$- Flue gas temperature in °C, $T_a$ - ambient air temperature)

## 3 RESULTS AND DISCUSSIONS

Tests are performed for 25%MSS/75%CSS and for 50%MSS/50%CSS in a 20kW bubbling fluidized bed combustor under A-F and O-E condition when steady state is reached to investigate the effect of MSS on the performance of a fluidized combustor and gas emission. The steady state is reached within 5 to 6 hours of operation. The oxygen is enhanced up to 8-10% for O-E conditions and a constant fuel feed rate of approximately 6 kg/hr for 25%MSS/75%CSS and 4 kg/hr for 50%MSS/50%CSS was maintained throughout each experiment.

### 3.1 Temperature profile & flue gas emission:

Both fuel blends are burned inside the combustor smoothly and efficiently. Temperature is recorded for the combustor bed, splash zone and freeboard along the height by employing thermocouple and data logger. Figure 4 shows the temperature variation along the height of combustor for both blends of MSS/CSS under both A-E and O-E conditions. Combustion for both the fuel blends is even and transition from the A-E to O-E is smooth. It observes that the maximum fuel blend is burned in the splash zone and freeboard for both fuels. This is because the density of fuel is low and fuel burn above the bed in splash zone and freeboard. A similar trend of temperature variation is observed for all cases. The maximum temperature in combustor is observed in the splash zone because most of the fuel burnt above the bed in splash zone and freeboard. Remaining fuel is burned in the freeboard. Combustion is improved inside the combustor under O-E condition when extra amount of primary oxygen introduced to bed through distributor and secondary oxygen to splash zone. Under O-E condition combustion is smooth and burning is improved in the freeboard. 50%MSS/50%CSS under O-E condition has the highest combustor temperature than others cases, because density of 50%MSS/50%CSS is more than 25%MSS/75%CSS, secondly under O-E condition extra amount of oxygen is available for complete combustion.

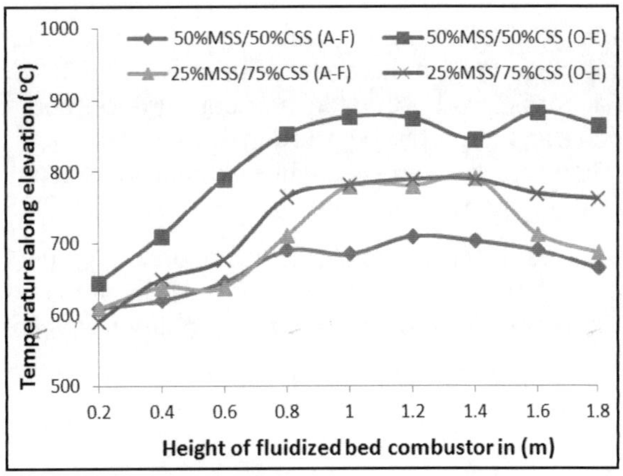


Figure 4: Temperature variation along height of combustor for both A-E and O-E condition.

### 3.2 Flue gas emission

The species concentration of gas emission is measured in the freeboard by using "testo 350" professional portable gas analyzer, when steady state is reached by. In Figure 5 CO concentration for MSS/CS under both A-E and O-E condition is shown. Under O-E condition combustion is improved and became more stable as the oxygen content is higher. It observed that the amount of carbon monoxide decreases with increase in oxygen concentration. There is no any remarkable difference was observed in CO concentration for both fuel blends.

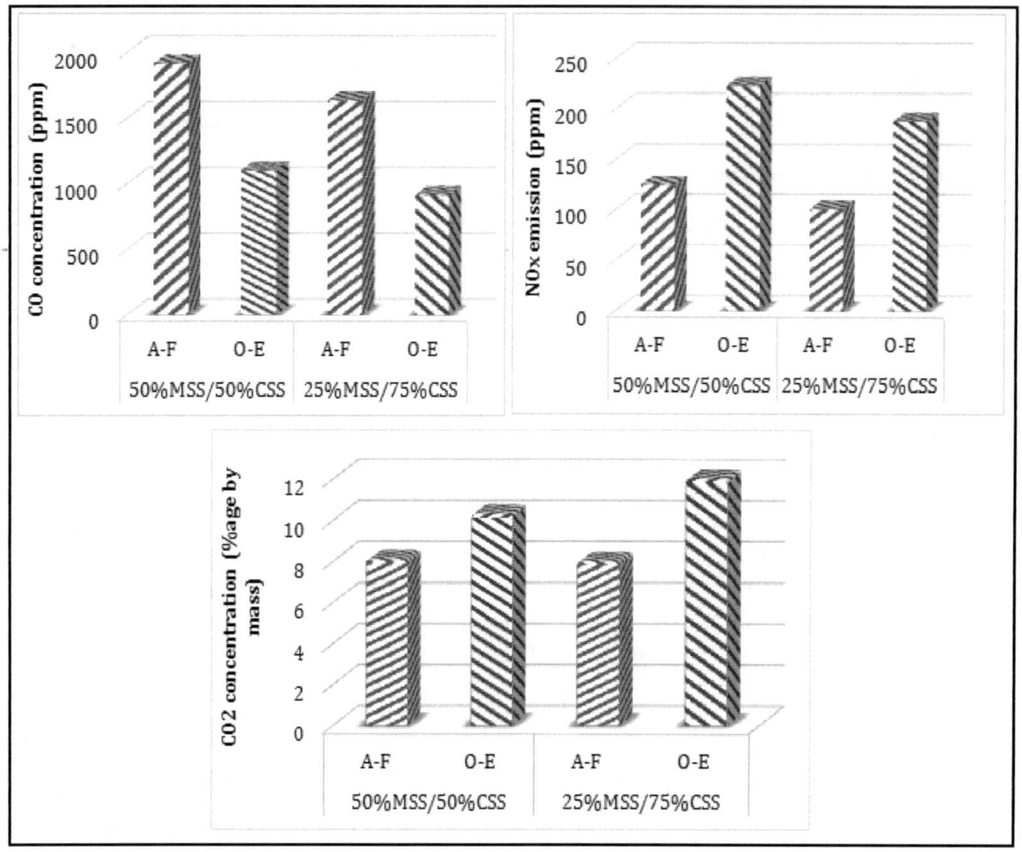


Figure 5: Gaseas emission under A-F and O-E condition

NOx concentration for both ratio of MSS/CS under steady-state A-E and O-E condition is shown in Figure 5. Generally, NO x production depends on nitrogen contents in fuels and it increases with increase in temperature. From Table 1 it observed very little amount nitrogen content for both fuel blends and hence NOx emission here much depend on temperature increase. With increase in oxygen concentration inside combustor, the combustion rate is increased which further increases the temperature and hence NOx increases. Maximum NOx emission was observed for 50%MSS/50%CSS under O-E condition.

$CO_2$ concentration by mass %age for both A-E and O-E condition is given in Figure 5. $CO_2$ concentration in emitted gases is increased with oxygen enhancement in the combustor, which is responsible for increase in combustion rate. In this work, maximum CO2 is observed for 25%MSS/75%CSS under O-E condition.

### 3.3 Combustion efficiency

Figure 6 shows gives combustion efficiency for both fuel blend under different condition, A-E and O-E and it observed that the combustion efficiency increases as the oxygen concentration is increased inside the combustor because the rate of the fuel oxidation is proportional to power of oxygen concentration. A maximum possible combustion efficiency of 96 % is achieved for 25%MSS/75%CSS under O-E condition.

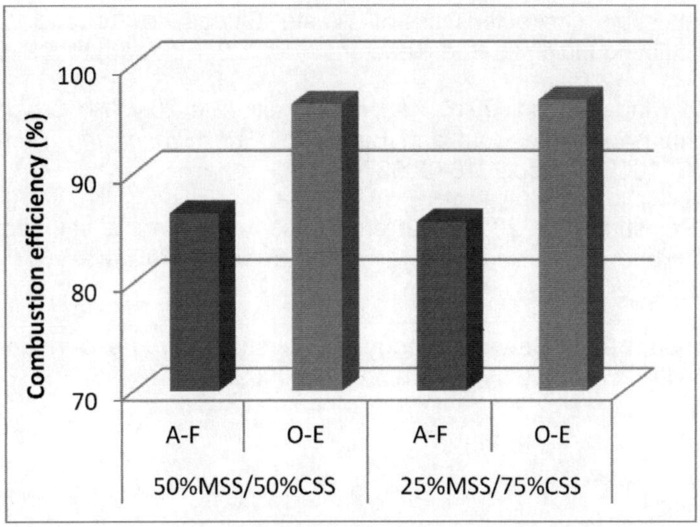


Figure 6: Combustion efficiency for 50%MSS/50%CSS and 25%MSS/75%CSS under A-E and O-E condition.

## 4 CONCLUSION

An experimental study of an oxygen-enriched (O-E) bubbling fluidized bed combustor has been carried out for co-combustion of MSS with CSS (for two ratio 25%MSS/75%CSS and 50%MSS/50%CSS) to examine the combustor zone temperature, emission, combustor performance under air-fired and oxy-fired condition. The following points are concluded from the above study.

1. MSS as a co-fuel burned successfully with CSS base fuel under bubbling fluidized bed combustor and the burnout of the blend is improved under O-E condition.

2. With the increase in oxygen concentration in supply gas combustion rate increases and the concentration of CO reduces and NOx concentration increases. The $CO_2$ concentration in flue gases is increased as a result of oxygen enhancement in supply gas.

3. The combustion inside the combustor was smooth. A maximum possible combustion efficiency of 96.5 % was achieved under O-E conditions for 25%MSS/75%CSS.

**Acknowledgements**

Authors are thankful to Department of Science and Technology, New Delhi, India for sponsoring a grant for this research work (Vide reference no. SB/FTP/ETA-0021/2014 dated July 17, 2014), authorities of BITS Pilani, Pilani campus, Rajasthan India for other administrative support required for this work and the farmer of Pilani region to provide fuel for this study.

**References**

"Biomass Fuel Supply Study ( Rajasthan ) for Rajasthan Renewable Energy Corporation Limited Prepared by ABI Energy Consultancy Services Private Limited." 2015.

Gupta, Neha, Krishna Kumar Yadav, and Vinit Kumar. 2015. "A Review on Current Status of Municipal Solid Waste Management in India." *Journal of Environmental Sciences* 37. Elsevier B.V.: 206–17.

Jang, Ha-na, Jeong-hun Kim, Seung-ki Back, Jin-ho Sung, Heung-min Yoo, Hang Seok, and Yong-chil

Seo. 2016. “Combustion Characteristics of Waste Sludge at Air and Oxy-Fuel Combustion Conditions in a Circulating Fluidized Bed Reactor.” *Fuel* 170. Elsevier Ltd: 92–99.

Kumar, Rajesh, and Ravi Inder Singh. 2016. “An Investigation in 20 kWth Oxygen-Enriched Bubbling Fluidized Bed Combustor Using Coal and Biomass.” *Fuel Processing Technology* 148. Elsevier B.V.: 256–68. doi:10.1016/j.fuproc.2016.02.037.

Singh, Ravi Inder, and Rajesh Kumar. 2016. “Current Status and Experimental Investigation of Oxy-Fired Fluidized Bed.” *Renewable and Sustainable Energy Reviews* 61. Elsevier: 398–420. doi:10.1016/j.rser.2016.04.021.

Werther, J., and T. Ogada. 1999. “Sewage Sludge Combustion.” *Progress in Energy and Combustion Science* 25 (1): 55–116. doi:10.1016/S0360-1285(98)00020-3.

Leadership in Sustainable Infrastructure
*Leadership en Infrastructures Durables*

**Case Study**

Vancouver, Canada
May 31 – June 3, 2017/ *Mai 31 – Juin 3, 2017*

# COMPARISON OF NITRILOTRIACETIC ACID AND [S,S]-ETHYLENEDIAMINE-N,N'-DISUCCINIC ACID IN UV-FENTON FOR THE TREATMENT OF OIL SANDS PROCESS-AFFECTED WATER AT NATURAL PH

Ying Zhang[1], Nikolaus Klamerth[1], Pamela Chelme-Ayala[1], and Mohamed Gamal El-Din[1,3]
[1]Department of Civil and Environmental Engineering, University of Alberta, Edmonton, Alberta, Canada, T6G 1H9
[2] Author Department of Biological Sciences, University of Alberta, 11455 Saskatchewan Drive, Edmonton, Alberta, Canada, T6G 2W2
[3] mgamalel-din@ualberta.ca

## 1 Introduction

Naphthenic acids (NAs) are a group of aliphatic and alicyclic carboxylic acids, which are primarily responsible for the toxicity of oil sands process-affected water (OSPW) generated in the bitumen extraction from oil sands and subsequent froth treatment process. Modified Fenton process at neutral pH with the addition of chelating agents or UV-Fenton have not been used for the treatment of OSPW yet. Therefore, the application of UV-Fenton processes with two chelating agents, nitrilotriacetic acid (NTA) and [S,S]-ethylenediamine-N,N'-disuccinic acid ([S,S]-EDDS), for the treatment of oil sands process-affected water (OSPW) at natural pH was investigated in this study. The efficiency of the UV-NTA/EDDS-Fenton processes was examined on the removal of acid extractable organic fraction (AEF), aromatics and the overall toxicity of OSPW towards *Vibrio fischeri*. The removals of classical naphthenic acid (c-NAs) and oxidized NAs (oxy-NAs) in the UV-NTA-Fenton, UV-$H_2O_2$ and NTA-Fenton processes were also compared.

## 2 Objectives

The objectives of this study were: (1) to explore the reduction mechanism of Fe(III)NTA/EDDS to Fe(II)NTA/EDDS; (2) to study the decomposition of Fe(III)NTA and Fe(III)EDDS under UV irradiation; (3) to compare the efficiency of the UV-NTA/EDDS-Fenton processes on the removal of acid extractable organic fraction (AEF), aromatics and the overall toxicity of OSPW towards *Vibrio fischeri*; (4) to investigate the toxicity of NTA, EDDS, Fe(III)NTA/EDDS and their degradation products towards *Vibrio fischeri*; and (5) to compare the removals of c-NAs and oxy-NAs in the UV-NTA-Fenton, UV-$H_2O_2$ and NTA-Fenton processes.

## 3 Methodology

UV irradiation experiments were conducted by placing a 100-mL beaker (5.4-cm diameter) with an 80 mL sample solution on a magnetic stirrer under a collimated beam UV apparatus equipped with a 1-kw medium-pressure (MP) Hg-lamp. $H_2O_2$, Fe, NTA, EDDS, metal ions in OSPW, Fourier transform infrared (FT-IR) spectra, synchronous fluorescence spectra (SFS), 1H nuclear magnetic resonance (NMR) analysis, NAs, toxicity toward Vibrio fischeri were analyzed by following standard analytical methods.

## 4 Results

The results indicate that UV-NTA-Fenton is more efficient in the reduction of acid extractable organic fraction (AEF) in OSPW compared with UV-EDDS-Fenton, primarily due to the lower scavenging effect of NTA on ·OH than that of EDDS. Both [1]H NMR and SFS spectra showed very high removal of aromatics in the UV-NTA-Fenton process, indicating high capability of this process to degrade aromatics. The toxicity test also showed that NTA is a much better chelating agent than EDDS due to its lower toxicity risk of the agent towards aquatic organisms. Figure 1 indicated that NA removal in the UV-NTA-Fenton process was much higher compared to the UV-$H_2O_2$ and NTA-Fenton processes.

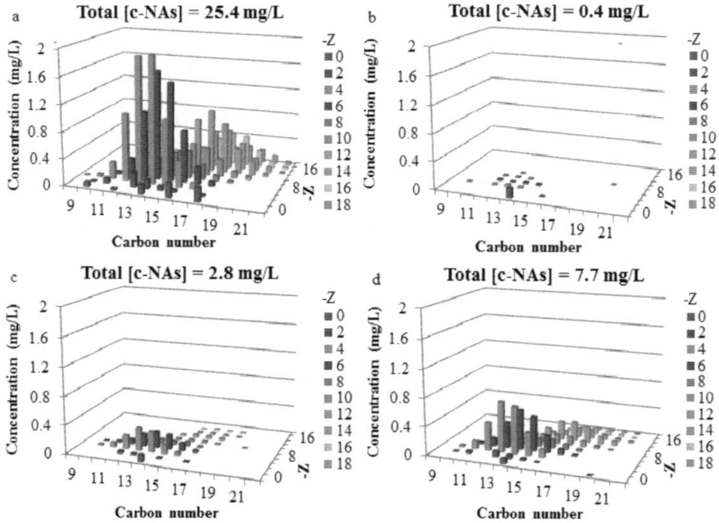


**Figure 1.** NA distribution in a) raw OSPW, and OSPW treated with b) UV-NTA-Fenton, c) UV-$H_2O_2$, and d) NTA-Fenton ($[H_2O_2]_0$ = 5.88 mM for all three processes. $[Fe]_0$ = 0.089 mM and $[NTA]_0$ = 0.72 mM for UV-NTA-Fenton and NTA-Fenton. 30 min UV irradiation for UV-NTA-Fenton and UV-$H_2O_2$) (Zhang et al., 2016).

## 5 Conclusions and Recommendations

Overall, NTA is a much better chelating agent than EDDS for the application of the UV-Fenton process on the treatment of OSPW at natural pH. The findings obtained in this study have significant impact for the further development of this process and OSPW remediation. The combination of UV-NTA-Fenton with biodegradation process might be an economical and efficient approach for the organic and toxicity removal in the OSPW remediation, which deserves more studies to confirm its effectiveness. The total cost should be considered for the application of UV-NTA-Fenton in the treatment of OSPW *in situ*, especially the electricity cost for UV irradiation.


### Acknowledgements

This research was supported by a research grant from a Natural Sciences and Engineering Research Council of Canada (NSERC) Senior Industrial Research Chair (IRC) in Oil Sands Tailings Water Treatment through the support by Syncrude Canada Ltd., Suncor Energy Inc., Shell Canada, Canadian Natural Resources Ltd., Total E&P Canada Ltd., EPCOR Water Services, IOWC Technologies Inc., Alberta Innovates - Energy and Environment Solution, and Alberta Environment and Parks. The financial supports provided by Trojan Technologies and an NSERC Collaborative Research and Development (CRD) grant, and the Helmholtz-Alberta Initiative are also acknowledged. The authors would like to thank the group of Dr. Ted Sargent from the Department of Electrical and Computer Engineering at the University of Toronto for the help in the measurements of half-wave potentials. Thanks also go to Dr.

Rongfu Huang for NAs analysis and Mr. Yuan Chen for his contribution in data mining, and Mr. Chengjin Wang for his help in the NMR analysis.

## Reference

Zhang, Y., Klamerth, N., Chelme-Ayala, P. and Gamal El-Din, M. (2016) Comparison of Nitrilotriacetic Acid and [S, S]-Ethylenediamine-N, N′-disuccinic Acid in UV–Fenton for the Treatment of Oil Sands Process-Affected Water at Natural pH. Environmental science & technology 50(19), 10535-10544.

Leadership in Sustainable Infrastructure
*Leadership en Infrastructures Durables*

**Case Study**

Vancouver, Canada
May 31 – June 3, 2017/ *Mai 31 – Juin 3, 2017*

# COMPARISON OF UV/HYDROGEN PEROXIDE, FERRATE(VI), AND OZONE IN OXIDIZING OIL SANDS PROCESS-AFFECTED WATER (OSPW)

Chengjin Wang[1], Nikolaus Klamerth[1], Selamawit Ashagre Messele[1], Arvinder Singh[1,2], Miodrag Belosevic[2], and Mohamed Gamal El-Din[1,3]
[1]Department of Civil and Environmental Engineering, University of Alberta, Edmonton, Alberta, Canada, T6G 1H9
[2] Author Department of Biological Sciences, University of Alberta, 11455 Saskatchewan Drive, Edmonton, Alberta, Canada, T6G 2W2
[3] mgamalel-din@ualberta.ca

## 1 Introduction

Oil sands process-affected water (OSPW) is the water produced in the bitumen extraction and the following upgrading processes. Its composition is very complex, especially the organic fraction. The organic fraction of OSPW contains classical naphthenic acids (NAs), sulfur and nitrogen containing species, and aromatic compounds including trace amount of polyaromatic hydrocarbon (PAH) (Jones et al. 2013). NAs refers to a mixture of alkyl-substituted acyclic and cycloaliphatic carboxylic acids, with a general formula $CnH_{2n+z}O_x$, where n indicates the number of carbon atoms, z specifies the hydrogen deficiency arising from double bonds and/or rings, and x is the number of oxygen atoms (Mohamed et al. 2013). The high concentration and low biodegradability of the organic matters in OSPW prompted extensive chemical oxidation researches in OSPW treatment. However, a well-defined performance comparison among the different oxidation processes, especially between selective oxidation and unselective oxidation, is currently not available, mainly due to varying raw OSPW quality among different researches and different experimental designs. To fill this gap, this research compared the oxidation performance by UV/$H_2O_2$, potassium ferrate(VI), and ozone (with and without radical scavenger).

## 2 Objectives

The specific objectives of this study were as follows: (1) to compare the oxidation processes including UV/$H_2O_2$, potassium ferrate(VI), and ozone (with and without radical scavenger TBA) in transforming NAs and aromatics in OSPW; (2) to investigate the toxicity of the OSPW treated by different processes; and (3) to explore the oxidation mechanisms behind the performance differences.

## 3 Methodology

Batch reactors with rapid mixing were used in this study. All tests were conducted in duplicates. The dosage of potassium ferrate(VI), ozone, and $H_2O_2$ investigated in this study ranged from 0.1 mM to 4.0 mM.

## 4 Results

The findings of this research can be summarized as follows.

- Ozonation was the most effective process in treating OSPW, achieving almost complete removal of fluorescing aromatics, sulfur-containing NAs (NAs+S), and classical NAs (97.1%) at the dose of 2.0 mM $O_3$. Both pathways via direct ozone reaction and •OH reaction played important roles in transforming the organics, as supported by ozonation performance with and without TBA.

- UV/$H_2O_2$ oxidation was a combination of direct UV photolysis and •OH reaction. The efficiency of this process was compromised by the low molar extinction coefficient of hydrogen peroxide. 42.4% classical NAs were removed at a 2.0 mM $H_2O_2$ dose and 950 mJ/cm$^2$ UV dose.

- Ferrate(VI) oxidation showed high selectivity in removing two-ring and three-ring fluorescing aromatics, NAs+S, and NAs with high carbon number and high hydrogen deficiency. 46.7% classical NA removal was achieved at 2.0 mM Fe(VI).

- Acute toxicity towards *Vibrio fischeri* was reduced, with the order of efficiency as "$O_3$ > ferrate (VI) > UV/$H_2O_2$" at the same molar doses of the oxidants. In addition, toxicity on goldfish PKMs was also mitigated after all the treatments, with the order of efficiency as "$O_3$ > ferrate (VI) ≈ UV/$H_2O_2$".

- The costs of treating 1 m$^3$ of OSPW at chemical dose of 2.0 mM were estimated to be \$0.5, \$1.8, and \$0.4 for UV/$H_2O_2$ oxidation (UV dose: 950 mJ/cm$^2$), ferrate(VI) oxidation, and ozonation, respectively. When NA removal is correlated to the chemical cost, it is apparent that ozonation is the most cost-efficient process in OSPW treatment.

## 5 Recommendations

Based on both the performance and the economic consideration, we recommend ozonation for OSPW oxidation. Based on the findings and unknowns regarding the mechanism study, we recommend the following for future studies. OSPW is a complex water matrix with various organic fractions. Understanding its composition is the precondition to explore the mechanisms behind its toxicity and to evaluate the performance of the water treatment technologies. Our NMR analyses indicated that the total concentration of aromatics in OSPW should be at the mg/L level, and very likely, these aromatics might be an important toxicity source in OSPW. It is advisable to work on exploring the concentrations, species, and the environmental impact of the aromatics in OSPW. Similarly, it is also essential for OSPW remediation to identify and quantify the sulfur and nitrogen containing species in OSPW with more analytical endeavors.

## Acknowledgements

This research was supported by a research grant from the Alberta Water Research Institute, the Helmholtz-Alberta Initiative, and a Natural Sciences and Engineering Research Council of Canada (NSERC) Senior Industrial Research Chair (IRC) in Oil Sands Tailings Water Treatment through the support by Syncrude Canada Ltd., Suncor Energy Inc., Shell Canada, Canadian Natural Resources Ltd., Total E&P Canada Ltd., EPCOR Water Services, IOWC Technologies Inc., Alberta Innovates - Energy and Environment Solution, and Alberta Environment and Sustainable Resource Development.

## References

Jones, D., Scarlett, A.G., West, C.E., Frank, R.A., Gieleciak, R., Hager, D., Pureveen, J., Tegelaar, E. and Rowland, S.J. (2013) Elemental and spectroscopic characterization of fractions of an acidic extract of oil sands process water. Chemosphere 93(9), 1655-1664.

Mohamed, M.H., Wilson, L.D., Peru, K.M. and Headley, J.V. (2013) Colloidal properties of single component naphthenic acids and complex naphthenic acid mixtures. J. Colloid Interface Sci. 395(0), 104-110.

Leadership in Sustainable Infrastructure

*Leadership en Infrastructures Durables*

Vancouver, Canada

May 31 – June 3, 2017/ *Mai 31 – Juin 3, 2017*

# COST FOCUSED EVALUATION OF ADVANCED OXIDATION PROCESSES TO REMOVE VENLAFAXINE IN WASTEWATER

Jordan Hollman [1, 6], John Albino Dominic [2], Gopal Achari [3], Cooper H. Langford [4] and Joo-Hwa Tay [5]

[1, 2, 3, 4, 5] University of Calgary, Canada

[6] jordan.hollman@ucalgary.ca

## ABSTRACT

Concern over the negative effects of discharging residual pharmaceuticals into public waters through wastewater effluent has led to new technologies to remove these contaminants. Advanced oxidation processes (AOPs) have been proposed as a strong option to remove pharmaceuticals, as they have been shown to effectively degrade a variety of recalcitrant contaminants and have the advantage of creating no additional waste stream. However, with many different chemical processes such as $O_3$, UV/$O_3$, UV/$H_2O_2$, Fenton reagent and photocatalysis, there are clearly a large variety of possible options for AOP treatment. This paper presents a laboratory study on the degradation of venlafaxine (VEN) using a variety of different AOPs. VEN is an antidepressant of high concern in wastewater discharge, chosen as a candidate compound to compare AOPs. To focus future research on the best possible AOPs, a cost analysis study was performed for the removal of VEN. The analysis looked at costs in terms of operating cost, capital cost, and required basin volume. Results indicated that UV/$H_2O_2$, UV/$O_3$ and $O_3$ are all cost competitive processes, while UV alone is not viable. Parametric analysis found that increasing $O_3$ dosage decreased both operating costs and basin volume. Increasing $H_2O_2$ dosage increases operating cost while decreasing required basin volume considerably. Increasing UV dose decreased basin volume in all cases, while increasing operating costs for UV treatment, and keeping operating costs stable in $H_2O_2$ based treatment.

Keywords: - Advanced oxidation process, ozone, hydrogen peroxide, UV, venlafaxine

## 1. BACKGROUND

Concerns over potential negative consequences of pharmaceuticals being discharged from municipal wastewater in recent years has become a topic of high concern in the advancement of wastewater treatment technologies. Biological studies on the negative impacts that low concentrations of pharmaceuticals can have on aquatic life are of primary concern. There are many potential negative effects, with endocrine disruption being of particular concern. Male fish have shown the development of female sex characteristics, with reproductive failure in fish populations as a potential outcome (Hamilton et al. 2016). While some compounds are partially removed by existing wastewater treatment processes, they are not currently brining concentrations below levels of environmental concern.

Venlafaxine (VEN) was selected as a candidate compound for this study to test the effectiveness of various advanced oxidation processes (AOPs). VEN is among the most highly prescribed antidepressants under the trade name Effexor®. In a study measuring the discharge concentration of antidepressants being discharged for WWTPs across Canada, VEN was found to be the most prevalent with discharge concentrations of up to 2.9 μg/L (Lajeunesse et al. 2012). VEN has been shown to have neuroendocrine disrupting properties at concentrations as low as 1 μg/L(Melnyk-Lamont et al. 2014). Based on high prevalence and low effect level, VEN is an antidepressant of high concern in municipal wastewater effluents. Based on the concentrations of discharge and of environmental effect levels found in literature, removal of about 65% would be required to bring discharge below levels of environmental effect. For this reason, a removal rate of 65% was targeted in this study as the level of treatment required in evaluation of various AOPs.

AOPs are a process with high potential to be used in municipal WWTPs to provide removal of pharmaceuticals from wastewater. AOPs function by creating highly reactive species such as hydroxyl radicals ($HO^{\bullet}$) that are able to chemically degrade recalcitrant contaminants. Typical AOPs consist of chemical reagents combined with UV light to create $HO^{\bullet}$, although processes utilizing photolytic materials may also be considered. One significant advantage that AOPs have over other treatment processes is that they do not create any additional waste streams, which are a major drawback of filtration or reverse osmosis (Subramani and Jacangelo 2014). An additional advantage of AOPs is that many existing WWTPs utilize UV for disinfection prior to discharge, allowing for considerable cost saving by simply modifying an existing process rather than installing a completely new treatment module.

Some literature exists showing that VEN can be successfully degraded by $HO^{\bullet}$, as presented by García-Galán et al. (García-Galán et al. 2016). However, a comprehensive study on the effectiveness of various AOPs to degrade VEN is lacking in literature. Various other studies have been published on the potential effectiveness of AOPs to degrade other pharmaceuticals in wastewater, demonstrating that AOPs have the potential to be a robust treatment technique to remove pharmaceuticals from wastewater (Ribeiro et al. 2015).

Evaluating the economic applicability of different AOPs can be a challenge, as studies typically provide results of a given method based on laboratory results measuring chemical and light dosages in the evaluation of different treatment methods, without translating those results to cost. In this paper, laboratory data is presented on a variety of AOP treatment options, with lab results being applied to estimate expected level of cost for each process. Laboratory data presented is based upon batch experiments on milli-Q water. It is noteworthy that significant uncertainty exists when scaling these experiments to full scale treatment. Some of the factors causing uncertainty are: experiments were conducted in milli-Q water rather than wastewater, concentrations tested were significantly higher than actual wastewater concentrations, and that tests were conducted in batch rather than flow through. The primary focus of the economic analysis is in this paper is for comparative purpose between different AOPs, with specific dollar amounts presenting high uncertainty. The level off accuracy in the values presented are acceptable for comparative purposes, providing a basis for which processes are worth further development.

When choosing a potential process, a holistic approach should be taken, with operating costs, capital costs, required basin volume, potential safety hazards, process adjustability and amount of maintenance required being some of the areas that should be considered when choosing the ideal process. For this paper, only areas that directly affect the costs of implementing each process are considered. By breaking down different potential treatment options directly into cost terms, choosing the ideal process for any given treatment facility becomes considerably easier.

## 2. MATERIALS AND METHODS

### 2.1 Chemicals

VEN was supplied by BOC Sciences at high purity (>98%). Methanol for High performance liquid chromatography (HPLC) analysis was provided by Sigma-Aldrich (>99% purity). Phosphoric acid for eluent acidification was provided by ACROS Organics (>85% purity). Hydrogen peroxide at 30% purity was provided by Fluka Analytical. Oxygen feed for ozone generator was provided by Praxair (99.99% purity).

Quenching solution was prepared using $Na_2SO_3$ at a concentration of 100 g/L in water.

### 2.2 Ozone Production

Ozone was produced using aa A2Z Ozone Inc. ozone generator (model 3GLAB) with ozone production by corona discharge tube. Feed gas was oxygen at 99.99% purity. Ozone was sparged into solution through a course gas diffuser.

Measurement of aqueous concentration of ozone was done using an indigo trisulfonate method (Bader and Hoigné 1981). Fine calibration of ozone dosage was performed before each set of experiments to provide consistency in ozone dosage.

### 2.3 Photoreactor

UV irradiation for all experiments was done in a Luzchem batch reactor (LZC-ORG, Luzchem Research Inc.). Low pressure mercury bulbs irradiating at a wavelength of 254nm. UV intensity was measured using ferrioxalate actinometry (Calvert and Pitts 1966), with light intensity adjusted by varying the number of active UV bulbs.

Experiments were conducted in a quartz beaker placed at the center of reactor containing 200mL of sample in all cases. Vessel was continuously mixed throughout all experiments.

### 3.4 Sample Analysis

Quantification of VEN in all samples was done by isocratic elution on HPLC with a UV-visible detector. All injections were performed in duplicate, with less than 1% variance noted between duplicate injections. The instrument used was a Varian Prostar 210 equipped with a Pinnacle DB C18 5μm column with dimensions of 150 x 4.6mm. A flow rate of 0.5mL/min was used with an eluent ratio of 30% water to 70% methanol. All eluents were acidified by phosphoric acid to a pH of 2.6. Calibration was performed using known standards, with a lower detection limit of 0.25 mg/L.

### 3.5 Experimental Procedure

Starting concentration for all experiments was selected at 10 mg/L of VEN in Milli-Q water, to accurately measure >95% degradation.

Prior to the beginning of experiments UV lamps and ozone generator were run for 10 minutes for warm up. VEN solutions equilibrated to room temperature prior to starting experiments. Before ozone experiments, calibration of ozone dosage was performed.

200 mL of VEN solution was placed into a quartz beaker. Sample volumes were 2 mL, with initial sample taken prior to starting experiments. UV irradiation, $O_3$ and/or $H_2O_2$ were added concurrently at the start of experiment. A small volume of quencher (<10 μL) was added to each sample vial, with exact amount determined by the concentration of $O_3$ or $H_2O_2$ in any given experiment. All samples were filtered using 0.45 μm polytetrafluoroethylene filters prior to analysis.

## 3. RESULTS AND DISCUSSION

### 3.1 Experimental Results

3.1.1 Degradation Kinetics

For comparative analysis of AOPs to degrade VEN in water, a variety of methods were tested. Results presented in this paper are degradation under UV, $O_3$, UV/$O_3$ and UV/$H_2O_2$. A sample of kinetics by different process are presented in Figure 1 for comparison. When reading the figure, it should be noted that the result for UV only follows a scale of minutes on the x-axis, while all other results are on a scale of seconds. Experiments on $H_2O_2$ alone, $H_2O_2$/$O_3$, as well as UV/$H_2O_2$/$O_3$ were also conducted, but are not presented here. Degradation by $H_2O_2$ alone was extremely slow to the point of being not competitive with other processes. $H_2O_2$/$O_3$ and UV/$H_2O_2$/$O_3$ experiments showed no note worthy improvement over AOPs with $O_3$ and UV/$O_3$, and were therefore not considered for analysis.

In processes using $O_3$, an ozone concentration of 14.4 mg/h and 28.8 mg/h were used, while $H_2O_2$ concentrations of 50 and 100 mg/L were used. For UV irradiation, light intensity ranging from 647 to 3502 W/M$^3$ were utilized for the comparison. Only one reagent dosage for different AOPs were presented in the figure for comparative purposes.

The different AOPs reported here were able to degrade VEN to the lower detection limit of detection by HPLC, over 95% degradation. However, degradation rates varied considerably between processes, with $O_3$, $O_3$/UV and $H_2O_2$/UV being the most rapid, degrading with a half life of 64s 63s and 77s for the parameters considered in Figure 1, respectively. Although UV alone could completely degrade VEN, it had a considerably higher half life of 43.4 min.

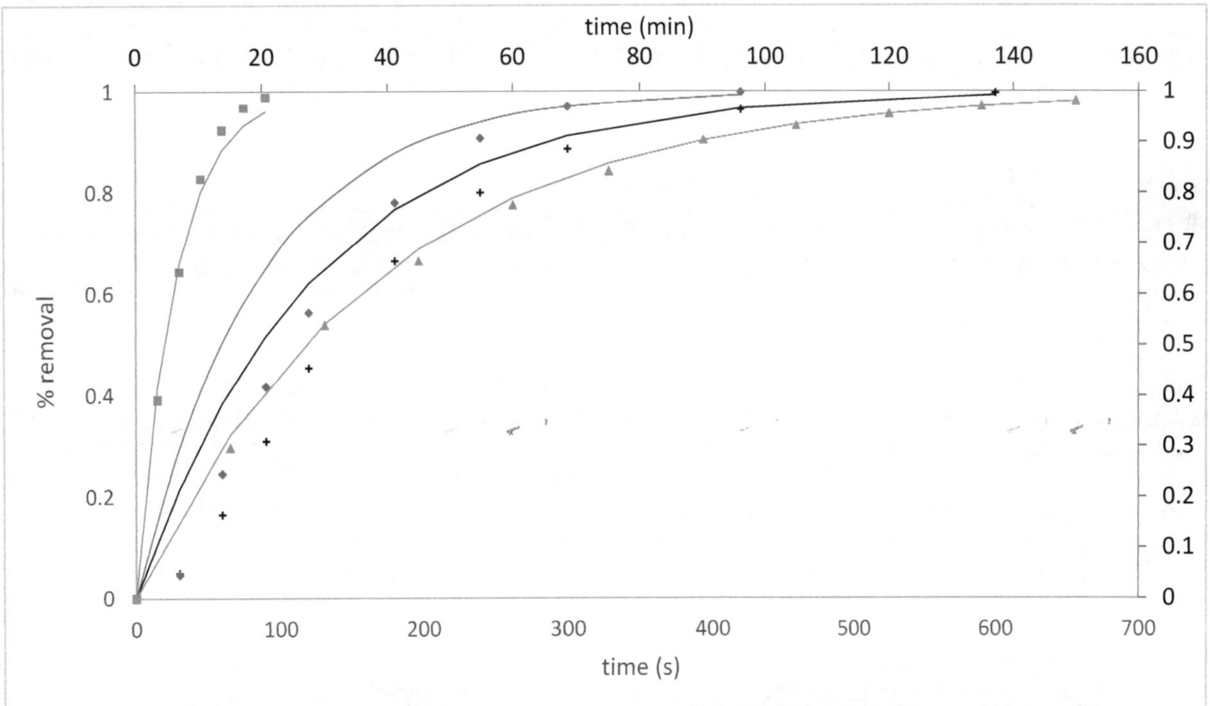


Figure 1: Removal kinetics for various AOPs
Note that UV alone follows a minute time scale

All other data follows second time scale

### 3.2.1 Parametric Analysis

Parametric analysis on the effect of UV, $O_3$, or $H_2O_2$ dosages have been discussed in other works currently seeking publication. In experiments using only UV, first order degradation rate constant was found to have a positive linear correlation with light intensity, with first order rate increasing by 0.1 $s^{-1}$ for every additional 1 $MW/m^3$ of additional light intensity. In experiments with $UV/H_2O_2$ and $UV/O_3$, first order degradation rate constant was again positively correlated with light intensity, with increases of approximately 7.0 $s^{-1}$ and 2.0 $s^{-1}$ for every 1 $MW/m^3$ increase in light intensity, respectively. When chemical dosages of $O_3$ or $H_2O_2$ were increased, an increase in degradation rate was seen, although rate increase was dependant on UV dosage and did not follow a linear trend.

The significant effect of light dosage on degradation of using $UV/H_2O_2$ compared to $UV/O_3$ indicates differences in breakdown mechanism between the two AOPs. In both $UV/H_2O_2$ and $UV/O_3$ processes, there is an expectation that in the presence of UV irradiation that HO• production will be of primary importance to the degradation of recalcitrant contaminants such as VEN. For $UV/H_2O_2$ processes, experiments utilizing $H_2O_2$ in dark conditions yielded extremely low degradation, indicating that when irradiated with UV the primary degradation mechanism is the formation of HO• that subsequently oxidizes VEN. Alternately, processes utilizing $O_3$ in dark conditions showed considerable degradation of VEN, with marginal improvement in degradation rate as UV intensity was increased. This demonstrates that $O_3$ is capable of degrading VEN alone, and that there is only partial HO• conversion as UV intensity is increased. Experiments to isolate the formation of HO• in solution have not been conducted for the purpose of this paper, leaving a gap in knowledge regarding how much degradation under $UV/O_3$ can be attributed to HO• formation as opposed to UV simply exciting VEN molecules making them more reactive to the $O_3$ present in solution. However, the practical implications of the experimental findings presented in this paper are clear regarding controlling design parameters in the development of AOP processes for large scale application; that increases in UV dosage provide much greater value in $H_2O_2$ AOPs compared to marginal improvement to $O_3$ processes.

## 3.2 Cost Analysis

When measuring the current discharge levels of VEN from WWTPs along with the lowest adverse effect levels seen in literature, it can be estimated that removal of roughly 65% of VEN must be accomplished during additional tertiary treatment if discharge levels are to be brought below levels of environmental concern.

Understanding the operational costs for each treatment process is of high importance in the determination of which process is likely to be applied on a large scale. Further, capital costs in the setup of a process are a major consideration. Basin volume requirements are directly related to the minimum retention time of each process and can indicate the amount of land space required for the process. As land costs can vary based on location, basic volume was taken as an economic consideration apart from capital costs. The relative cost increase to increase basin volume can vary dependant on WWTP location. It should be noted, that although plant size is an important factor in determining both operating and capital costs, it has been excluded from this analysis for simplicity.

To provide a general overview of how the different AOPs presented relate in terms of applicability, they were each ranked from 1-4 in each of the above categories for quick comparison.

3.2.1 Operating Cost

Calculation of estimated operating cost was performed considering parameters that would be required to achieve 65% removal of VEN, based on lab studies. As all AOPs considered followed pseudo first order degradation kinetics, removing the desired percentage of VEN should be comparable.

Pricing should be considered rough values, as costs for energy and chemicals can vary considerably between regions. Dollar values are presented in American dollars, as pricing information for more readily available. Data on chemical costs and process efficiencies were sourced from specifications and prices publicly listed by manufacturers and suppliers.

Some of the input data used for cost analysis is as follows:

- Energy costs were estimated at $0.12/kWh as an average value in the United States.
- $H_2O_2$ pricing estimated at $0.345/lb at 50% purity ($0.157/kg). With 100 mg/L of $H_2O_2$ being considered, pricing is approximately $0.03/m$^3$ of water treated. Pricing assumes large quantity of $H_2O_2$ and does not include freight costs USP Technologies, 2017).
- $O_3$ production was assumed to be performed using a compressed air to $O_3$ setup. Estimated energy costs are based on power requirements of 50 kW for feed gas compressor and 150 kW for ozone generator. Generator considered produces 10 kg of $O_3$ per hour (Baratharaj, n.d.).
- Energy efficiency of low pressure mercury UV bulbs is 40% conversion to light at the required 254nm (Haraeus, 2017).

Table 1: Estimated operating costs to remove 65% venlafaxine from 1.0 m$^3$ of water

| Process type | Chemical Dosage | No UV | 647 W/m$^3$ UV | 1847 W/m$^3$ UV | 3502 W/m$^3$ UV |
|---|---|---|---|---|---|
| Ozone | 14.4 mg/h | $0.0062 | $0.0112 | $0.0133 | $0.0148 |
| Ozone | 28.8 mg/h | $0.0093 | $0.0112 | $0.0121 | $0.0120 |
| UV alone | NA | NA | $0.2954 | $0.2459 | $0.2828 |
| $H_2O_2$ | 100 mg/L | NA | $0.0335 | $0.0345 | $0.0344 |
| $H_2O_2$ | 50 mg/L | NA | $0.0183 | $0.0171 | $0.0163 |

When considering the operating costs of the different processes analysed, it is apparent that the two processes which provide the strongest value for treatment are $O_3$ and $H_2O_2$ processes. Operating cost of UV alone becomes very high due to the long retention times needed.

When considering the operating costs for $O_3$ AOPs, operating cost increases as UV intensity is increased. With only marginal improvements in degradation rate by adding UV to these processes, the faster degradation obtainable with UV is not enough to offset the additional energy costs required to power the UV bulbs. When $O_3$ dosage is increased for $O_3$ alone, operating cost increases to obtain the same level of treatment. However, processes with UV/$O_3$ show lower operating cost as $O_3$ dosage is increased.

In general, when considering $O_3$ AOPs, there is clearly some trade off that must be made between cost efficiency and degradation rate.

In $H_2O_2$ AOPs, there is little impact on operating cost as UV dose is adjusted. Degradation rates under high intensity increase enough to offset the additional energy cost of using a higher UV intensity. Therefore, designing a process with very high UV intensity can provide significantly faster degradation for the same operational cost. However, increasing $H_2O_2$ dosage considerably increases operational costs, with doubling the $H_2O_2$ dosage from 50 mg/L to 100 mg/L roughly doubling costs at all UV intensities tested. With the significant impact that $H_2O_2$ dosage has on operating costs, care should be taken in selecting the minimal dosage required for the required task, with higher dosages useful for applications that require short retention time.

Ranking based on operating costs:
1. $O_3$ alone
2. UV/$O_3$
3. UV/$H_2O_2$
4. UV alone

3.2.2 Capital Cost

There can be significant capital costs when setting up an AOP process on large scale for wastewater treatment. These costs can vary significantly depending on what processes are currently in place. For instance, adding an UV/$H_2O_2$ process to a facility that already utilizes UV disinfection can be done at considerably lower cost than a completely new system. For this analysis, it will be assumed that there is no existing infrastructure in place, to provide a fair comparison of the technologies for new facilities. Analysis presented is of a basic nature, and consists primarily of major costs obtained from literature only, to allow for a general comparison.

Considerable literature has been published on the costs of implementation of various treatment processes, with costs can varying considerably based on many factors. Influent quality, desired effluent quality, site specifics, required capacity and variability in manufacturer quotes are only a few of the factors that can impact capital costs. Due to the complexity of the issue, a review of literature was performed to provide a general estimate of how costs compare between different treatment processes.

Plumlee et al. (2014) provided the most comprehensive available cost data on the processes considered, and was therefore used as the primary resource in ranking capital costs of treatment equipment (Plumlee et al. 2014). For UV/$H_2O_2$ based systems, total capital costs for various system capacities were determined, with average total unit cost estimated in terms of million dollars spent per mega gallon per day (MGD) of treatment capacity. For this analysis, an average plant capacity of 50 MGD was chosen, unit total unit cost was estimated to be 0.21 $M/MGD. It should be noted that $H_2O_2$ dosage considered for the estimate was considerably lower than that considered for this paper, but change in $H_2O_2$ dosage will have only a small impact on total capital cost.

Ozone based systems, estimated by Plumlee et al. (2014) showed that at roughly 50 MGD, that the total capital costs are estimated at 0.29 $M/MGD. It should be noted that estimates for capital costs of ozone based systems show considerably more variance that $H_2O_2$ based systems, with costs per treatment decreasing considerably as plant capacity increases.

A paper by Cotton et al. provided a detailed breakdown of costs associated with implementing UV treatment systems (Cotton et al. 2001). Based on their estimation, at a design for of roughly 50 MGD, a capital cost of roughly $72 000 per MGD were expected. These costs can vary considerably based on required UV intensity, with the upper range of UV intensity of this study being roughly 5 times higher than in standard practice for disinfection. Further, the very high contact times for UV treatment alone would necessitate a drastically larger number of lamps, ranging from 20 to as high as 100 times the amount of UV lamps required compared to other systems considered based on retention time requirements. An estimate for the amount of UV scale-up required was taken to be on the low end at 20 times traditional UV disinfection treatment, considerably increasing the capital costs to 1.4 $M/MGD.

A combination of the above costs was considered for estimating the capital costs of a UV/$O_3$ treatment system. A combination of the estimated capital costs for an $O_3$ system along with a standard UV disinfection system yielded an estimated total capital cost of 0.36 $M/MGD.

Ranking based on capital cost:
1. UV/$H_2O_2$
2. $O_3$ alone
3. UV/$O_3$

4. UV alone

3.2.3 Basin volume requirements

When considering any potential new process, retention times are of high importance. With the large water volumes being treated daily at municipal wastewater treatment plants, even small increases in retention time can lead to very large increases in required basin volume. The significance of these requirements will vary depending on the location of a given facility. Therefore, estimated volume requirements for each process will be determined by tank volume needed per 1000 $m^3$/d of flow input to the facility.

Table 2: Estimated basin volume needed to remove 65% venlafaxine from a 1000 $m^3$/d influent

| Process type | Chemical Dosage | No UV | 647 W/$m^3$ UV | 1847 W/$m^3$ UV | 3502 W/$m^3$ UV |
|---|---|---|---|---|---|
| Ozone | 14.4 mg/h | 1.50 $m^3$ | 1.87 $m^3$ | 1.40 $m^3$ | 1.04 $m^3$ |
| Ozone | 28.8 mg/h | 1.13 $m^3$ | 1.10 $m^3$ | 0.88 $m^3$ | 0.65 $m^3$ |
| UV alone | NA | NA | 158.5 $m^3$ | 45.6 $m^3$ | 28.0 $m^3$ |
| $H_2O_2$ | 100 mg/L | NA | 1.35 $m^3$ | 0.64 $m^3$ | 0.34 $m^3$ |
| $H_2O_2$ | 50 mg/L | NA | 1.51 $m^3$ | 0.87 $m^3$ | 0.42 $m^3$ |

UV/$H_2O_2$ processes provide a significant advantage over all other processes when considering required basin volume. Ozone based processes are competitive with $H_2O_2$ based processes, particularly when UV irradiation is utilized. In general, the addition of UV provides the ability to significantly reduce the volume required in any of the AOPs discussed.

Considering treatment with UV alone, the volume requirement to treat VEN are high enough to make it non-viable.

Ranking based on required basin volume:
1. UV/$H_2O_2$
2. UV/$O_3$
3. $O_3$ alone
4. UV alone

3.2.4 AOP Comparison

When selecting an AOP for a given process, there are multiple factors to consider, the most significant of which is cost. In comparing the potential costs of AOPs to degrade VEN, cost considerations were broken down into three categories: operating cost, capital cost, and required basin volume. As research on new technologies for the removal of pharmaceuticals from water emerge, it is essential that engineers keep a perspective on which processes are the most viable financially, to ensure that research is focused on areas with the highest potential.

As detailed above, operating costs and basin volumes were based on extrapolation of data obtained in lab scale experiments to degrade VEN. Capital costs were based on a review of literature on existing processes that are similar to the proposed AOPs. Trying to estimate costs for a process to remove VEN or other pharmaceuticals at this early point of technology refinement does not allow for high accuracy in cost analysis. However, data was sufficient to do a ranking based comparison, with differences in cost between different process being large enough to do so confidently.

It is immediately clear that attempting to degrade VEN by UV alone would not be a cost-effective method of treatment. Existing WWTPs that utilize UV disinfection are typically orders of magnitude below the UV dosage that would be required to sufficiently remove VEN from wastewater effluents. The massive volume requirements to obtain sufficient removal, along with the energy requirements to reach sufficient UV dosage makes UV alone a process that is not viable for removing VEN in wastewater.

When considering $O_3$ and UV/$O_3$ processes, both processes are competitive in terms of pricing analysis. On site ozone generation, along with contact tanks and ozone diffusers come with an increase in capital costs, these costs can be offset by reducing the required basin volume. When choosing to add UV along with an ozone based process, the primary trade-off is a reduction in required contact time versus the additional operational cost required to power the UV system. If ozone based AOPs are ultimately found to be the most prominent technology finding implementation in coming years, there will likely always be a balance between the design levels of UV intensity and cost. Urban environments where space is at a premium may find that using a high UV intensity allows them to use considerably smaller contact tanks, while small rural WWTPs may choose to utilize UV alone.

AOPs using UV/$H_2O_2$ present an interesting case, because they allow for a very high amount of control over degradation parameters. When applied with $H_2O_2$, UV irradiation showed a very large impact on degradation rates of VEN, with the affect being particularly pronounced at higher concentrations of $H_2O_2$. In terms of operating costs, $H_2O_2$ based AOPs showed a trend that as UV dosage was increased, operating costs either remained constant or decreased. Although still behind $O_3$ processes for operating costs, this trend presents an interesting opportunity in the cost optimization of $H_2O_2$ processes, where increasing UV intensity allows for a decrease in not only retention times, but also in operating costs. It is possible that with a high level of optimization, $H_2O_2$ processes may prove to be ideal for many applications. The relative simplicity of applying $H_2O_2$ AOPs allows for lower capital costs, with further a further advantage being that existing WWTPs with UV disinfection systems could be easily retrofitted to utilize $H_2O_2$. When optimizing $H_2O_2$ AOPs, there appears to be opportunity to create a process that is highly efficient in terms of all the cost categories considered if UV intensities can be increased by a factor of 5 or more.

## 4. CONCLUSIONS

A laboratory study was conducted to determine the effectiveness of various AOPs to remove the pharmaceutical VEN. The methods compared in this study were: UV alone, UV/$H_2O_2$, UV/$O_3$, and $O_3$ alone. All methods can achieve greater that 95% degradation of VEN. A cost analysis was then performed based upon energy consumption and chemical requirements for each process.

The major conclusions of this study are:

- UV/$H_2O_2$, UV/$O_3$ and $O_3$ are all potentially cost competitive options for the removal of VEN. Large retention time and energy requirements makes UV an non-viable choice for the treatment of VEN.
- $O_3$ proved to be the best option in terms of operating cost. The addition of UV to $O_3$ treatment showed a decrease in required retention time, but the increased energy requirement makes the process less cost efficient.
- UV/$H_2O_2$ was found to be more expensive in terms of operating costs. However, increasing UV intensity allows for decreased retention times, while remaining neutral in terms of operating costs.
- In terms of basin volume, UV/$H_2O_2$ was the best, with UV/$O_3$ second. Required basin volume can be significantly reduced by increasing UV dosage in $H_2O_2$ processes, and marginally improved by increasing UV dosage in $O_3$ processes.
- UV/$H_2O_2$ was determined to have the lowest capital costs, followed by $O_3$ then UV/$O_3$.

Based on this study, all the considered processes except for UV have potential to be effective to remove VEN. Determining the most effective process may come down to a case by case basis. It is significant however that higher UV dosages can provide substantial improvements to AOP processes. Particularly for $H_2O_2$ based processes, drastic reduction in the required basin volume along with no increase in operational costs makes finding methods to increase UV intensity a strong consideration.

## 5. REFERENCES


Bader, H., and J. Hoigné. 1981. "Determination of Ozone in Water by the Indigo Method." *Water Research* 15 (4): 449–56. doi:10.1016/0043-1354(81)90054-3.

Baratharaj, V. n.d. "How to Evaluate and Select an Ozone Generator.". Ozone Technologies & Systems (India) Pvt. Ltd. Accessed February 7, 2017. http://www.otsil.net/articles/HOW%20TO%20EVALUATE%20AND%20SELECT%20AN%20OZONE%20GENERATOR.pdf

Calvert, JG, and JN Pitts. 1966. "Photochemistry." http://www.bcin.ca/Interface/openbcin.cgi?submit=submit&Chinkey=136979.

Cotton, C. A., Douglas M Owen, G. C. Ćline, and Timothy P Brodeur. 2001. "UV Disinfection Costs for Inactivating Cryptosporidium." *Journal / American Water Works Association* 93 (6): 82–94.

García-Galán, Jesús J., Alba Anfruns, Rafael Gonzalez-Olmos, Sara Rodríguez-Mozaz, and Joaquim Comas. 2016. "UV/H2O2degradation of the Antidepressants Venlafaxine and O-Desmethylvenlafaxine: Elucidation of Their Transformation Pathway and Environmental Fate." *Journal of Hazardous Materials* 311. Elsevier B.V.: 70–80. doi:10.1016/j.jhazmat.2016.02.070.

Heraeus "UV Low-Pressure Lamps.". Accessed February 7, 2017. https://www.heraeus.com/en/hng/products_and_solutions/uv_lamps_and_systems/uv_low_pressure_lamps.aspx.

Hamilton, Patrick B, Ian G Cowx, Marjorie F Oleksiak, Andrew M Griffiths, Mats Grahn, Jamie R Stevens, Gary R Carvalho, Elizabeth Nicol, and Charles R Tyler. 2016. "Population-Level Consequences for Wild Fish Exposed to Sublethal Concentrations of Chemicals - a Critical Review." *Fish and Fisheries* 17 (3): 545–66. doi:10.1111/faf.12125.

Lajeunesse, A., S. A. Smyth, K. Barclay, S. Sauvé, and C. Gagnon. 2012. "Distribution of Antidepressant Residues in Wastewater and Biosolids Following Different Treatment Processes by Municipal Wastewater Treatment Plants in Canada." *Water Research* 46 (17): 5600–5612. doi:10.1016/j.watres.2012.07.042.

Melnyk-Lamont, Nataliya, Carol Best, Manuel Gesto, and Mathilakath M. Vijayan. 2014. "The Antidepressant Venlafaxine Disrupts Brain Monoamine Levels and Neuroendocrine Responses to Stress in Rainbow Trout." *Environmental Science and Technology* 48 (22): 13434–42. doi:10.1021/es504331n.

Plumlee, Megan H., Benjamin D. Stanford, Jean-François Debroux, D. Cory Hopkins, and Shane A. Snyder. 2014. "Costs of Advanced Treatment in Water Reclamation." *Ozone: Science & Engineering* 36 (5). Taylor & Francis: 485–95. doi:10.1080/01919512.2014.921565.

Ribeiro, Ana R., Olga C. Nunes, Manuel F R Pereira, and Adrián M T Silva. 2015. "An Overview on the Advanced Oxidation Processes Applied for the Treatment of Water Pollutants Defined in the Recently Launched Directive 2013/39/EU." *Environment International* 75. Elsevier Ltd: 33–51. doi:10.1016/j.envint.2014.10.027.

Subramani, Arun, and Joseph G. Jacangelo. 2014. "Treatment Technologies for Reverse Osmosis Concentrate Volume Minimization: A Review." *Separation and Purification Technology* 122. Elsevier B.V.: 472–89. doi:10.1016/j.seppur.2013.12.004.

USP Technologies "How Much Does H2O2 Cost?." . Accessed February 7, 2017. http://www.h2o2.com/faqs/FaqDetail.aspx?fId=25.

Leadership in Sustainable Infrastructure
*Leadership en Infrastructures Durables*

Vancouver, Canada
May 31 – June 3, 2017/ *Mai 31 – Juin 3, 2017*

# CROSS-CORRELATING LANDFILL GAS QUALITY WITH CLIMATIC FACTORS AT THE CITY OF REGINA LANDFILL

Richter, Amy[1] and Ng, Kelvin T. W.[2]
[1,2] Faculty of Engineering and Applied Science, University of Regina, Canada
[2] kelvin.ng@uregina.ca

**Abstract:**

Landfill gas is a by-product of the anaerobic decomposition of organic waste, and consists mainly of $CH_4$ (methane) and $CO_2$ (carbon dioxide), both of which are important when considering discussions on climate change and global warming. Five years of landfill gas data at the Regina landfill was collected to study the impacts of selected climatic factors on landfill gas collection using cross-correlation analysis. Cross-correlation is used to estimate the degree of correlation between two sets of time-series data, and returns the lag, or delay, between each series. Cross-correlation has been used successfully in a number of different fields, including seismic refraction, finance, and climate; however, its use in the field of solid waste management is extremely limited. The objective of this study is to examine five different climatic factors (rainfall, snow cover, temperature, pressure and relative humidity) on landfill gas quality ($CH_4$:$CO_2$) and residuals (the gas collected that is not $CH_4$ or $CO_2$), of which precipitation will be discussed in this paper. Surprisingly, a consistent lag time was not observed at the Regina landfill during the study period. Results suggested that a number of reasons could be behind this phenomenon. Understanding the impact that these climatic factors may have on landfill gas collection could help to increase efficiency of collection, especially in harsh climates. As well, information on the relationships in between climatic events and increases or decreases in landfill gas quality may help landfill designers and operators to optimize the landfill system and improve environmental compliance at the landfill.

## 1 Introduction

Canadians are one of the largest producers of solid waste in the world, generating 965 kg/cap in 2010 (Richter et al., 2017; Bruce et al., 2016). The majority of Canadian waste is sent to landfills, due to the high availability of un-developed land (Bruce et al., 2016). Saskatchewan and western Canada have grown substantially, making efficient management of solid waste more difficult (Wang et al., 2016).

Landfills contribute a significant amount of anthropogenic greenhouse gases to the atmosphere (Park and Shin, 2001; Fourie and Morris, 2004; Di Bella et al., 2011; Spokas et al., 2005; Themelis and Ulloa, 2007). In Canada, landfills account for approximately 20% of all national methane emissions (Environment Canada, 2014); while in the United States, landfills are the third largest source of methane emissions (USEPA, 2015). Landfill gas consists primarily of methane ($CH_4$) and carbon dioxide ($CO_2$), the two most important greenhouse gases, along with other trace gases. $CH_4$ is 23 times more potent than $CO_2$ on a molar basis over a 100 year time frame (Spokas et al., 2005).

The City of Regina landfill is located in a semi-arid climate, with average precipitation of about 308.9 mm per year (Statistics Canada, 2016). The landfill has two peaks, one of which has been closed and covered with final cover. The landfill also has a landfill gas (LFG) collection system, which collects and flares LFG.

Precipitation data and LFG composition data were examined in this study to explore their inter-relationships.

In their 1973 study, Farquhar and Rovers found that moisture content of waste plays an important role in supporting maximum LFG production. The moisture content of waste entering the majority of US landfills at the time of the study was much lower than the 60-80% wet weight suggested for maximum LFG generation (1973).

Munasinghe (1997) studied how hydraulic retention time (HRT) at a landfill was related to changes in landfill gas and leachate quality. The HRT is the time that water spends suspended with the solid waste in a landfill. Munasinghe (1997) found that as the HRT reduces (from increased precipitation) $CH_4$ concentration is increased because of the dissolution of $CO_2$ into the water that enters the landfill, as well as enhanced methanogenesis because of decreased $CO_2$ partial pressure.

Wreford et al. (2000) measured and analyzed gas composition from 8 wells at an active municipal solid waste (MSW) landfill in Vancouver. Field evidence suggested that $CH_4$ production was affected by precipitation 14 days prior to sampling, and that high moisture input coincided with peak $CH_4$ production during the study period (Wreford et al., 2000). According to their study, this may have been due to a number of factors, including organic content (which would have been placed at the top of the landfill) leaching down as a result of precipitation to increase microbial activity in other areas in the landfill, and thus increase the amount of $CH_4$ being produced. As well, a reduced HRT may have caused pH to rise, which then favours methanogenesis. LFG quality ($CH_4$:$CO_2$) also increased dramatically following periods of intense rainfall (7 days prior to testing), which may have been due to dissolution of $CO_2$ (Wreford et al., 2000).

Park and Shin (2001) studied surface effluxes of methane from an MSW landfill in Korea, receiving 1318 mm of rain per year. During summer, the efflux rate rapidly increased compared to other seasons because of higher precipitation. This caused the pores to close off on the surface of the landfill and thus act as a barrier for the landfill gas to efflux from the landfill; however, once the sun came out, the surface would dry and surface emissions would increase. This Korean landfill experiences Regina's yearly precipitation in a 4 month timeframe.

However, contradictory field observations were reported in literature. Farquhar and Rovers found that large infiltrations of water into the test cells disrupted $CH_4$ production via inhibition of methanogenic bacteria in their study (1973). Peer et al. (1992) collected data including landfill geometry, well depth, and climate variables from 21 US landfills that were either optimizing or attempting to optimize gas recovery, and found that the best predictive variables for methane emissions using a regression model were depth and waste mass. Climatic variables such as precipitation, temperature, and dew point, however, were not found to be statistically significantly related to annual methane recovery.

Relationships between climatic variables and landfill moisture content are still not clear. The environment in which the landfill is located will determine the amount of precipitation that the landfill receives, although the amount of water that physically percolates into the landfill may vary greatly depending on the physical and operational conditions at the landfill (such as the site topography, cell geometry and design, final cover and daily cover practices at the landfill). Sadek et al. (2007) found that in semi-arid climates, dessication cracks and burrowing animals along with dry periods tended to cause clay covers to meet their plastic limit quickly. Their study found that after the first wetting and drying cycle, there was a significant increase in water infiltration through the cover due to cracking.

Based on the conflicting literature, the objective of this study is to use cross-correlation to determine if precipitation has any effect on LFG quality at the City of Regina landfill. Unlike most closed landfills, the City of Regina landfill is still operating, making it more susceptible to moisture changes within the landfill. Cross-correlation studies have been used to detect the similarity of two series as a function of the lag (delay) in between the two. It has been applied successfully in a number of different fields; however, its use in the field of waste management is limited.

# 2 Methodology

## 2.1 Study Site

The City of Regina landfill is located in the northeast quadrant of the city, with a total area of 97 ha, with refuse occupying 43 ha (Lefevbre, 2006). Cut-fill waste disposal practices occurred until 1963, and since then, the site has operated as a monolithic unlined cell. The landfill serves the City of Regina and surrounding area, and accepts a variety of waste including: municipal solid waste, building demolition waste, fill dirt, recyclable concrete, asphalt and other materials (Gallant, 2011). The average typical composition of waste at the City of Regina landfill includes about 34.5% organic waste (organics and leaf and yard waste), as well as about 20% paper and cardboard waste, with plastics making up about 7.5% of waste. The remaining waste (glass, metal, and household hazardous waste) made up about 10%, and the remaining waste (26.5%) was categorized into the other category (City of Regina, 2009). It is estimated that 3.1 Mt of waste had been placed at the City of Regina landfill in between 1961 and 2011. The LFG well field consists of 27 extraction wells (Latoski, 2014), and covers approximately 38% of the total area of the landfill and was covered in 2007 with 1 m of low permeability soil with 150 mm of vegetated top soil (Lefevbre, 2006).

The LFG collection system at the City of Regina landfill is an active system, with a blower used to collect LFG from vertical wells. LFG was analyzed by a real-time SCADA (Supervisory Control and Data Acquisition) system, and data for every single minute (except during scheduled maintenance and unexpected outages) was recorded. For the purpose of this research, the per-minute readings of $CH_4$ and $CO_2$ percentages were consolidated into daily averages. During the analysis of the LFG data, it became clear that a small number of LFG measurements were considerably lower than others. An investigation of these abnormal readings was carried out. A normal distribution was used to classify anomalous data points as outliers, and for any data point that fell below three standard deviations (encompassing 99.7% of the data) from the mean, the daily SCADA files were examined. The normal distribution was used because large data sets tend towards a normal distribution, and the application of the central limit theorem, If the daily readings dropped to a negative value, the entire set of averaged values was removed. After carrying out these steps, a total of 55 data points (about 2.8% of the total) were removed. Most outliers originated from scheduled downtimes and maintenance works.

Data was collected in between August 2008 and January 31$^{st}$, 2014, with the exception of $CO_2$ percentages, which were only recorded for the period of August 2008 to January 31$^{st}$, 2013. Precipitation data (daily rainfall and total precipitation) was gathered from Environment Canada Gilmour Weather Station, 50.667°N, 104.833°W, approximately 25.4 km northwest of the landfill. Rainfall was measured by a standard Type B rain gauge, measuring in millimeters. From 1981-2010, Regina experienced an average annual rainfall of 308.9 mm per year (Statistics Canada, 2016).

## 2.2 Data Analysis

Unlike LFG quantity, quality or composition is less sensitive to site conditions and waste characteristics. In this study, $CH_4$ and $CO_2$ composition are analyzed as a ratio ($CH_4$:$CO_2$) along with percentage gas residuals (=100%-($CH_4$%+$CO_2$%)). Key parameters in gas composition such as $CH_4$:$CO_2$ and percent residual were selected to facilitate comparisons with previous work (Wreford et al., 2000).

Seven study periods were chosen for this study, all of which occurred during the spring and summer (no events were considered where there was snow or freezing temperatures) of 2010, 2012, and 2013. Each study period was categorized based on the ratio of days with any amount of precipitation to the total number of days in the study period, hereafter referred to as the 'precipitation ratio'. The precipitation ratio (PR) was chosen so that there were periods where less than, equal to, and greater than 50% of the days had precipitation. Data for the intensity (mm/hr) of each individual rainfall event was not available, but study periods were chosen so that either there was a day in the study period with a significant amount of rain (greater than 50 mm) or there were a number of concurrent days that had significant amounts of rain (multiples days with precipitation greater than 15 mm). Despite having 6 years of data available, data was only chosen for 3 years because of the long sections of complete data for both precipitation and LFG quality. These events were chosen because they are believed to reflect typical rainfall events that occur in Regina.

Cross-correlation studies have been used to detect the similarity of two series as a function of the lag (delay) in between the two. It has been applied successfully in a number of different fields, including seismic refraction, finance, physiology, climate, and genomics (Barnston, 1993; Podobnik, 2010; Mustafa and Hakan, 2014).

Cross correlation is used to estimate the degree of correlation of two series as a function of a shift of one of the curves. The cross-correlation function, for a pair of time series data can be given by the following equation:

[1] $C_{xy} = (t+\tau,t) = \frac{1}{T}\int_0^T x(t+\tau)y(t)dt$

Where $\tau$ is the lag (or delay) is between two data sets, T is the duration, and t is the time increment (Chen et al., 2007). The maximum value of the cross-correlation coefficient (y-axis in Figure 1) is ±1 (±100%) and denotes that both sets of data are exactly the same (positive) or exactly dissimilar (negative) at the same time ($\tau = 0$). If two sets of time series data are similar, but not exactly the same, the value of the cross-correlation coefficient will in between 0 and ±1 (Vorburger et al., 2011). In order to ensure fair comparison between study periods, all data was normalized. The confidence interval was set to 95% a typical value for data analysis in this field.

## 3 Results and Discussion

A typical cross-correlation plot developed in this study is shown in Figure 1, which corresponds to row 3 of Table 1, below. The lag in the figure is shown in units pre-determined by the software used, and is translated to days for the purpose of analysis. For a PR of 0.450, there is a strong positive correlation (+0.530) in between precipitation and an increase in $CH_4$:$CO_2$ at approximately 4.62 days after the initial rainfall occurred, shown as a lag of "-3" in Figure 1.

Please note that cross-correlation analysis can be subjective, as the researcher is required to select the study period for each precipitation event. The results that occur for lag greater than zero (on the right hand side of Figure 1) imply that first there is an increase in the LFG ratio, and then precipitation occurs, a finding that is not consistent with observations in the field. Results from the statistical analysis also support this, as most of results on the right hand side were less than the 95% confident interval. As such, data analyses and discussion will only focus on the left-hand side of the figure (lag equal to or less than 0).

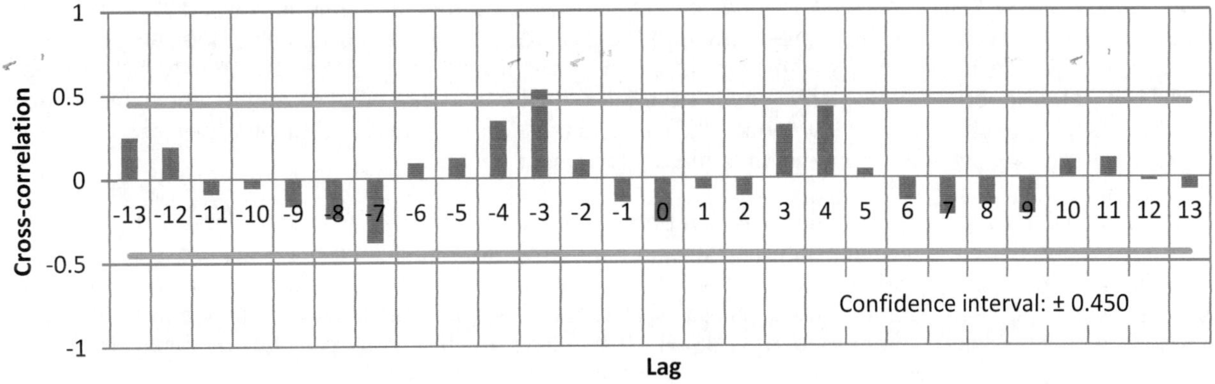


Figure 1: Example of Cross-correlation between precipitation and $CH_4$:$CO_2$ (PR of 0.450)

From Table 1, the lag varies significantly, with a minimum lag period of 0 days and a maximum lag period of 13.85 days with a mean lag period of about 4.9 days. Most cross-correlation coefficients were positive, indicating that $CH_4$:$CO_2$ increased as precipitation increased. Wreford et al. (2000) reported a dramatic increase in $CH_4$:$CO_2$ after 7 cumulative days of rainfall leading up to sampling, citing dissolution of $CO_2$ as a likely cause for this increase. Using a confidence interval of 95%, 3 positive correlations (PR = 0.450, 0.500, 0.563) and 1 negative correlation (PR = 0.588) were observed.

Two (PR=0.533 and PR=0.588) of the seven cross-correlation coefficients were found negative with values of -0.509 and -0.492, respectively. These negative coefficients are considered less reliable for the following reasons: (i) When compared to the averaged values in Table 1, we can see that the maximum daily rainfall and cumulative precipitation over the specific study period associated with PR=0.533 and PR=0.588 are considerably lower. Because of the lower amount of water involved in the PR=0.533 and PR=0.588 rain events, it is believed that actual infiltration rates and their effects are also lowered due to absorption of water from wetting and surface ponding, (ii) the negative coefficients are statistically less significant as the results are below (PR=0.533) or marginally above (PR=0.588) the 95% interval.

The precipitation is also lower than average for PR=0.650, but the cross-correlation coefficient is found to be positive. It should be noted that the cross-correlation coefficient is much lower than the 95% confidence threshold (=±0.450). The corresponding lag (13.9 days) is also much higher than the average value. According to our results, precipitation may or may not be a factor in the quality of collected landfill gas, as further discussed below.

Table 1: Summary of Cross-correlation between precipitation and $CH_4$:$CO_2$

| Precipitation Ratio | Max Rainfall (mm) in study period | Cumulative Precipitation (mm) | Lag (days) | Cross-correlation Coefficient | 95% Confidence Interval |
|---|---|---|---|---|---|
| 0.450 | 19.6 | 71.8 | 4.62 | 0.530 | ±0.450 |
| 0.500 | 63.0 | 145.8 | 0.00 | 0.409 | ±0.352 |
| 0.533 | 15.2 | 40.8 | 1.36 | -0.509 | ±0.524 |
| 0.563 | 26.1 | 74.7 | 0.00 | 0.523 | ±0.506 |
| 0.566 | 18.0 | 61.6 | 4.50 | 0.315 | ±0.475 |
| 0.588 | 12.2 | 54.7 | 9.92 | -0.492 | ±0.490 |
| 0.650 | 6.8 | 34.0 | 13.85 | 0.287 | ±0.450 |
| **AVERAGE** | **23.0** | **69.1** | **4.90** | | |

The City of Regina landfill has final cover on the closed peak, which encourages surface runoff and limits infiltration rate. However, dessication and cracking of the cover can occur in arid climates (Daniel and Wu, 1993; Sadek et al., 2007), and via thermal contraction (Andersland and Al-Moussawi, 1987), both of which would allow water to seep more easily into the covered portion of the landfill. Other areas of the landfill are still operational during the study period, and so water may be able to seep laterally through the landfill. As Farquhar and Rovers (1973) reported, a large increase in infiltrations into the landfill disrupted $CH_4$, a possible reason why results are inconclusive. The fact that Regina is located in a semi-arid climate and is a partially closed landfill contributes to the complexity of analysis.

Direct relationships were not observed between precipitation ratios and lag time from Table 1. In order to see how well related various outcomes (lag and cross-correlation coefficient) from the cross-correlation analysis were plotted against the precipitation factors (precipitation ratio, maximum precipitation and cumulative precipitation) over the study period. The results suggest that the lag is to a certain degree related to all precipitation factors ($p<0.15$, $0.35<R^2<0.45$), whereas the cross-correlation coefficient is not ($p>0.15$, $R^2<0.2$ in all cases). A weak positive correlation is found between precipitation ratio and lag, as shown in Figure 2. The rainfall regime and pre-existing moisture conditions at and within the landfill will also likely affect how precipitation affects LFG quality changes. As mentioned previously, Regina only receives about 308.9 mm of rainfall per year. The work done by Wreford et al. (2000) was for a coastal area in western Canada that received 1,167 mm of annual precipitation. Because Regina receives much less precipitation than Wreford et al.'s study (about 26% of precipitation), the moisture content of the buried waste in Regina may be much lower than in their study. It is hypothesized that the lower moisture saturation of waste at the Regina landfill may be partially responsible for a lack of strong correlation between the lag time and the quality of LFG, since any moisture received during drier periods may saturate the waste first, before other mechanisms help to increase production of $CH_4$.

The semi-arid climate at the City of Regina landfill is likely a factor in the inconsistent results observed when comparing precipitation and $CH_4$:$CO_2$. Data in the present study is taken from multiple years, and conditions at the landfill prior to the rainfall events were not considered, so it is possible that even the amount of rainwater received may not significantly affect the LFG quality. Munasinghe (1997) found that the HRT is lower when there is more precipitation. A lower HRT will result in higher $CH_4$ concentration (observed in our case as an increase in $CH_4$:$CO_2$) because of the dissolution of $CO_2$ into the water, as well as increased methanogenesis because of a decrease in $CO_2$ partial pressure. Blight et al. (1992) studied water storage of landfills in semi-arid climates and found that it was difficult to accurately quantify in these areas since it depended heavily on the properties and disposition of intermediate cover layers. Peer et al., (1992) also found that the best predictive variables for $CH_4$ production were depth and mass of waste at a landfill. Information on landfill conditions (in semi-arid climates) and other landfill properties are recommended for future studies.

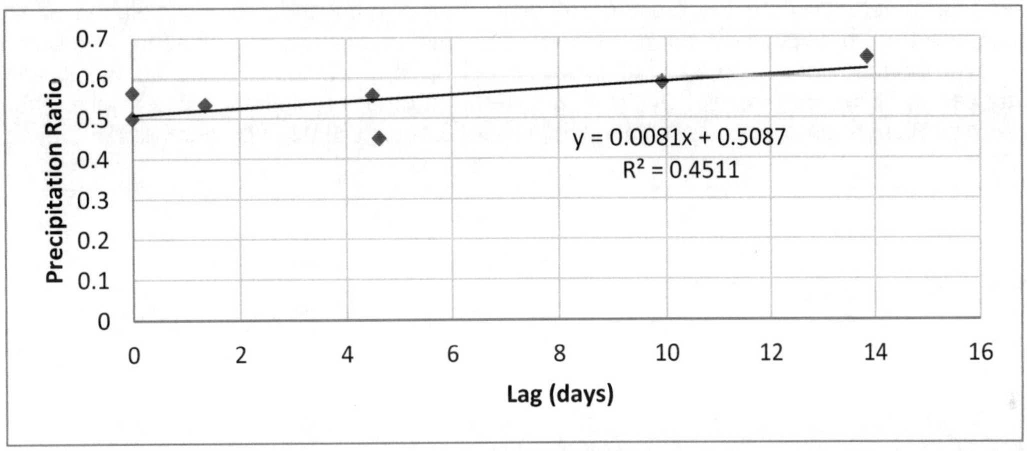


Figure 2: Relationship between precipitation ratio and lag period for $CH_4$:$CO_2$

From Table 2, the average lag observed in between precipitation and changes in LFG residual was about 5.3 days, with a minimum lag of 0 days and a maximum lag of 12.31 days observed. In 4 cases (PR=0.500, 0.533, 0.563, 0.588) the cross-correlation coefficient was positive, implying that precipitation caused an increase in the amount of residuals present. In the other cases (PR=0.450, 0.566, 0.650) the relationship was negative, meaning that precipitation caused a decrease in LFG residuals. Using a 95% confidence interval, results suggested that there is a positive relationship between PR and lag (see, for example PR = 0.500, 0.533, 0.563). Direct relationships were not observed between precipitation ratios and lag time for residual gas.

It is hypothesized that precipitation may adversely impact the operation of the landfill gas collection system, and thus cause an increase in residuals. For example, the readily degradable organics located at the top of the landfill may leach out during heavy precipitation. The LFG collection pipes placed in the landfill have a porous filter, which could be clogged by microbes and fines due to this washing action. This phenomenon has been well reported in the literature with regards to leachate collection systems (LCS). Microbial activity including the formation of multiple types of slime, or biologically induced precipitation of minerals may form 'biorock', which reduces void spaces in filter media in LCS systems (Fleming and Rowe, 2004). Although there is no research showing that this phenomenon might also occur for LFG collection systems, water infiltration after precipitation may cut the vacuum from the gas management system., This likely will lead to air intrusion and an increase of residual gases (mostly air, in this case). More research is required before definite conclusions can be made.

Table 2: Summary of Cross-correlation between precipitation and LFG residuals

| Precipitation Ratio | Max Rainfall (mm) in study period | Cumulative Precipitation (mm) | Lag (days) | Cross-correlation Coefficient | 95% Confidence Interval |
|---|---|---|---|---|---|
| 0.450 | 19.6 | 71.8 | 4.62 | -0.373 | ±0.450 |
| 0.500 | 63.0 | 145.8 | 5.83 | 0.500 | ±0.352 |
| 0.533 | 15.2 | 40.8 | 0.00 | 0.889 | ±0.524 |
| 0.563 | 26.1 | 74.7 | 2.67 | 0.621 | ±0.506 |
| 0.566 | 18.0 | 61.6 | 4.50 | -0.380 | ±0.475 |
| 0.588 | 12.2 | 54.7 | 7.08 | 0.482 | ±0.490 |
| 0.650 | 6.8 | 34.0 | 12.31 | -0.355 | ±0.450 |
| **AVERAGE** | **23.0** | **69.1** | **5.3** | | |

Again, the various outcomes from the cross-correlation analysis (lag and cross-correlation coefficient) were plotted against the precipitation factors (precipitation ratio, maximum precipitation, and cumulative precipitation) over the study period. Results shows that lag is poorly related to precipitation ratio ($p=0.29$, $R^2=0.22$), as seen in Figure 3. The relationship in between the other precipitation factors (maximum and cumulative precipitation) had non-existent relationships ($p>0.8$ and $R^2<0.02$ in both cases).

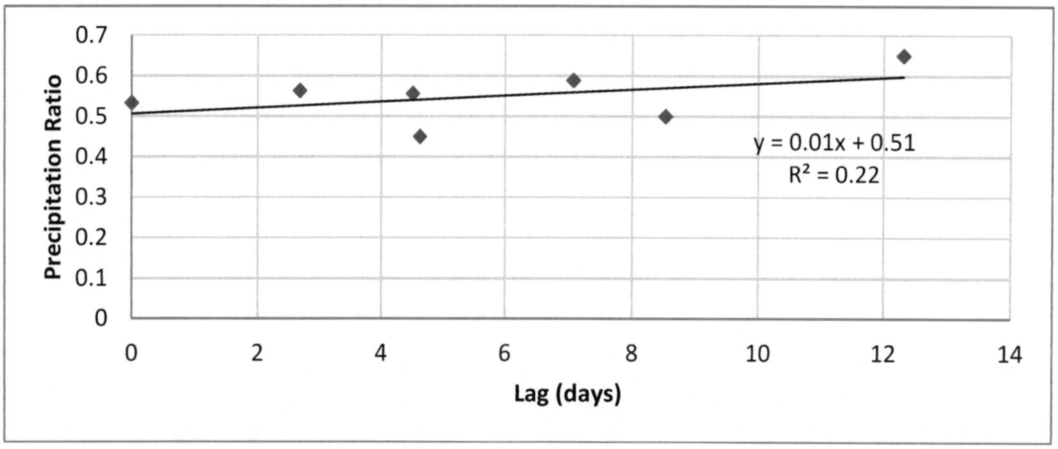


Figure 3: Relationship between precipitation ratio and lag period for residual gases

Comparing Figures 2 and 3 we can see that it appears as if $CH_4$:$CO_2$ is more closely related to precipitation than residuals when comparing $R^2$ values ($R^2_{ratio}=0.4511$ vs. $R^2_{residual}=0.2229$) and p-values ($p_{ratio}=0.0984$ vs. $p_{residual}=0.2847$). This follows from the work of Wreford et al. (2000) and Munasinghe (1997) which discuss the role of precipitation on reduction of $CO_2$ due to dissolution in water. Also, by using the confidence interval as a benchmark, we can see that about 57% of the time, cross-correlation analysis of precipitation and $CH_4$:$CO_2$ yielded results that extended past the 95% confidence limits; whereas this was only the case 43% of the time for residuals. Because of this, it is suggested that $CH_4$:$CO_2$ be used in future studies.

Research in this area is still ongoing. Selection of a lower confidence interval and different durations of rain-events are also being investigated in this study, as well as using differing definitions of continuous rainfall events. Furthermore, taking into account the previous conditions of the landfill (dry or wet years, for example) may help to explain inconsistent results. Possible areas of future work including looking at other climatic factors, such as snow cover and temperature. As well, auto-correlation, a statistical tool similar to cross-correlation where time series data is compared to itself to find periodicity, could be used to see if there is any fluctuation in LFG quality from day to night. Further to this, it may be interesting to look more in depth at the actual constituents of LFG residuals, that is – their chemical composition. Without knowing the exact composition of residuals gases, it is difficult to come to any conclusions about chemical and biological activity in the landfill that may relate LFG residuals to precipitation.

## 4 Conclusions

The purpose of this study was to use cross-correlation to determine if there was any relationship in between precipitation and changes in LFG quality at the City of Regina landfill. Results show that there is not a consistent lag period observed in between $CH_4$:$CO_2$ and LFG residuals and precipitation. An average lag period of about 5 days was observed in between precipitation and changes in $CH_4$:$CO_2$. In five of the seven cases studied, there was an increase in $CH_4$:$CO_2$, while in 2 cases, a negative cross-correlation was observed. This could be the case for a number of reasons, including much less precipitation occurring on average in Regina (308.9 mm per year), as well as having drier and wetter years. Results showed that $CH_4$:$CO_2$ was relatively well related to all precipitation factors (precipitation ratio, maximum precipitation, and cumulative precipitation). During dry years, an increase in HRT may not be observed, simply because increased precipitation does not increase HRT as much. More research is required in order to come to a more definite conclusion.

The relationship in between precipitation and LFG residual was only somewhat relatively related to the precipitation ratio only (unlike the ratio, which was relatively related to all precipitation factors used in this study). The average lag observed for residuals and precipitation was 5.3 days. It is likely that the relationship in between precipitation and residuals is more related to operational characteristics of the landfill gas management system. For example, microbial activity, which has been reported to cause clogging of leachate collection systems, may potentially also do the same to landfill gas collection wells, due to the downward movement of highly leachable organic material during rain events. Precipitation may thus cover the pores in the gas collection filter material, and cause the blower to work harder, causing the collection of residual gases.

As mentioned previously, research using this method is ongoing, and currently work is being carried out to see if there is any relationship in between LFG quality and snow cover. Other areas of study will include temperature, relative humidity, and pressure. With respect to this work, research could be done investigating rainfall regimes and intensity, and their effect on LFG quality, especially in arid and semi-arid climates. Rainfall intensity data could not be gathered for the study periods used in this paper. The position of the landfill and wind direction were not studied in this part of the project, but may be studied as an area of future work.

The authors believe that landfill operators would benefit greatly from this work, since knowing how climatic factors affect LFG quality could help to develop more efficient maintenance schedules. As well, developing an understanding of how these climatic factors affect LFG may help to improve the design of landfill, and systems within the landfill.

## 5 Acknowledgement

This work is financially supported by a research grant (RGPIN-385815) from the Natural Sciences and Engineering Research Council of Canada (NSERC). The authors are grateful for their support. Special acknowledgment goes to the team at Regina Solid Waste Department, who supported the data collection. The views expressed herein are those of the writers and not necessarily those of our research and funding partners.

## 6 References


Andersland, O.B., Al-Moussawi, H.M. 1987. Crack Formation in Soil Landfill Covers due to Thermal Contraction. *Waste Management and Research,* **5**, 445-452.

Barnston, A.G. 1993. Atlas of frequency distribution, auto-correlation and cross-correlation of daily temperature and precipitation and stations in the U.S., 1948-1991. Camp Springs, MD. U.S. Department of Commerce, National Oceanic and Atmospheric Administration, National Weather Service.

Blight, G.E., Ball, J.M., Blight, J.J. 1992. Moisture and suction in sanitary landfills in semi-arid areas. *Journal of Environmental Engineering,* **188**(6), 865-877.

Bruce, N., Asha, A., and Ng, K. T. W. 2016 "Analysis of Solid Waste Management Systems in Alberta and British Columbia using Provincial Comparison". Canadian Journal of Civil Engineering, **43**(4): 351-360,

Chen, Z., Haykin, S., Eggermont, J.J. Becker, S. 2007. *Correlative Learning: A Basis for Brain Adaptive Systems*, John Wiley and Sons, Hoboken, New Jersey, USA.

City of Regina. 2009. Waste Plan Regina Report. Prepared by Genivar Consultants LP for City of Regina, Public Works Division.

Daniel, D.E., Wu, Y-K. 1993. Compacted Clay Liners and Covers for Arid Sites. *Journal of Geotechnical Engineering,* **199**(2), 223-237.

Di Bella, G., Di Trapani, D., Viviani, G. 2011. Evaluation of methane emissions from Palermo municipal landfill: Comparison between field measurements and models. *Waste Management,* **31**, 1820-1826

Environment Canada. 2014. National Inventory Report 1990-2012. Greenhouse Gas Sources and Sinks in Canada. Part 2. *The Canadian Government's submission to the UN Framework Convention of Climate Change.* Statistics Canada, Ottawa, ON, Canada.

Farquhar, G.J., Rovers, F.A. 1973. Gas Production During Refuse Decomposition. *Water, Air, and Soil Pollution,* **4**, 483-495

Fleming, IR., Rowe, R.K. 2004. Laboratory studies of clogging of landfill leachate collection and drainage systems. *Canadian Geotechnical Journal,* **41***,* 134-153.

Fourie, A.B., Morris, J.W.F. 2004. Measured gas emissions from four landfills in South Africa and some implications for landfill design and methane recovery in semi-arid climates. *Waste Management and Research*, **22**, 440-453

Gallant, K. 2011. Annual Fleet Street Landfill Report. Year: 2011. City of Regina. Regina, SK, Canada.

Latoski, M. 2014. City of Regina Fleet Street Landfill Annual Report. Year: 2013. City of Regina. Regina, SK, Canada.

Lefebvre, M. 2006. Landfill gas Collection Pre-Design Brief. Conestoga-Rovers and Associates.

Munasinghe, R. 1997. Effect of Hydraulic Retention Time on Landfill Leachate and Gas Characteristics. PhD Thesis, University of British Columbia.

Mustafa, S., Hakan, K. 2014. A Semi-Automatic Approach to Identify First Arrival Time: the Cross-Correlation Technique (CCT). *Earth Science Research Journal,* **18**(2), 107-113.

Park, J., Shin, H. 2001. Surface emissions of landfill gas from solid waste landfill. *Atmospheric Environment,* **35**, 3445-3451

Peer, R.L., Epperson, D.L., Campbell, D.L., von Brook, P. 1992. Development of an Empirical Model of Methane Emissions from Landfills. *US Environmental Protection Agency*, Research Triangle Park, NC.

Podobnik, B., Wang, D., Horvatic, D., Grosse, I., Stanley, H.E. 2010. Time-lag cross-correlations in collective phenomena. *A Letters Journal Exploring the Frontiers of Physics*, **90**, 1-6.

Richter, A., Bruce, N., Ng, K. T. W., Chowdhury, A. and Vu, H. L. "Comparison between Canadian and Nova Scotian waste management and diversion models – A Canadian Case Study" 2017. Sustainable Cities and Society, **30**, 139-149

Sadek, S., Ghanimeh, S., El-Fadel, M. 2007. Predicted performance of clay-barrier landfill covers in arid and semi-arid environments. *Waste Management,* **27**(4), 572-583.

Spokas, K., Bogner, J., Chanton, J.P., Morcet, M., Aran, C., Graff, C., Moreau-Le Golvan, Y., Hebe, I. 2005. Methane mass balance at three landfill sites: What is the efficiency of capture by gas collection systems? *Waste Management* **26**, 516-525.

Statistics Canada. 2016. Canadian Climate Normals 1981-2010 Station Data. Statistics Canada, Ottawa, Ontario.

Themelis, N., Ulloa, P.A. 2007. Methane Generation in Landfills. *Renewable Energy,* **32**, 1243-1257.

USEPA. 2015. Inventory of U.S. Greenhouse Gas Emissions and Sinks (1990-2013). *United States Environmental Protection Agency.*

Vorburger, T.V., Song, J.F., Chu, W., Ma, L., Biu, S.H., Sheng, A., Renegar, T.B. 2011. Application of cross-correlation functions. *Wear,* **271**, 529-533.

Wang, Y., Ng, K. T. W., and Asha, A. 2016. Non-hazardous Waste Generation Characteristics and Recycling Practices in Saskatchewan and Manitoba, Canada. *Journal of Material Cycles and Waste Management.* **18**(4): 715-724.

Wreford, K.A., Atwater, J.W., Lavkulich, L.M. 2000. The effects of moisture inputs on landfill gas production and composition and leachate characteristics at the Vancouver Landfill Site at Burns Bog. *Waste Management and Research,* **18**, 386-392.

Leadership in Sustainable Infrastructure
*Leadership en Infrastructures Durables*

Vancouver, Canada
May 31 – June 3, 2017/ *Mai 31 – Juin 3, 2017*

# DESIGN AND OPERATION OF ACTIVELY-AERATED FIELD-SCALE METHANE BIOFILTERS

Gunasekera, Samadhi A. [1,5], Hettiaratchi, Joseph P. A [2], Bartholameuz Eranda M [3], and Irvine, E. [4]
[1,2,3,4] Centre for Environmental Engineering Research and Education (CEERE), Schulich School of Engineering, University of Calgary, Calgary, Alberta, Canada
[5] samadhi,gunasekera@gmail.com

## ABSTRACT

Methane is the second most prevalent contributor to global warming after $CO_2$, with a global warming potential 34 times higher than $CO_2$. 60% of global methane emissions are a result of human activities such as the oil and gas industry, livestock farming, and landfilling. In the oil and gas industry, excess or unwanted flammable gases are often disposed by flaring or venting. Methane biofiltration is an economically feasible, low maintenance, biological method that can remediate low volume point source methane emissions from sources such as oil wells, and landfills. An active aeration closed methane biofilter (MBF), first of its kind, was designed, constructed and installed at a single well battery site in Hannah, AB. The MBF was packed with 100% compost into a 4m$^3$ conical frustum shaped tank at a density of 800kg/m$^3$. The source methane supply consists of 92% $CH_4$ and the inlet $CH_4$ flowrate varying between 10-25m$^3$/day. Heated air was mixed with methane and injected into the bottom of the MBF to diffuse uniformly across the MBF. Inlet and outlet methane flowrate, air flowrate, inlet temperature and the temperature inside different locations of the MBF will be monitored throughout the lifetime of the MBF. The efficiency of the MBF will be calculated using this information, and will determine the carbon offsets achieved. This paper presents preliminary results during the operation of the MBF and the study of the temperature profiles.

Keywords – Methane, Biofiltration, Active-Aeration, Field-Scale, Design

## 1. INTRODUCTION

Methane ($CH_4$) is 34 times more effective at absorbing radiation than $CO_2$, even though it has a short lifetime (~10 years) in the atmosphere. Therefore, over a 100-year period $CH_4$ has a global warming potential 34 times higher than $CO_2$. The average global methane concentrations increased by a factor of 2.5 since pre-industrial times, from 722 parts per billion (ppb) to 1800 ppb by 2014, a concentration unprecedented within the last 800,000 years (IPCC, 2014). 60% of these emissions are a result of human activities such as natural gas and petroleum systems, enteric fermentation, solid waste landfills, and coalmines. Waste landfills produce biogas as a result of anaerobic degradation of waste, contributing towards 17% of the worldwide anthropogenic $CH_4$ emissions (Wuebbles and Hayhoe, 2002). When the $CH_4$ concentration in biogas is above 30% (v/v), the collected biogas can have economic value and practical use. In such instances, biogas is burnt in boilers for heating and other purposes. However, when the $CH_4$ concentration is lower than 30% (v/v), collected biogas is burned on site in order to reduce the risks of explosion associated with $CH_4$, or vented to the atmosphere (Streese et al., 2003).Methane biofiltration technology can be used remediate such low volume, point source, methane emissions.

## Methane Biofiltration (MBF) Technology

Treatment of $CH_4$ through methane biofiltration technology aims at the aerobic conversion of $CH_4$ to $CO_2$ by methanotrophic bacteria. Methanotrophs are capable of converting $CH_4$ to $CO_2$ without producing unwanted by-products, making this a very attractive solution compared to any other method of handling $CH_4$. Methanotrophs are found in natural environments such as tropical forests, grasslands and meadows, landfill cover soils, deserts, and agricultural soils (Nikiema et al., 2007). The simplified biological degradation reaction of methane is given by eq.1

[1] $CH_4 + 2O_2 \rightarrow CO_2 + 2H_2O$ + bacterial biomass (-890 KJ/mol)

A methane biofilter is a box like structure designed and constructed to achieve maximum methane oxidation by utilizing this phenomenon. Solid biofiltration medium, nutrients and water for growth of microorganisms and sufficient porous space for gas exchange, are the three essential constituents of a functional methane biofilter. MBFs can be classified into two main configurations, closed system and open system (Nikiema et al., 2005). Open systems are mostly found in landfill sites and utilizes a passive aeration system. Gas collected through a landfill gas collection system is directed to the bottom of the landfill, thus the flow proceeds upwards through the filter bed, while oxygen diffuses downwards from the ambient air into the bed. However, open systems are highly susceptible to environmental conditions, and pose problems in field scale MBFs as its filter media is constantly subjected to rainfall, snow and wind. Closed systems are more desirable as it is more controlled and allow for better estimation of the conditions inside the system. Since closed systems do not receive ambient oxygen from the top, they are aerated actively using an air distribution system. As can be seen from the equation above methane oxidation is an exothermic reaction producing heat energy (Reible, D. D., 1998). Therefore MBFs are capable of retaining its temperature above zero degrees even during cold winter months

## Application of MBF technology in the oil and gas industry

In the oil and gas industry, excess or unwanted flammable gases are often disposed by flaring or venting. Flaring is the process of combusting the gases in an open atmosphere, to dispose flammable gases such as methane cost-effectively. However, if the emission flowrates are too low or too intermittent, or the heating value of gas is too low to sustain combustion, it is not economically feasible to burn surplus gases in a stable method. These low volume point source emissions could be directed into a methane biofilter instead of venting it directly to the atmosphere.
An active aeration field scale methane biofilter was designed, constructed and installed at a single well battery site in Hanna, AB. A battery is a facility where production fluid from a well is separated into gas, oil and water. A storage tank is used as a treater which separates the fluid into these three layers; salt water at the bottom, crude oil emulsions in the middle and natural gas at the top. The gas layer at the top may also contain solution gas which gets released when the fluid pressure in the treater is reduced. The methane biofilter discussed in this paper is designed to receive a line of natural gas from the storage tank as the source of methane and would receive 92% CH4 with some impurities consisting of salt water and crude oil.

# 2. MATERIALS AND METHODS

## Apparatus

Currently all field scale MBFs are open biofilters employing a passive aeration system. A closed MBF with active aeration was proposed for this site due to anticipated extreme weather conditions with high winds and high snowfall. Having the MBF closed, protects the media from rainfall, snowfall, and provides better control over the MBF.

As can be seen in figure 1, the system was designed to consist of a conical frustum shaped tank accompanied by a housing structure, built using 2"x4" wooden blocks on the wooden skid next to the tank.
The curved walls of the tank offer less wall resistance during gas migration as opposed to a box like structure with sharp corners. The housing structure stores all monitoring equipment and the air blower to pump air into the MBF. This provides a confined space to heat thus ensuring that the air sent into the tank is at a desired temperature.

The use of the wooden skid has two main advantages. It provides a stable structure to transport the setup and possible relocations at a later time. In addition, it ensures that the structure remains levelled over time. If some settling was to occur due to the weight of the tank, this would still be uniform.

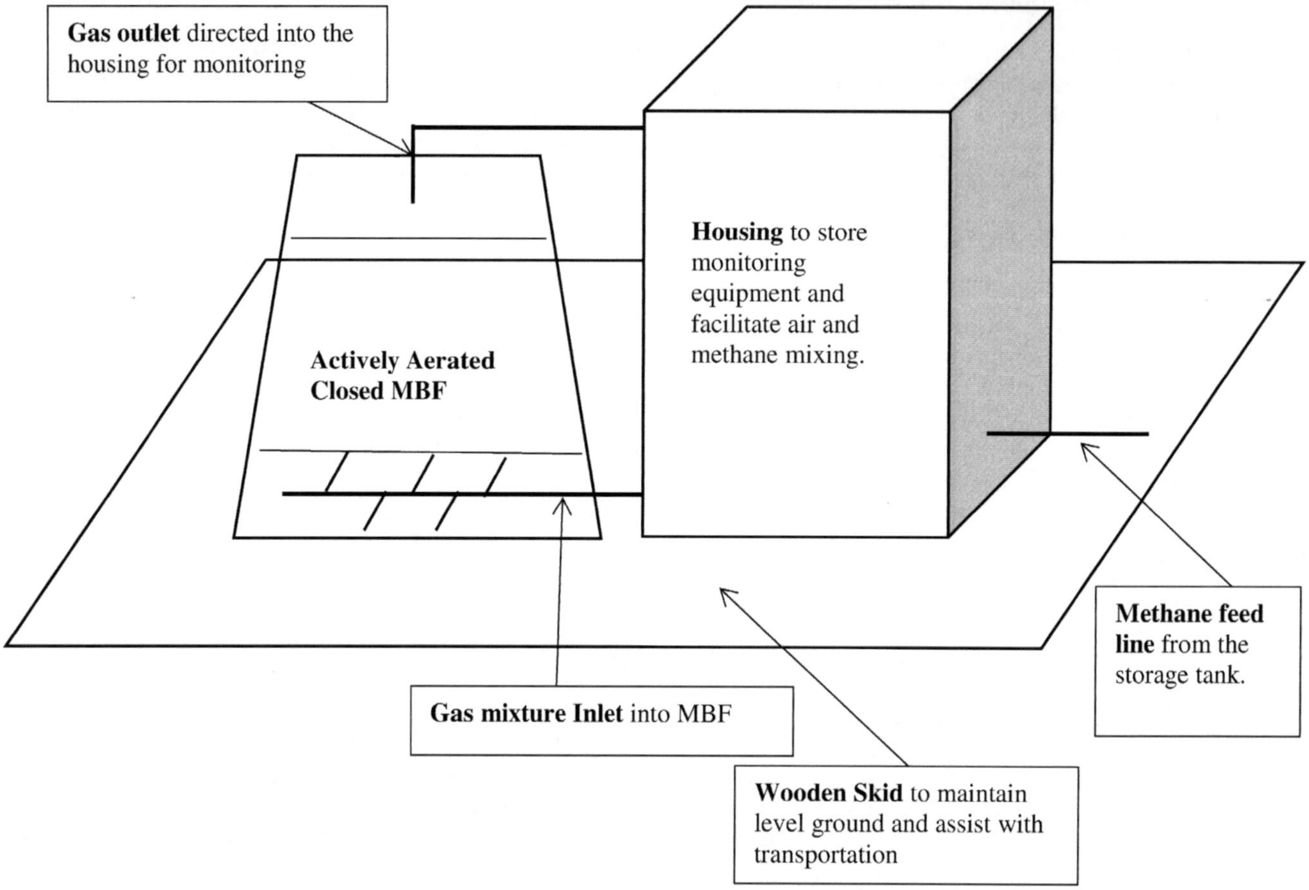


Figure 1: Schematic of MBF

An active MBF has an expected efficiency three times higher than a passive MBF. Therefore, a higher concentration of methane could be treated within a smaller volume. The current MBF has a volume of $4m^3$ and a maximum intake capacity of about $40m^3$/day of $CH_4$. This would provide an empty bed retention time of 2.4 hours. The actual retention time will be lower when the tank is packed with compacted media thus providing sufficient time for the microbes to act on the methane. A commercially available HDPE water tank was chosen to serve as the MBF which saves the additional cost of a custom made tank. Figure 2 shows the exact dimensions of the tank.

The internal walls of the housing and the external walls of the tank were insulated using 2" closed cell spray foam insulation providing an R value of 12.5. In order to protect the insulation foam from UV damages it was painted with a double coating. The lid of the tank was sealed by weather stripping the top perimeter of the tank and then wrapping coper pipe insulation around the edge. Silicone was passed inside the insulation to tighten the seal ensuring there were no leaks through the lid of the tank.

*Gas Distribution System*

At design capacity, the MBF would receive $40m^3$ of methane and $200m^3$ of air, adding up to a total of $240m^3$ of air-methane mixture per day. This gas should be well mixed and uniformly distributed across the cross section of the MBF. The gas distribution system consists of a piping network, gravel, and a geotextile to undertake this task. The distribution system was fabricated using 2" ABS pipes, with holes drilled into its surface to distribute the gas. The hole sizes were calculated to increase in diameter systematically, thus ensuring that the rear end of the MBF will receive equal amounts of gas as the front end. The distribution system was tested by connecting an air blower into the inlet to confirm that air was passing through all sides of the distribution system uniformly.

The holes were drilled at 45 degree angles facing the bottom to allow the gas to first reach the bottom of the tank, and seep up uniformly. 20mm crushed concrete gravel was filled up to about 10cm. Next, the gas distribution system was placed on top of the gravel and levelled using small wooden blocks. More gravel was filled on top of the distribution system to tightly secure it's positioning, and finally, the geotextile was placed on top of the gravel layer to prevent compost from settling down and clogging the gravel layer. A compressed $CO_2$ cylinder was used to send gas through the distribution system, and flux measurements were taken at the surface of the geotextile to confirm uniform distribution throughout the perimeter of the tank.

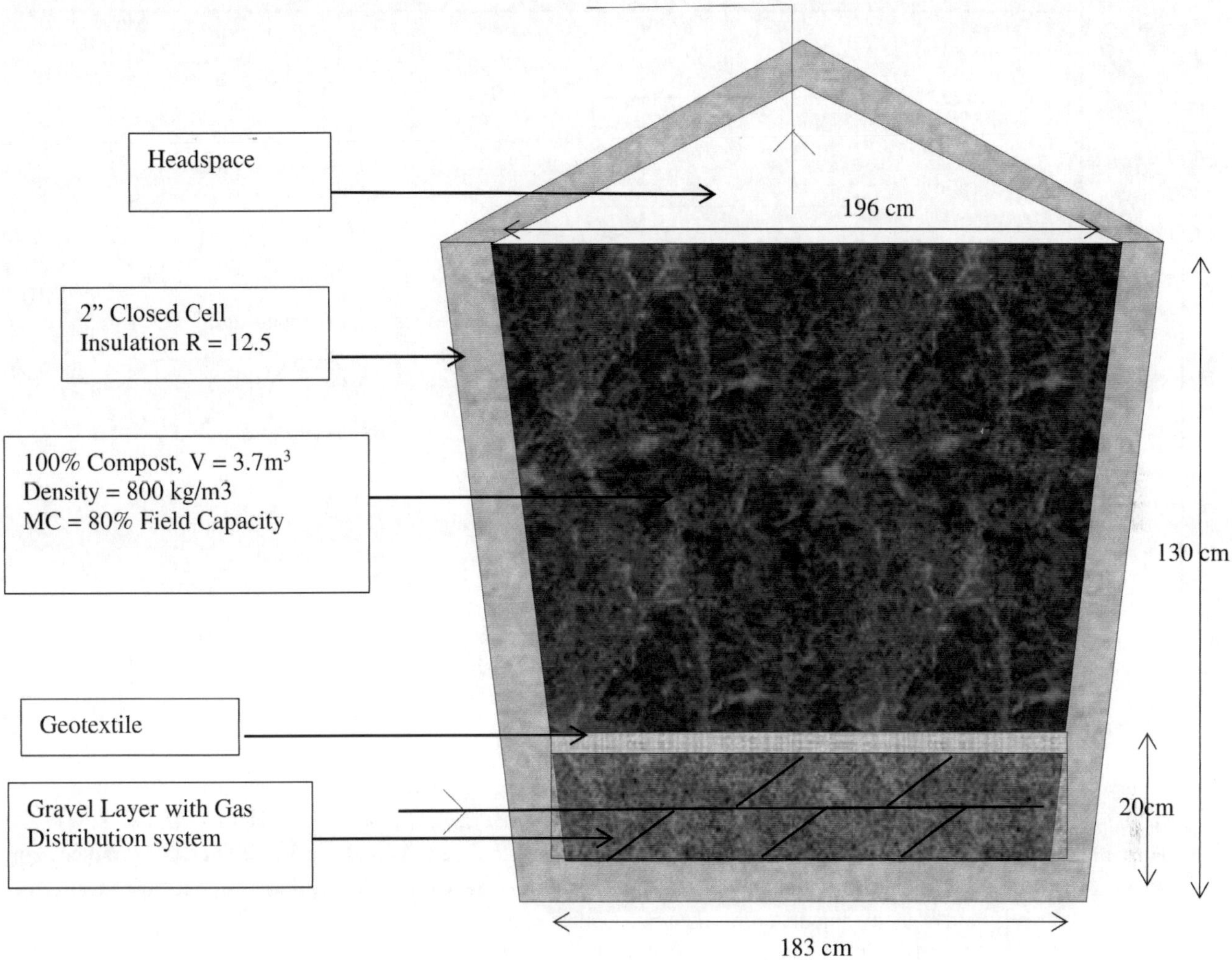


Figure 2: Detailed Schematic of MBF

*Active Aeration System*

According to stoichiometry (eq. 1), an air to methane ratio of 10:1 (v/v) is optimum for methane oxidation. However, increasing the air flow would result in higher flowrates resulting in a reduced retention time. If the time of contact between methane and the MBF media is limited, the methantrophs would not have sufficient time to oxidise the methane. Therefore an air to methane ratio of 5:1 was chosen. Thus, the system is designed to receive 200m3/day of air, with the option to increase if need be.

A 24 V DC, air blower which could send in up to 650m3/day (16 CFM) is used to pump air. This blower is powered by two adjacent 100W, 12V solar panels, connected to two 12V,100Ah AGM Batteries. A 12V catedyne heater, which uses natural gas as its power source, will be installed to heat the housing structure, was installed and set to heat the housing structure to about 15°C thereby heating the air that will be pumped into the MBF. Figure 3 shows the details of the air supply system.

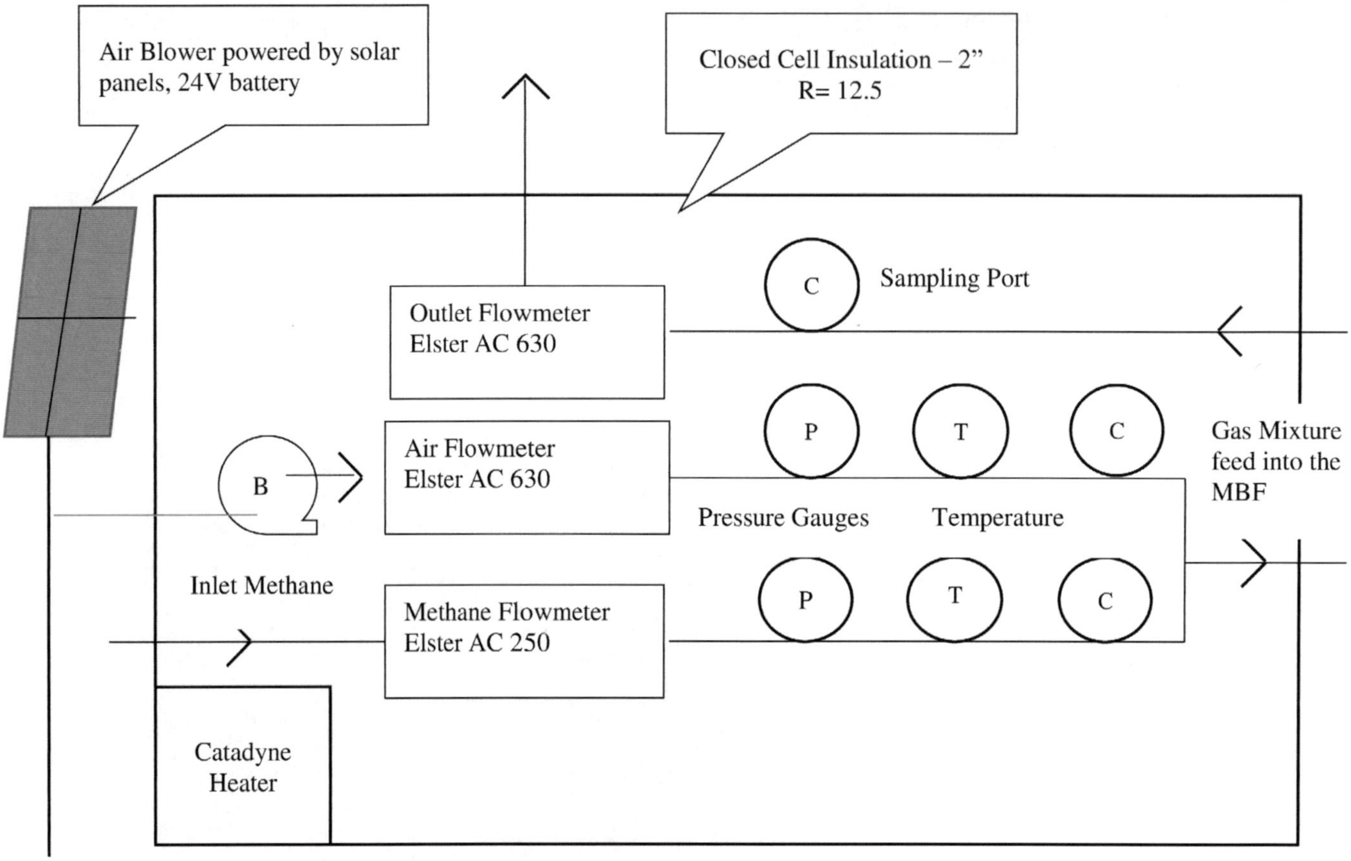


Figure 3: Detailed Schematic of Housing

**Materials**

Compost has a high organic content, nutrients and water holding capacity, providing an ideal environment for methanatrophs and yields the best oxidation efficiencies during laboratory experiments (Hettiaratchi, J. P. and O. Hurtado, 2010). It also has known parameters such as the thermal conductivity and specific heat capacity rather than when using a mixture of media. Compost was collected from the East Calgary Landfill, and tested to measure the moisture content, field capacity and dry solids content. Table 1 shows the media properties of the compost.

*Moisture content (MC), dry solids (DS)*

The Moisture contents (MC) of compost samples were determined by drying 10 g of in an oven (model 1510 E, Sheldon Manufacturing Inc.) at 105 ◦C for ~12 h until a constant weight was obtained. The weight of the sample before and after drying was measured (scale model: METTLER AT 250, Fisher Scientific). The weight lost during drying was the MC and the weight remained was the DS content.

*Field capacity (FC)*

Funnel experiments were conducted to determine the field capacity (FC) of the compost. First, the material was oven-dried at 105 °C for 2 h. A filter paper was wetted, drained and fitted into a 500 mL funnel. The funnel was filled with the compost samples and the bottom tip of the funnel was closed. Known volume of water was added slowly to the compost until it was saturated. A beaker was placed at the bottom of the funnel, the bottom tip was opened, and the drained water was collected. The water was allowed to drain for ~1 h, until no water drained into the beaker. The field capacity was calculated using the difference between the added water and the collected water volume.

*Table 1: Media properties*

| Media | Field Capacity (FC) | Moisture Content (MC) | Dry Solids (DS) |
|---|---|---|---|
| 100% Compost | 50% | 40% | 60% |

The compost was packed into the tank systematically to achieve the desired density i.e. 800kg/m$^3$. $CO_2$ was passed through the MBF using a compressed cylinder after every 20cm of packing, to ensure equal distribution of gas until it was confirmed that the gas reached the top of the filter in all sides of the tank.

## Methods

*Methane Removal Capacity/Efficiency*

During operation, the performance of the MBF is measured by its methane removal capacity. Inlet and outlet methane flowrates and air flowrates are measured using flowmeters as can be seen in Figure 3. Inlet gas concentration and outlet gas concentrations are also measured through sampling ports as seen in Figure 3. If there are no leakages in the system, the efficiency of the MBF can be calculated using the flowrates and concentration measurements.

*Operation and Maintenance*

Controlling the back pressure is a major concern during operation of the MBF. Since the tank is closed, the pressure inside the tank could build up. This MBF, when packed to design criteria will have a pressure of 8.6KPa (P = 1.1m*800kg/m3*9.81m/s2). Therefore, the methane and air entering the MBF should have a pressure greater than this value in order to reach the top of the MBF. Furthermore, methane and air together is flammable, and its contact with electricity could be very dangerous. Therefore it is important to have control over the air to methane ratios at all times, as well as the pressures within the gas lines. Three way valves are installed to divert the flow and take pressure, temperature and concentration measurements from the corresponding gas lines as seen in figure 3.

*Temperature modelling*

Methane oxidation is an exothermic reaction and generates heat. Therefore, there is a direct correlation between the temperature rise inside an MBF and the methane oxidation that took place during a given period time. 15 Temperature sensors were installed in three levels as shown in figure 4 to monitor the temperature profiles inside an MBF.
One sensor will be kept outside to record the ambient temperature data, and the temperature inside the housing structure is recorded. The 15 sensors are connected to two data loggers, and stored inside the housing structure.

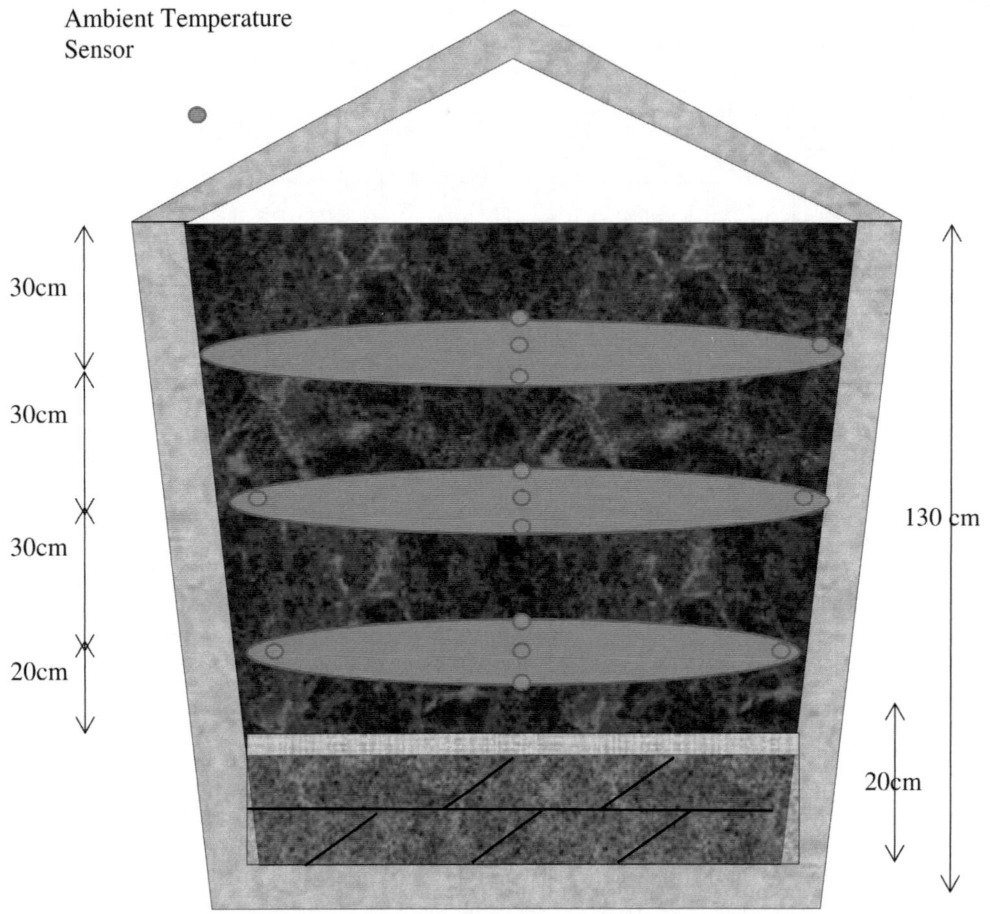


Figure 4: Temperature Sensor Locations

## 3. PRELIMINERY RESULTS

Baseline data of the temperature variation inside the MBF was obtained before the methane supply was connected to the MBF. This would show how to the solar radiation, wind and soil insulation affects the temperature inside the MBF, before methanatrophic activity takes place. It is also a good indicator of whether there are other bacterial activities acting as a heat source inside the MBF.

Figure 5 shows how the average temperature inside the MBF changes with the change in ambient temperature during the month of September. As can be seen, the temperature inside the MBF is consistently lower than the outside temperature. This difference in temperature is due to the thermal resistance caused by the HDPE walls, insulation and compost of the MBF. Therefore, these preliminary results can quantify to which degree the atmospheric temperature influences the temperature inside the MBF.

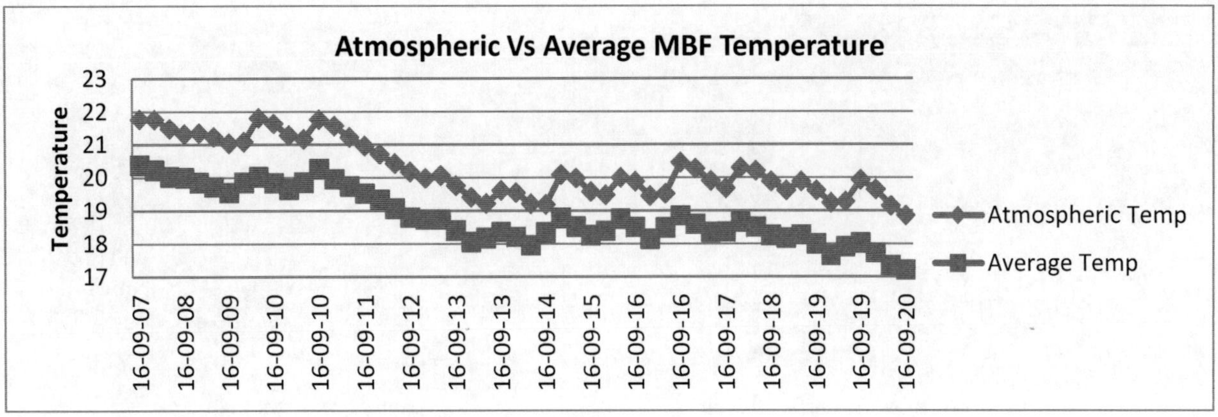


Figure 5:Change in MBF temperature with ambient temperature

The $CH_4$ supply from the storage tank was connected to the MBF on 20th of September 2016. Table 2 tabulates the source $CH_4$ composition.

Table 2: Source Methane Composition

| **Source Methane Composition (%)** | |
|---|---|
| **$CH_4$** | 91.7 |
| **$CO_2$** | 0.15 |
| **$O_2$** | 1.85 |
| **$N_2$** | 7.63 |

During the seeding period of an MBF, methanatrophic activity will be low. Therefore the MBF will not be able to perform at a high efficiency. The temperature fluctuations inside the MBF give a good indication of the amount of $CH_4$ converted to $CO_2$ by methanotrophs. Figure 6 shows the average temperature change inside the MBF from November 2016 to March 2017. The temperature inside the MBF reduces to about 10°C between December to January (the coldest months) indicating the seeding period, but gradually increases to about 40°C by the beginning of March.

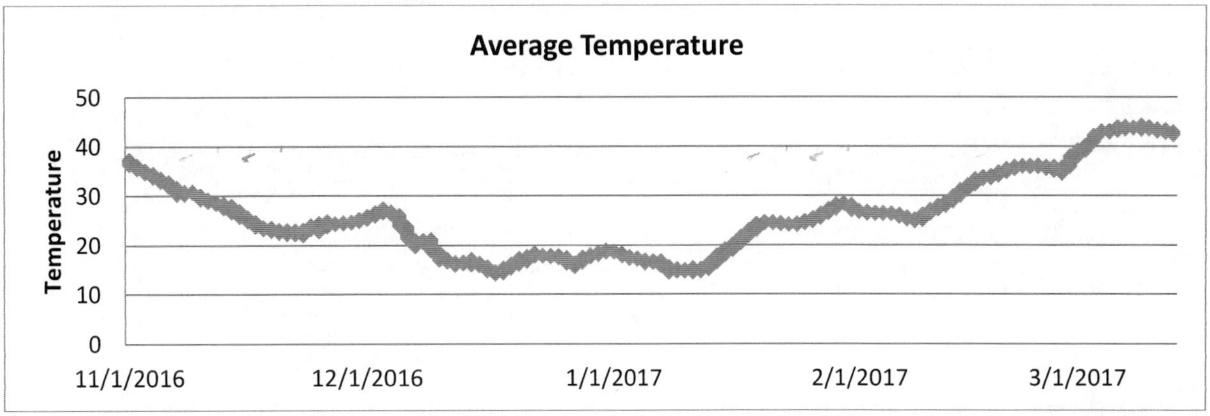


Figure 6: Change in average MBF temperature over time

The contour plots in Figure 7 and 8 show the distribution of heat across the cross section of the MBF at 60cm and 90cm on 1st of December.

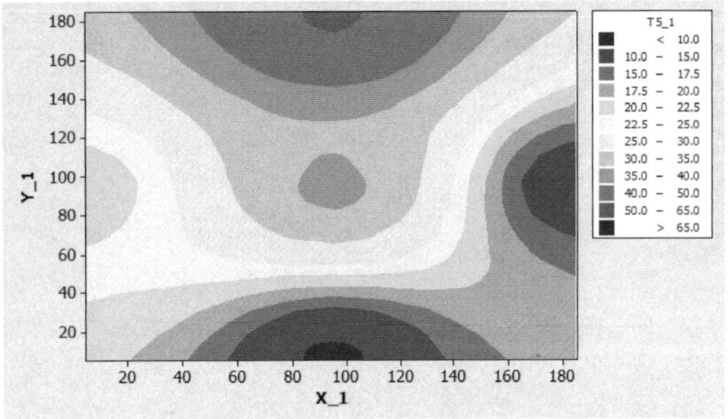

Figure 7: 60cm below surface level

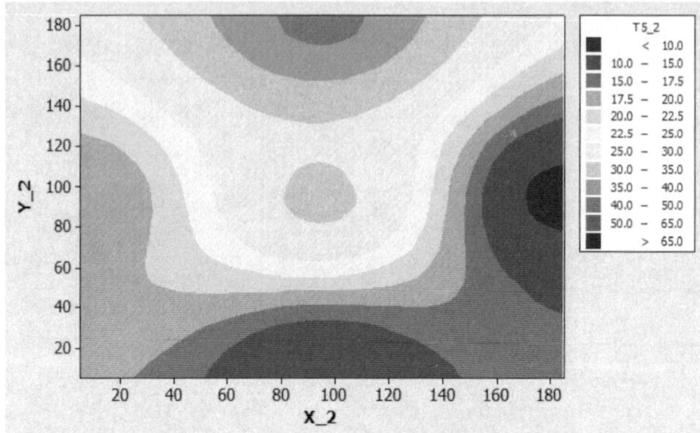

Figure 8: 90cm below surface level

As can be seen the temperature inside the MBF was higher than 10°C, which was much greater than the recorded average atmospheric temperature (-5°C) on 1st of December confirming bacterial activity. It is also evident that the temperature follows the exact pattern at both the surface levels although there is a slightly higher average temperature at 60cm. Since the two opposite sides of the tank have the highest and lowest temperatures, this could be due to the wind direction and solar radiation falling directly onto one side. Compost sample from various locations of the MBF should be analysed for its nutrient content and bacterial population to confirm findings.

The methane and air flow-rates were manipulated based on the temperature indications of the MBF performance. Composition of the inlet and outlet gas mixtures were measured using a Gas Chromatograph (GC) and the efficiency of the MBF was calculated based on equation [2]. Equation [2] includes a $N_2$ component, which accounts for dilution that takes place through the addition of air. Table 3 shows how the efficiency of the MBF improved over time.

[2] $Efficiency = (CH_{4,in} - CH_{4,out} * (N_{2,in}/N_{2,out}))/CH_{4,in} * 100$

Table 3: MBF performance

| Date | Gas | Flowrate (m3/day) | Inlet Conc. (%) | Outlet Conc. (%) | Efficiency (%) |
|---|---|---|---|---|---|
| 27/02/2017 | $CH_4$ | 23 | 11.52 | 7.38 | 38.73 |
| | $O_2$ | 121 | 19.32 | 19.34 | |
| | $N_2$ | | 72.98 | 76.31 | |
| | $CO_2$ | | 0 | 0 | |
| 13/03/2017 | $CH_4$ | 23 | 11.11 | 6.5 | 42.75 |
| | $O_2$ | 171 | 18.89 | 20.9 | |
| | $N_2$ | | 70.85 | 72.4 | |
| | $CO_2$ | | 0.2 | 0.2 | |
| 10/4/2017 | $CH_4$ | 21 | 64.2 | 2.5 | 98.45 |
| | $O_2$ | 5 | 8.18 | 20.92 | |
| | $N_2$ | | 30.74 | 77.47 | |
| | $CO_2$ | | 0.11 | 0 | |
| 10/4/2017 | $CH_4$ | 8.15 | 26.88 | 0.49 | 98.65 |
| | $O_2$ | 15 | 15.82 | 21.41 | |
| | $N_2$ | | 58.68 | 79.15 | |
| | $CO_2$ | | 0.03 | 0 | |

## 4. CONCLUSION

Methane biofiltration is an economically feasible, low maintenance, biological method that can remediate point source methane emissions from sources such as oil wells, and landfills. An active aeration closed methane biofilter (MBF) was constructed and installed at a single well battery site in Hannah, AB during the summer of 2016. The MBF will receive source methane consisting of 92% $CH_4$ with a design capacity of oxidizing 40$m^3$/day of methane. 100% compost, with moisture content at 80% field capacity is used to pack the MBF to a density of 800kg/$m^3$. It is insulated to achieve a 12.5 R-value, and minimize heat loss during cold months. Flow rates of inlet and outlet methane, flowrate of the air, inlet temperature and the temperature inside different locations of the MBF are monitored throughout the lifetime of the MBF. Initially the methane supply was connected to the system at a flow rate of 20$m^3$/day, and will be increased gradually as the MBF starts to oxidize methane at its full capacity Preliminary temperature profiles inside the MBF show that there is bacterial activity inside the MBF and is rapidly increasing. This inference is also confirmed by the increasing efficiency calculated using inlet and outlet methane concentrations. A 98% efficiency was obtained at the lower flow rate of 21$m^3$/day with minimal air supply, suggesting that the MBF is now ready to accept a higher load of $CH_4$. However, temperature inside the MBF is not uniform across its cross section, suggesting presence of hot spots with higher bacterial activities, or a non-uniform effect of ambient conditions on the MBF. Further analysis on compost at different locations of the MBF is required to confirm the bacterial population at different areas of the MBF.

## 5. REFERENCES


IPCC AR5, 2014. “Climate Change 2014: A synthesis report”.

Wuebbles, D., Hayhoe, K., 2002. Atmospheric methane and global change. Earth-Science

Streese, J. and Stegmann, R. 2003. Microbial oxidation of methane from old landfills in biofilters. Waste Management, 23, 573-580.

Nikiema, J., Brzezinski, R., Heitz, M. 2007. Elimination of methane generated from landfills by biofiltration: a review. Reviews in Environmental Science and Bio/Technology, 6 (4), 261–284.

Nikiema, J., Bibeau, L., Lavoieb, J., Brzezinski, R., Vigneux, J. and Heitz, M. 2005. Biofiltration of methane: An experimental study. Chemical Engineering Journal , 113, 111–117.

Reible, D. D. 1998. Fundamentals of environmental engineering. CRC Press.

Hettiaratchi, J. P., and O. Hurtado. 2010. "Compost as a Growth Medium of Methanotrophic Bacteria." *Journal of Solid Waste Technology & Management* 36.1

Leadership in Sustainable Infrastructure

*Leadership en Infrastructures Durables*

Vancouver, Canada

May 31 – June 3, 2017/ *Mai 31 – Juin 3, 2017*

# DESIGN WASTE GENERATORS IN CONSTRUCTION PROJECTS

M.Osmani [1, 2]

[1] School of Civil and Building Engineering, Loughborough University, UK

[2] m.osmani@lboro.ac.uk

**ABSTRACT**

The construction industry in the UK is by far the greatest waste producer among all industries, being responsible for 32% of total waste generated, which equates to three times the combined waste produced by all households. Consequently, the last few years saw the development of several government-driven waste related regulations, policies and consultation documents to reduce construction waste production and increase reuse and recycling rates. Equally, global research on construction waste has been conducted over the last decade ranging from 'soft' onsite waste management tools to 'hard' material and recycling technologies. However, there is an insufficient effort and no structured approach to examine the underlying generators of design waste. Hence, this research engaged the top 100 UK contractors and architectural practices through a questionnaire survey and 24 follow up interviews to investigate the direct and indirect root origins and causes of construction waste across all stages of the building design process that culminated in the development of a set of design waste maps. Respondents reported that designing out waste has never been a priority in the design process. Moreover, results reveal that design waste is affected by a wide practice of not embedding waste reduction in briefing, no waste reduction target setting, and lack of designers' understanding of design waste causes; hindered by incoherent coordination and communication between project members; and impeded by time constraints

Keywords: Construction waste; design waste mapping; architects; contractors; UK.

## 1. INTRODUCTION

The construction industry makes a vital contribution to the competitiveness and prosperity of the UK economy; with an annual turnover in excess of £100 billion, contributes 9% of GDP, and provides employment for over three million people. However; it is by far the greatest consumer of resources and waste producer of all industries in the UK being responsible for 32% of total waste generated, which equates to three times the combined waste produced by all households (DEFRA, 2007). Additionally, an estimated 25 million tonnes of construction and demolition waste ends up in landfill without any form of reuse or recovery, costing the industry £1 billion per year in disposal costs (Moon, 2008). Consequently, construction waste management and minimisation became a priority in the UK environmental policy programmes such the UK Strategy for Sustainable Construction-SSC (BERR, 2008), which contains a target to halve waste to landfill by 2012. A zero waste target was debated, but concerns regarding industry fragmentation and poor engagement led to its omission. More than 500 companies have signed the 'Halving Waste to Landfill' commitment, including 185 contractors and 64 designers, in addition to clients, professional institutions and government agencies. Additionally, a host of existing and forthcoming environmental legislation is expected to drive the resource efficiency momentum, and the issue of waste reduction is likely to rise in importance among all construction stakeholders. The SSC recognised the importance of considering construction waste minimisation during design; hence the notion of 'designing out waste', as espoused in recent guidance for building projects (WRAP, 2008; WRAP, 2010). Despite international academic endeavours over the last 15 years, design waste reduction research is limited and piecemeal; and as such "more work is essential to investigate C&D (construction and demolition) waste issues in project design (Lu and Yuan, 2010). Hence, the aim of this paper is to investigate the root causes construction waste during the design process. Building 'design waste' is defined in this paper as waste arising from building sites as a direct and indirect result of the design process, including opportunities to reduce waste by all project's stakeholders throughout the design stages.

## 2. INSIGHTS INTO CURRENT CONSTRUCTION WASTE MANAGEMENT RESEARCH

An ever-increasing construction waste related research has been conducted ranging from 'soft' mapping, management, and reduction tools and methodologies to 'hard' material and recycling technologies Yuan and Shen (2010) presented insights into the recent construction and demolition (C&D) waste management research trends. The study showed that there was no clear signposting on a single research direction by reporting that the bulk stream of publications was devoted to broad brush topics such as environmental regulations and cost benefit analysis. Similarly, Lu and Yuan (2010) developed a framework for understanding global construction waste management research.. They indicated that the majority of published papers were related to the construction and demolition stages, with very few attempts were devoted to investigate design waste. That said, Yuan and Shen (2010) and Lu and Yuan (2010) reviews were limited to a number of journals on the one hand; and they did not cover other academic work, such as doctoral studies; and industry and government publications on the other. For example, Gamage et al. (2009) and Gamage (2011) assessed the relationship between construction procurement systems and construction waste generation, which led to the development of a procurement waste minimisation framework for construction projects. As such, Osmani (2011) revealed that the current and on-going research in the field of construction waste management and minimisation can be broadly categorised into 13 clusters, including procurement waste. In terms of government publications, WRAP (2008) introduced a new guide to help architects reduce the amount of construction waste sent to landfill in the UK. The guide comprises five principles: design for reuse and recovery; design for offsite construction; design for material optimisation; design for waste efficient procurement; and design for deconstruction and flexibility. The latter is a step forward to engage architects in designing out waste; however, it did not associate the proposed principles with all parameters of the design process environment; including design stage sequencing and stakeholders' interactions and roles. More importantly, the guide failed to conduct a waste diagnosis across all design stages to map out the direct and indirect design waste sources and origins that are critical in informing and implementing designing out waste principles and strategies.

Emerging information technologies, bar coding systems, and GPS, GIS and the wide area networks (WAN), are being introduced into construction waste management and minimisation research (Cheng et al. 2011). Most recently; saw the emergence of Building Information Modelling (BIM) techniques (RIBA, 2012), which can be adopted to assist architects minimising waste in their design projects (Liu and Osmani, 2011). Similarly, a steering committee consortium, chaired by the author, is in the process of developing a new British Standard, BS 8895,that aims to provide principles and an implementation framework for waste prevention and minimisation during the design briefing stage of building projects. BS 8895 Part 1 forms the foundation standard for a suite of future British Standards and Codes of Practice intended to address specific and interrelated issues and processes of designing out waste in building projects (BSI, 2013).

The extant of literature suggests that all construction stages directly or indirectly contribute to construction waste generation. However, the level and severity of waste production varied from stage to stages depending on a number of variables that include type of procurement, project brief, stakeholders' engagement and commitment, etc. That said, it is widely argued that future waste efforts should focus on designing waste (Baldwin et al, 2006; Poon et al., 2004; Poon, 2007; Osmani et al, 2008; Osmani, 2011; Chen et al, 2011; Lu and Yuan, 2010).

## 3. METHODOLOGY

The research adopted a triangulated methodological approach, consisting of a desk study, quantitative and qualitative research methods. The quantitative research took the form of two sets of postal questionnaire surveys distributed to top 100 UK architectural practices and contractors respectively. The aim of the twofold questionnaire survey was to establish a general industry-wide perspective on causes and origins of construction waste. The qualitative data collection comprised 24 follow up semi-structured interviews that were conducted with 12 architects and 12 contractors to provide the qualitative research for the research.

The Statistical Package for Social Science (SPSS) software was used to analyse questionnaires' results. The qualitative data from the interviews were firstly recorded, which enabled full transcripts, and finally coded and analysed using NVivo.

.

## 4. RESULTS

The response rates from the questionnaire were 40% and 49% from participating architects and contractors respectively; representing a 44.5% mean response rate. Both questionnaires' results led to a rating signposting of broad waste related causal parameters. However, there was a need to comprehensively map the root origins and underlying causes of design waste and establish correlation between onsite waste generation and the building design process. As such, interviewees were asked to identify the potential causes of design waste during each briefing, design development and detailing and specification. Results are summarised and discussed below

### 4.1 Construction waste sources

Respondents were asked to rate from 1 (lowest) to 5 (highest) waste sources in construction projects. Questionnaire results indicate that responding architects rated 'unused materials' as the highest waste generator with a mean value of 4.30; followed closely by 'offcuts' (4.15) and 'client-led changes' (4.00). Similarly, responding contractors accorded the highest and same mean importance rating (3.57) to both 'offcuts' and 'client-led; followed by an equal rating of 3.51 for 'design changes' and 'improper onsite material storing'.

Figure 1. Sources of waste across a construction project life cycle

In-depth investigation was needed to examine the underlying design waste causes and sources during the design stages. As such, 24 follow up semi-structured interviews were conducted with 12 architects and 12 contractors from the responding architectural practices and contracting firms respectively. The results of the interviews are reported and discussed in the sections below.

### 4.2 Construction waste causes during design stages

4.2.1 Construction waste causes during project brief development

The interviews identified a number of design waste causes during the briefing stage. These are summarised in Figure 2 and discussed below.

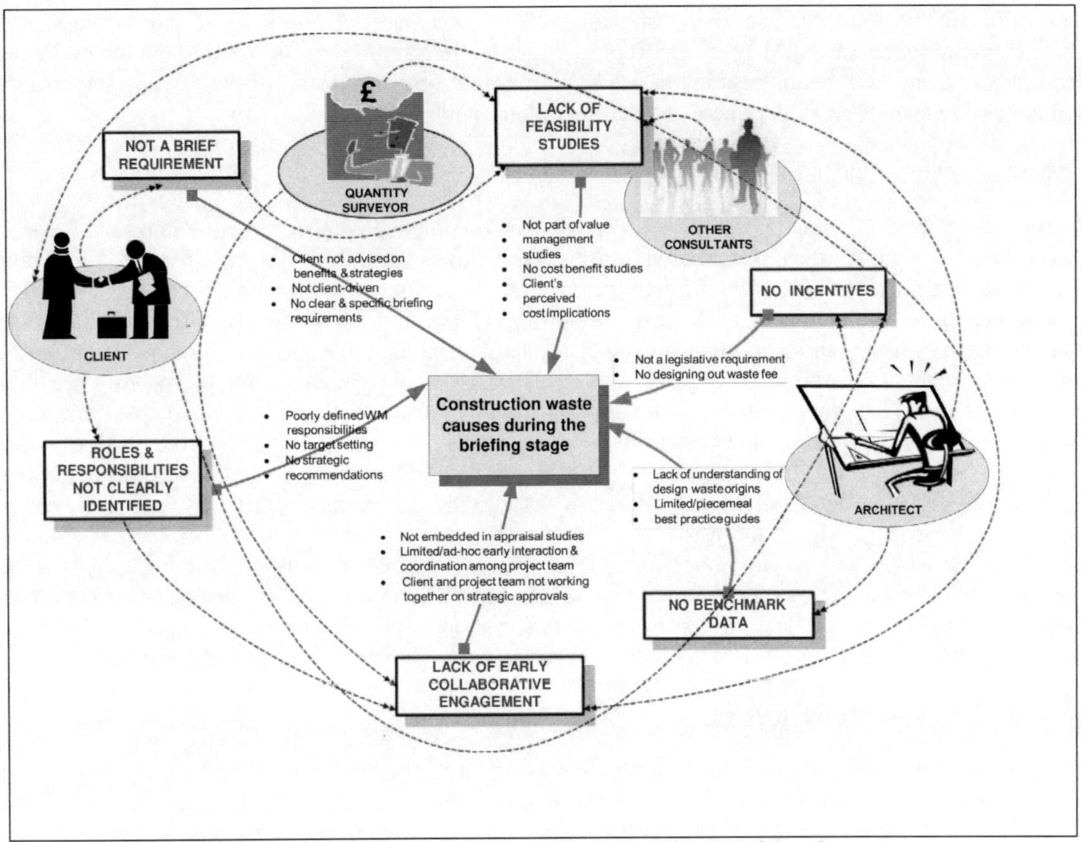


Figure 2: Construction waste causes during project brief development

4.2.1.1 Designing out waste is not a brief requirement

There was a consensus among participating contractors and architects that waste minimisation is generally not a brief requirement in most projects. The client being unaware of the benefits, especially associated costs with waste reduction measures, was raised by the majority of interviewees. It was emphasised by both architects and contractors that "this is the responsibility of the quantity surveyor to identify potential benefits and communicate this to the client", who should be taking the lead in making sure that all project stakeholders are informed about the importance and impact of waste minimisation throughout the project life cycle. When probed on reasons for the lack of architect, consultants, and contractors' engagement in advising the client regarding waste control and management; contractors were of the opinion that this is mainly due to time constraints imposed by clients. Similarly, architects reported that they work in accordance with a tight time schedule from start to end and there is "no time to look at best design"; and encapsulated their thoughts by stating that they "are not influencing the fundamentals of design after all". On the other hand, contractors argued that the client and design team "should be in a position to identify a waste minimisation target during the briefing stage, that's as far as they can go". On the other hand, architects rejected the argument by stating that architects "don't even know at that point what the building is going to be made of as the design brief is a broad brush stage".

4.2.1.2 Roles and responsibilities not clearly identified

Participants were asked whether the client should endeavour to define and allocate responsibilities for waste minimisation at the briefing stage. A representative response was forwarded by one contactor: "If waste minimisation is driven from the project outset and written into contracts, then responsibilities were made through the key stages of design that would help enormously".

4.2.1.3 Lack of early collaborative engagement

Little interaction among architects, consultants and contractors was a factor identified by all interviewees. The need for a whole team approach was considered important if waste minimisation was to filter through the entire process, as one contractor put it: "waste minimisation has to be introduced right at the start of the briefing stage so that the client and the design team have all got to collaborate to implement it".

4.2.1.4 No waste minimisation feasibility studies and benchmarking data

Most contractors were of the view that the briefing stage should comprise detailed research into how waste can be minimised through design. As such, one contractor argued that this is partially the client's responsibility; however, most of the time "the client will just give the architect an embryonic brief with broad requirements". This was seen as an opportunity for architects to conduct a waste minimisation feasibility study whereby information is assembled, waste reduction target is set, and a mechanism is put in place to monitor the process throughout the project life cycle. This should include working out rough ideas on materials, assess their resource efficiency suitability, and develop an initial cost plan with quantity surveyors. On the other hand, architects reported that currently waste minimisation is not a design priority and "feasibility studies during the briefing stage of a construction project will be looking at big issues on sites". One architect went further by stating that "the extent at which waste minimisation will be considered at these stages will depend on how it fits into the most critical design issues". However, by and large architects acknowledged that if the waste issue is not addressed at the project brief, then there is a potential to create a framework which will go on being wasteful all the way through the project life cycle. That said, both participating contractors and architects concurred that at present, waste minimisation endeavours are not considered during feasibility studies and the lack of waste related information, especially benchmarking data, makes it even more difficult to pragmatically assess the potential for waste reduction during the design and construction process.

4.2.1.5 No financial incentives for designers

The lack of financial incentives was raised as another key issue for the passive motivation by architects in investing extra time and effort required to design out waste. As such, one architect pointed out that, "the client should perhaps set aside an additional fee; for the architect to consider waste minimisation in the design process". Another architect stated that waste minimisation "is often a moral expectation from the architect", and argued that this is often not sufficient pressure to consider it in design. Conversely, most contractors agreed that design waste considerations should be part of architects' conventional work process without being paid extra for it. However, architects argued that financial incentives could objectively drive forward the waste agenda during the design process; and proactive actions in this matter will need architects to go "above and beyond their legal requirements". On the other hand, a responding architect compared the waste issue to health and safety considerations by stating that 10 years ago, architects did not think about health and safety too much because it was assumed to be the contractor's responsibility but now if it is routinely considered part of CDM Regulations (HM Government, 2007); and as such architects eventually will get at that stage with waste minimisation. He argued; however; for the moment "the idea has not permeate the architectural offices".

4.2.2 Construction waste causes during design development

Figure 3 summarises the interviewees' insights into the salient design waste origins and causes during design development, which are discussed below.

4.2.2.1 Lack of partnering commitment and coordination

Both respondents agreed that wastage can indirectly occur because of lack of coordination between members of the design team on one hand, and the architect and the client on the other. Interviewees went further by associating waste generation with weak linkages between designers and manufacturers and poor coordination and communication among designers.

4.2.2.2 Lack of architects' engagement

Contractors commented that waste minimisation is not considered a priority by architects. Responding architects They opined that even though design waste will be an initial consideration, other issues soon distract architects from implementing it, such as "getting through planning and other regulatory approvals" that require more urgent thoughtful attention.

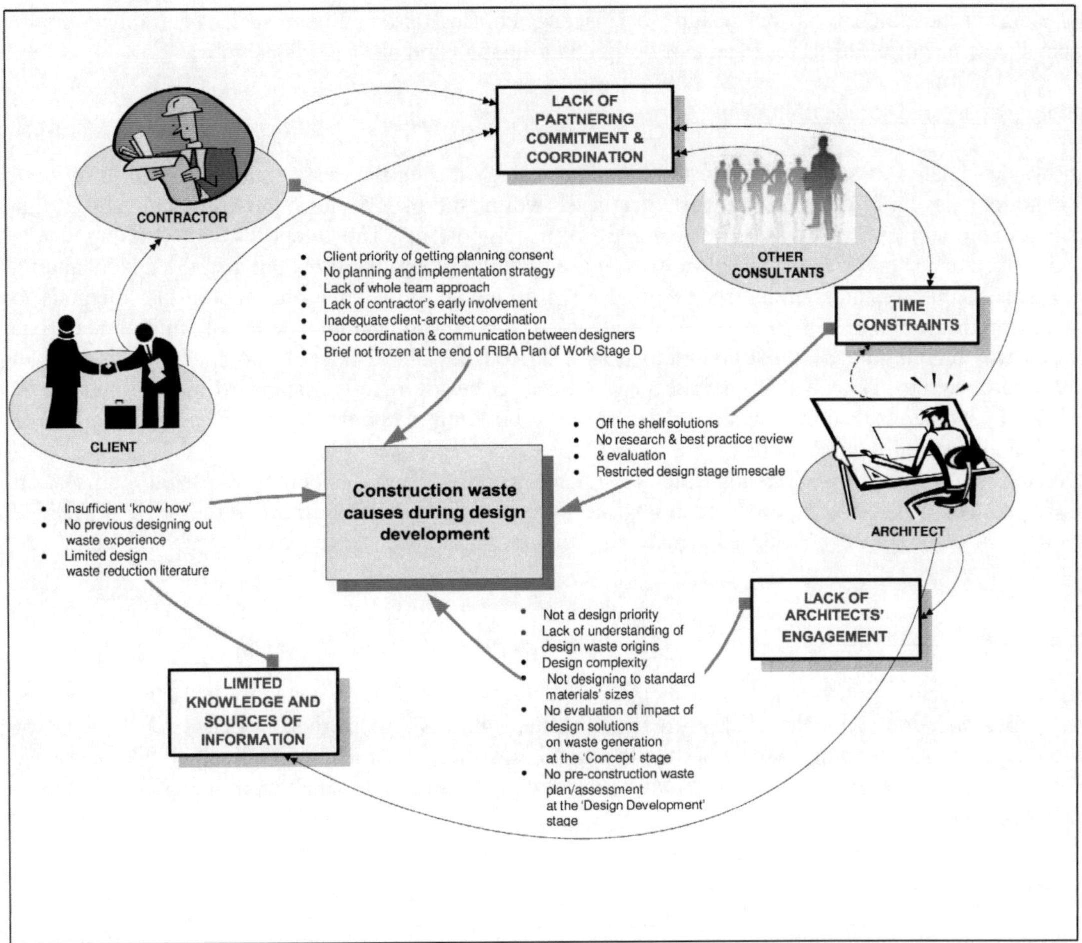


Figure 3: Construction waste causes during design development

The issue of design complexity fuelled greater debate among the respondents. This was particularly emphasised by contractors as a key source of waste resulting in unconventional shapes that are beyond traditional site personnel buildability skill sets. This often results in abortion of work or trimming off materials to fit their specific applications. Indeed, offcuts and rework are by and large indirect consequences of design waste as an architect stated: "every time more complexity is added, there is potential for more waste". For contractors, the key action for architects is to produce a more integrated and informed design process to avoid offcuts. However, they argued that the overtly complex design makes it difficult to specify materials to standard dimensions. Nonetheless, another architect identified the issue of design complexity as an opportunity to explore and promote the use of modern methods of construction: "when a certain level of complexity is a concern then there is a need to explore offsite fabrication and manufacturing". Responding architects commented that design waste can be reduced without compromising creativity and innovation: "if architects standardised everything they would automatically eliminate design, which is actually the most important element of architecture". An architect also added that it is possible to "manage the process to have a complex design but still minimise waste within the parameters of that particular project".

Architects attributed their lack of engagement in designing out waste to their limited involvement in design development. This is due to an increasing amount of specialist design, particularly in Design and Build (D&B) procurement. The restricted contribution of architects in the whole process especially in a supervisory role was also raised as a factor that could lead to design waste: "our role in onsite supervising work is diminished; therefore the level of experience that our technicians are getting, going out and seeing what is going on onsite is getting less". They concluded that current and up-coming architects will have little or no practical knowledge of site activities and will be unable to relate the impact of their design on what happens onsite. The prevalent practice that architects are not responsible for the entire design was raised. One contractor agreed that "architects rarely produce a design for

100% of the building" as most the specialised design work packages in D&B procurement, such as curtain walling are done by specialist contactors. This leads to lack of design coordination and ultimately design waste.

#### 4.2.2.3 Limited knowledge and source of information

The lack of knowledge related to an effective waste reduction strategy during the design stage was raised by both architects and contractors as an indirect design waste origin. However, the majority of contractors pointed out that "having that background information of where waste comes from is important". They went further by stating that "a focussed education for architects needs to take place first, not only giving them the tools but also the understanding of how to use them". Similarly, most architects acknowledged a lack of knowledge and appropriate methods to minimise waste during the design process. Furthermore, the lack of robust technical knowledge of building materials and site processes was highlighted by most contractors as a serious impediment facing architects to design out waste. As such, one contractor opined that "architects will need to be aware of all standard sizes produced by manufacturers"; while another argued that "poor understanding of building materials" by designers tends to lead to uninformed design and material specification.

One of the recurring issues raised by all respondents was related to limited and incoherent accessible information that contributes to the lack of knowledge on designing waste. Accurate and available information, i.e. best practice guides, was seen as an enabler in design decision making, which could consequently lead to reduction in design waste.

#### 4.2.2.4 Lack of early team interaction

Both architects and contractors consented that efficient design waste minimisation can only be achieved if it is pursued by all parties and driven by the client. Architects opined, however, that this "depends on the kind of contractor and the type of procurement route". On the other hand, contractors maintained that this should be firmly set during the early design stages and argued that "if waste is not considered or known about at the briefing stage, it will be extremely difficult to carry it further".

#### 4.2.2.5 Time constraints

Insufficient time to consider waste minimisation is a common thread in most of the work stages. Due to time constraints, architects argued that they cannot adequately explore individual solutions and often make use of design and specification data from past projects. This was reinforced by one participating architect: "if there is no time to research systems, architects will keep defaulting and probably pull off what they have used or heard of before". Another architect suggested that if there was no sufficient time, issues are considered in order of importance. He went further to explain that "time is always short to do everything perfectly but it doesn't mean that one does not have a go at doing it, if it is considered important enough". Contractors also commented that clients want "buildings designed, built and occupied as quickly as possible". They added, however, that if longer periods were allowed for pulling project details together, then issues such as waste and alternative methods can be realistically considered. They went further by acknowledging that if designers are up against very tight deadlines, "they will go with what they know".

## 5. CONCLUSIONS

Quantitative and qualitative data collection that were captured from the major UK architectural practices and contracting companies allowed a holistic and detailed mapping of waste causes across design stages in construction projects. Design waste during the briefing stage is attributed to the lack of project strategic waste minimisation commitment driven by misinformed clients; absence of benchmarking data and legislative requirements for designers; and limited early interaction of project team members. The key waste causes during the design development stage are associated with limited knowledge on the underlying design-related waste origins and designing out tools and methodologies; and stilted client-architect, architect-material manufacturers; and architect-consultant coordination and communication.

Results suggest that design waste is affected by a wide practice of not having waste minimisation as a brief requirement, no baseline setting, and lack of designers' understanding of design waste origins; hindered by limited 'know-how' and incoherent coordination and communication between project members; and impeded by time constraints and disjointed design information; which cumulatively disallow due waste minimisation consideration, implementation and monitoring during the design stages.

## REFERENCES


Baldwin, A., Poon, C., Shen, L., Austin, A., Wong, I., 2006. *Designing out waste in high-rise residential buildings: analysis of precasting and prefabrication methods and traditional construction*. In: Runming, Y.,Baizhan, L., Stammers, K. (Eds.), International Conference on Asia-European Sustainable Urban Development, Chongqing, China, 5–6 April 2006, Centre for Sino-European Sustainable Building Design and Construction, Biejing (ISBN 0-903248-03-04).

BERR -Business, Enterprise and Regulatory Reform, 2008. *Strategy for Sustainable Construction*, BERR, London.

Bossink, B.A.G., Brouwers, H.J.H., 1996. Construction waste: quantification and source evaluation. *Journal of Construction Engineering and Management*, 122: 55-60.

Cheng, X., Chen, J., Chen M. 2011. Construction Waste Management: Current Status and Directions for Further Study. *Applied Mechanics and Materials*, 71-78: 4628-4633.

DEFRA- Department of Environment, Food and Rural Affairs, 2007. *Waste Strategy for England 2007*. The Stationery Office, London.

Ekanayake, L.L., Ofori, G., 2000. Construction material source evaluation. *Proceedings of the 2nd Southern African Conference on Sustainable Development in the Built Environmen*t, Pettoria, 23-25 August 2001.

Gamage S.W., Osmani, M., Glass, J., 2009. An Investigation into the impact of procurement systems on waste generation: the Contractors' perspective, *Proceedings of the Association of Researchers in Construction Management (ARCOM) Conference*, Nottingham, UK, 103-104.

Gamage S.W., 2011. *A waste minimisation framework for the procurement of design and building construction projects*. PhD thesis, Loughborough University, UK.

Gavilan R.M., Bernold, L.E., 1994. Source evaluation of solid waste in building construction. *Journal of Construction Engineering and Management* 120(3): 536-552.

HM Government, 2007. *Health and Safety: the Construction (Design and Management) Regulations*. The Stationary Office Ltd, London.

Liu, Z. and Osmani, M. 2011. The Potential use of BIM to aid construction waste minimisation. *Proceedings of the Joint CIB W78-W102 International Conference on Computer Integrated Construction*, 26-28 October 2011, Sophia Antipolis, France.

Lu, W., Yuan, H., 2010. Exploring critical success factors for waste management in construction projects of China. Resources, *Conservation and Recycling*, 55(2): 201–208.

Osmani, M., Glass, J., Price, A.D. 2006. Architect and contractor attitudes towards waste minimisation. *Waste and Resource Management*, 59: 65-72.

Osmani, M., Glass, J., Price, A.D., 2008. Architects perspectives on construction waste minimisation by design. *Waste Management*, 28: 1147-1158.

Osmani, M., 2011. *Construction Waste,* in Letcher T.M. and Vallero D. A. (eds) Waste: A Handbook for Management, Elsevier, 207-218.

Poon C.S., 2007. Reducing construction waste. *Waste Management,* 27: 1715-1716.

Poon, C.S., Yu, A.T., JAILLON, L., 2004. Reducing building waste at construction sites in Hong Kong. *Construction Management and Economics*, 22(5): 461-470.

RIBA -Royal Institute of British Architects, 2009. *Plan of work: multi-disciplinary services.* RIBA, London.

RIBA -Royal Institute of British Architects, 2012. *BIM overlay to the RIBA outline plan of work*. RIBA, London.

WRAP - Waste and Resources Action Programme, 2008. *Designing out waste: a design team guide for buildings*. WRAP, Banbury, UK.

WRAP - Waste and Resources Action Programme, 2010. *Designing out waste: a design team guide for civil engineering*, WRAP, Banbury, UK.

Yuan, H., Shen, L., 2010. Trend of the research on construction and demolition waste management. *Waste Management*, 31(4): 670–679.

Leadership in Sustainable Infrastructure
*Leadership en Infrastructures Durables*

Vancouver, Canada
May 31 – June 3, 2017/ *Mai 31 – Juin 3, 2017*

# DIFFERENT ELECTRICAL REGIMES TO DEMULSIFY OIL SLUDGE - WETTABILITY INDICATOR

Esmaeel Kariminezhad [1, 3], Maria Elektorowicz [2]
[1, 2] Department of Building, Civil and Environmental Engineering, Concordia University, 1455 Boul. de Maisonneuve W, Montreal, PQ, Canada
[3] es_karim@encs.concordia.ca

**Abstract:** Extraction and refinery of oil by oil-producing industries result in the production of a huge amount of oil sludge which disposal might present a threat to the environment. The treatment of oil sludge is challenging due to its physical and chemical properties. Among these properties, a stable emulsion is one of the main obstacles to an effective treatment. Adsorption of asphaltenes and resins on the surface of solids drastically increases the hydrophobicity of the solid particles and produces meta-stable water-in-oil emulsions. In this study, an electrokinetic (EK) approach was employed to decrease the double layer thickness and destabilize emulsions to facilitate the separation of water from oil phase. Four different types of electrical fields were applied to test their impact on the stability of water-in-oil emulsions. They are: constant direct current (CDC), pulsed DC (PDC), incremental DC (IDC) and decremental DC (DDC). Furthermore, the effectiveness of DDC and IDC methods remarkably improved demulsification compared to the sequential solvent extracted oil sludge with toluene and tetrahydrofuran. Wettability alternation analyses showed an increase in the population of fine solids in the aqueous part following EK treatment which in turn led to the higher hydrophilicity of the solids. The applied DDC and IDC fields enhanced the wettability towards water-wet solids and were better than the other tested configurations, which can result in better solid-oil phase separation.

## 1 INTRODUCTION

The environmental challenges regarding the treatment and disposal of waste generated by oil industries are not restricted to the processing stages of extraction and separation only. For example in upstream, oil extraction conducted from geological strata produces a complex mixture of sand, clay, minerals, water and oil. The resulting oil sludge represents a serious hazard to the environment if not properly treated and disposed. Oil sludge might be comprised of 30-50% oil, 30-50% water and 10-12% solids, and forms a complex matrix which is difficult to dewater (Mikula et al., 1996;Jeeravipoolvarn et al., 2009;Da Silva et al., 2012). The high water content contributes to the large volume of material to be disposed of and renders the geotechnical properties of the oil sludge more intricate (Snars and Gilkes, 2009).

Oil sludge is generally stored in tailing ponds, below or above ground with the help of perimeter dikes or lagoons. The storage of huge volumes of oil sludge represents a significant technical challenge on one hand, and their treatment prior to disposal underlies another. Inadequately designed storage structures (insufficient drainage, piping, and overtopping) can significantly affect operation and incur damage to the surrounding environment (Rico et al., 2008;Blight, 2009). This problem can be further exacerbated during other activities such as transport and disposal of oil sludge thereby increasing the risk of severe contamination of water bodies, the demise of aquatic life and heavy metal contamination (Ayotamuno et al., 2007).

Chemically, oil sludge is a very stable colloidal suspension system as a result of electronegativity and hydration (Guolin et al., 2009). The presence of oil and water, two immiscible liquids, in oil sludge wastes, results in the formation of water-in-oil emulsions. These emulsions are meta stable due to the presence of a protective layer of fine clay which prevents the emulsions from coagulation or coalescence. Petroleum hydrocarbons (PHCs), such as, asphaltenes and resins act as natural emulsifiers in these protective layers. There are also other natural emulsifiers in these layers such as fine solids, oil-soluble organic acids and other fine materials that contribute to the stabilization of water-in-oil emulsions (Yang et al., 2008). In addition, asphaltenes and resins contain hydrophilic functional groups that act as lipophilic emulsifiers. The structure of the water-in-oil emulsions is further stabilized by the pH of oil sludge which usually ranges from 6.5-7.5 (Elektorowicz et al., 2006). The formation of these stable water-in-oil emulsions renders the whole separation and treatment process of oil sludge to be very challenging (Heidarzadeh et al., 2010;Hu et al., 2013).

Surface properties of solids in oil sludge play important roles on the stability of the emulsion. Chen et al. (2017) show the effect of ununiform particles with varying hydrophobicity on the emulsion stabilization that can be an example of solids on the oil sludge. The ratio of asphaltenes to resins plays a very important role in the stability of the emulsions and adsorption on the solids (Sullivan and Kilpatrick, 2002;Sparks et al., 2003). Mineral particles show a significant effect on the emulsion stability, while resins and asphaltenes associate to form colloidal aggregates. Asphaltenes interact through hydrogen bonding and π-bond overlap. Through polar functional group interactions, resins solvate asphaltene aggregates to each other. However, asphaltene colloidal aggregates associate to form an interfacial film. Primary asphaltene aggregates cross-link to form a rigid, viscoelastic structure at the oil-water interface. The process is further compounded by the adsorption of resin/asphaltene aggregates to hydrophilic particles. As more aggregates adsorb, the particles become preferentially wetted by the oil phase ($\Theta > 90°$). Ionic and π-bondings have been found between asphaltene/resin and clay surface (Sullivan and Kilpatrick, 2002;Sparks et al., 2003).

Electrokinetic (EK) method show the ability to remediate organic and inorganic contaminated soil and recently much attention had been shifted on the usage of EK remediation for oil contaminated soil (Lim et al., 2016). For the first time, the EK method was developed to separate the different phases (water, oil and solids) in oil sludge through the application of low-intensity direct current across a pair of electrode causing electro-osmosis of liquid phase, movement of ions and electrophoresis of charged particles (Kuo et al., 2011). This method can dewater oil sludge by nearly 63% and reduce the light hydrocarbon content by about 43% (Elektorowicz et al., 2006). The effect of EK separation, however, has not been studied on the fine solid particles in oil sludge.

The aim of this project was to further explore the effect of EK processes on the surface properties of oil sludge conglomerates. Thus, wettability alternation of solid particles is possible due to electro-demulsification processes, the wettability was examined as an indicator of destabilization of oil sludge. Study was conducted by the application of different direct current (DC) operation regimes to the sludge originated from a refinery disposal pond. Changes induced on the surface properties of solids, wettability alternation and contact angle were studied before and after application of the electric fields. It was assumed that the study outcomes would help to improve management of oil sludge which can be beneficial to environment, soil and water pollution.

## 2 MATERIALS AND METHODS

### 2.1 Setting-up of an electrokinetic reactor

To achieve the above-mentioned objectives, the 1-L electrokinetic (EK) reactor was set up by supplying an electric current across two electrodes by a DC power supply (BK Precision 1902, USA) which converts an AC to a DC. This unit enabled the supply of a constant direct current (CDC). Incremental or decremented DC was applied by adjusting the voltage on a daily basis. In order to create pulsed DC (PDC), a function generator (BK Precision 4001A, USA) was used to induce a DC offset. The resulting power was increased by an amplifier (model 7224 AE Techron, USA), which was connected to a 30 Hz Dual Trace Analog Oscilloscope (BK Precision 2120B, USA) to track the electrical function. A Digital Multimeter (Mastech MAS830, USA) was connected to the probes which were strategically positioned in the EK reactor, and used to monitor and measure the voltage supplied by the different electrical configurations across the electrodes.

#### 2.1.1 Application of different electrical regimes

EK studies were conducted using lagoon sludge sampled from an oil refinery located in Montreal, Quebec, Canada. A control reactor was set up with no potential gradient. A Constant DC (CDC) reactor was designed by applying a potential gradient of 1 V/cm across the electrodes. This gradient of voltage was chosen based on Elektorowicz and Habibi (2005), who found its suitability for very low permeable sludge, which was the case for the lagoon sludge used in this current study. For both Incremental and Decremented currents (IDC and DDC), a potential gradient ranging from 0.25 to 1.75 V/cm was applied while for the Pulsed DC (PDC), a 0.5 V/cm of DC was applied in combination with a 0.5 V/cm of AC pulse at squared frequencies of 60 Hz with a 1/5 pulse period.

### 2.2 Characterization of solids in oil sludge

#### 2.2.1 Sequential solvent extraction solids

A two step solvent extraction has been used to evaluate the efficiency of EK systems (Mochida et al., 1986;Wang et al., 2010). The first extraction was performed using 25 g of oil sludge with 100 mL of toluene, while the mixture was boiled under reflux for 15 hrs in a Dean-Stark apparatus (Kimble KIMA Soxhlet Apparatus, Fisher Scientific). The second extraction was conducted with 100 mL more polar tetrahydrofuran (THF) over a period of 15 hrs. The extracted solids from solvent extraction was subjected to different examinations as described below to compare with other EK methods.

#### 2.2.2 Wetting properties of solids

Oil sludge solids are rich in organics, which plays important roles in the sorption of oil particles to the surface of the solid matrix. Depending on the solids and organic content in the solid aggregates, the particles can be classified as completely hydrophobic (Hydrophobic Aggregate, HAGG) or biwettable (Biwettable Aggregate, BAGG). However, it is not uncommon to have some hydrophobic materials in variable degrees for the BAGG-type solids. In addition, ultra-fine clays (UFCs) may also be present and they are an additional contributor in intractable solids for stable emulsions in oil sludge.

In order to assess wettability, a mixture of 2 mL toluene and 8 mL of water was added to 200 mg of dried solids, then vortexed for 2 minutes at a maximum speed of 1500 rpm to have a homogenous mixture, and incubated overnight in the fumehood at room temperature (Sparks et al. 2003). Dried sludge in this study was prepared by using a vacuum oven to reach a constant weight at 38°C (Ukwuoma and Ademodi, 1999). Three kinds of sludge were dried. First, original sludge is the sludge that has not been treated. Second are sludge samples treated with different EK techniques. Third is the extracted sludge with toluene and THF to have comparison between EK and solvent extraction. All three kinds of sludge were dried by this technique. The wettability of the solids was determined as the percentage of solids present in the aqueous phase by weighting the aqueous phase solids (Somasundaran and Zhang, 2006;Jiang et al., 2011). In addition, the change of solids wettability over a short period (30 min) and a long period (3 days) of time was investigated.

The degree of biwettability is variable, thus, it is expected that some hydrophobic materials will also be present in the hydrophilic phase and vice versa. Therefore, the percentage weight of the solids that were partitioned to the aqueous phase which show the water-wet ratio of the sample was evaluated. These tests have been applied twice to each sample for more representative results. The emulsions stability could be evaluated by the results from this analyse.

### 2.2.3 Contact angle measurement

The surface hydrophilicity and surface energy change in each sample after EK treatments were assessed through contact angle measurements of DI water using the sessile drop method (VCA, video contact angle system, AST Products, Inc., Billerica, MA, USA). The right and left angles of the water drop were measured using the system software (VCA optima XE). 10 mm pellets were prepared with laboratory press (Model C12 Ton Benchtop, Carver, Inc., USA). 0.5 g of dried sample was added to the die, and 25 MPa pressure was applied to prepare the pellets (Wang et al., 2015). At least three points from the pressed sample were selected for contact angle measurements; an average of three measurements was reported. The purpose of this test is to have a comparison and a evaluation on the solids surface wettability and emulsion destabilization.

## 3 RESULTS AND DISCUSSION

All EK regimes, as defined in section 2.1.1, were able to destabilize oil sludge but at different extents. Figure 1 shows an example of a demulsified oil sludge sample prepared for further wettability tests. The samples from different reactors have been prepared like the one in Figure 1 to study the effects of solids wettability on the emulsion stabilities. Though this analyse, the comparison can be done on the destabilization ability of different EK technique and solvent extraction.

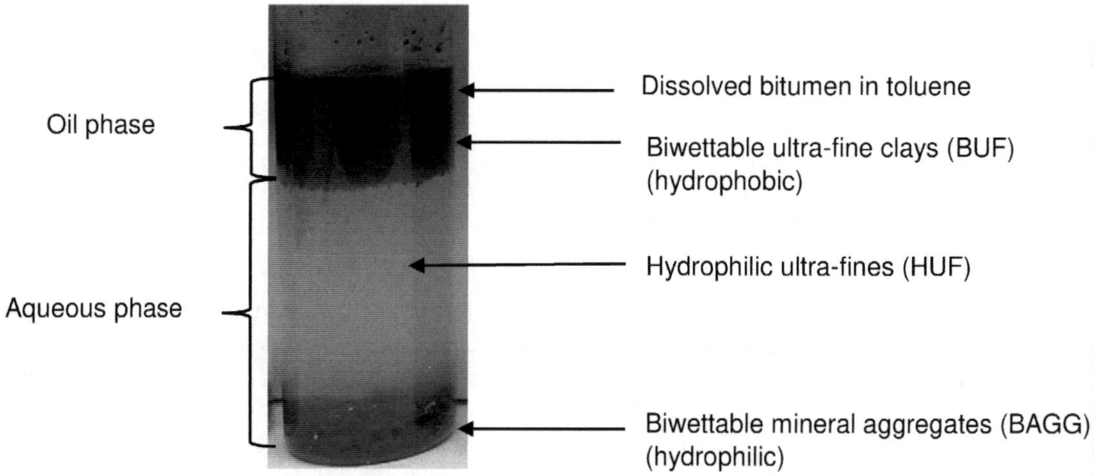


Figure 1: Fractionation of oil sludge solids into organic and aqueous phases

### 3.1 Wetting properties comparison between original, solvent extracted and EK-treated oil sludge solids

To evaluate the wettability alteration of solids with EK treated sludge, wettability analyses have been applied. The results from wettability analyses have been compared with two samples; original (control) sludge solids and solvent extracted sludge solids that have been extracted by toluene and THF. Solids from original sample showed no significant changes over short and long periods of time. By evaluating the changes through time the stability of emulsions was found to decrease. This was mainly due to the effect of the applied electrical field on the ion concentration on the surface of solids.

For the long-time testing, when the separation of phases was complete, solids from different sections of the reactor showed different behaviours. At the bottom anode, which included most of the solids, the wettability showed clear changes of solids wettability to water-wet compared to the original control sludge. Under the PDC, IDC and DDC electrical regimes, the water-wet portion of solids increased more than sequential extracted oil sludge solids. Most of the solids at the bottom cathode showed better water-wet properties compared to the original control. The DDC reactor showed the most alternation in water-wet conditions at the bottom and middle parts of the reactor. For all the reactors, the water-wet percentage increased to more than 60%. The top points of reactors had low amounts of solids that were mostly oil wet. The IDC reactor exhibited an interesting change in the water-wet behavior at the bottom section of the reactor where it had the highest movement of solids to the anode side. Subsequently, it was concluded that both horizontal and vertical mobilisation of particles vs. water takes place under EK treatment (Figure 2).

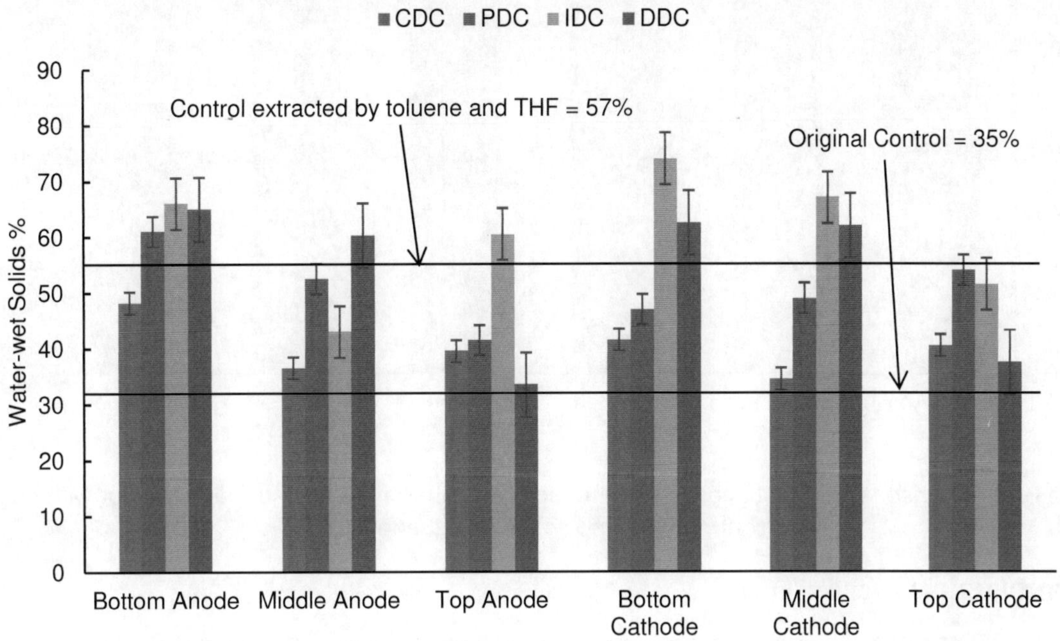


Figure 2: Percentage of solids that migrated to the aqueous phase in the four reactors following application of different electrical regimes.

### 3.2 Contact angle

The contact angle measurements of the dried solids confirmed the wettability alternation of the solids from oil-wet to water-wet for all four electric regimes. As mentioned before, most of the solids moved to the bottom anode side in the IDC and DDC reactors (Figure 2). Thus, the wettability alternation (contact angle) changes were significant for the bottom anode solids (Figure 3). PDC reactor configuration showed the most alternation on the solid surface properties. As indicated before, the wetting properties of solids were observed for the IDC and DDC reactors followed by the PDC reactor, which all of them had more decrease on contact angle degree compare to the toluene/THF extracted control.

Hence, the current study shows that EK treatment can have a significant impact on the wettability alternation of fine solids in oil sludge. These changes might occur due to changes in functional groups at the surface of the solids as well as the bonding formation between organic compounds, mainly asphaltenes and resins. Subsequently, wettability analyse can serve as an adequate indicator of demulsification and dewatering of oil sludge. Comparison between contact angle and wetting properties of solids showed that decrease on the hydrophobicity of surface destabilize the emulsions and make more water wet solids. The minor change on the contact angle can make a marked difference in the solids and emulsions stability.

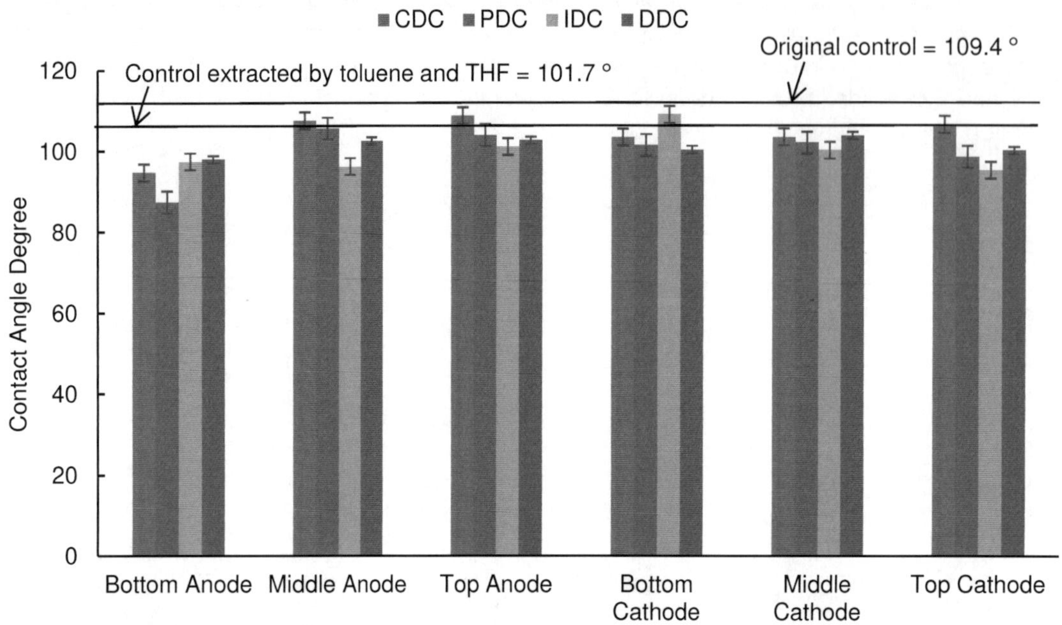


Figure 3: Comparison of contact angle degree of solids in different section of each reactor with original and control extracted solids by toluene and THF.

## 4 CONCLUSION

In this study, different configurations of DC electrical regimes were applied to oil sludge treatment. It was found that they all induce changes to the surface properties of the fine solids. For all the tested EK systems, the solids wetting properties changed to water-wet and the changes even surpass the sequential solvent extracted solids. It was concluded that wettability is a valuable indicator of water-in-oil emulsion destabilisation that can play an important role in oil sludge treatment and volume reduction. The results have been confirmed by contact angle analyse that almost in all sections of reactors the contact angle of the solids decreased even lower than the solids that have been solvent extracted. The DDC and IDC electrical regimes induced the highest extent of solids movement to the anode side and wettability alternation to water-wet. Hence, they can be applied as advanced treatment technology to destabilise oil sludge emulsions. In comparison between contact angle and wetting properties it can be determined that minor change on the contact angle can make a marked difference in the solids and emulsions stability.

### Acknowledgements

The authors would like to express their gratitude to NSERC for supporting this research through Discovery Grant (156161) awarded to Dr. Elektorowicz.

### References

Ayotamuno, M., R. Okparanma, E. Nweneka, S. Ogaji and S. Probert (2007). Bio-remediation of a sludge containing hydrocarbons. *Applied energy*, **84**(9): 936-943.

Blight, G. (2009). Geotechnical engineering for mine waste storage facilities, CRC Press.

Chen, Q., I. Stricek, M. R. Gray and Q. Liu (2017). Influence of hydrophobicity distribution of particle mixtures on emulsion stabilization. *Journal of Colloid and Interface Science*, **491**: 179-189.

Da Silva, L. J., F. C. Alves and F. P. de França (2012). A review of the technological solutions for the treatment of oily sludges from petroleum refineries. *Waste Management & Research*, **30**(10): 1016-1030.

Elektorowicz, M. and S. Habibi (2005). Sustainable waste management: recovery of fuels from petroleum sludge. *Canadian Journal of Civil Engineering*, **32**(1): 164-169.

Elektorowicz, M., S. Habibi and R. Chifrina (2006). Effect of electrical potential on the electro-demulsification of oily sludge. *Journal of colloid and interface science*, **295**(2): 535-541.

Guolin, J., L. Ming and Q. Shaopeng (2009). The Use of Nonionic and Anionic Surfactant in the Treatment of Oily Sludge. Energy and Environment Technology, 2009. ICEET'09. International Conference on, IEEE.

Heidarzadeh, N., S. Gitipour and M. A. Abdoli (2010). Characterization of oily sludge from a Tehran oil refinery. *Waste Management and Research*, **28**(10): 921-927.

Hu, G., J. Li and G. Zeng (2013). Recent development in the treatment of oily sludge from petroleum industry: A review. *Journal of Hazardous Materials*, **261**(0): 470-490.

Jeeravipoolvarn, S., J. Scott and R. Chalaturnyk (2009). 10 m standpipe tests on oil sands tailings: long-term experimental results and prediction. *Canadian geotechnical journal*, **46**(8): 875-888.

Jiang, T., G. J. Hirasaki, C. A. Miller and S. Ng (2011). Wettability Alteration of Clay in Solid-Stabilized Emulsions. *Energy & Fuels*, **25**(6): 2551-2558.

Kuo, C.-H., H.-Y. Chang, C.-P. Liu, S.-H. Lee, Y.-W. You and J.-J. Shyue (2011). Effect of surface chemical composition on the surface potential and iso-electric point of silicon substrates modified with self-assembled monolayers. *Physical Chemistry Chemical Physics*, **13**(9): 3649-3653.

Lim, M. W., E. V. Lau and P. E. Poh (2016). A comprehensive guide of remediation technologies for oil contaminated soil — Present works and future directions. *Marine Pollution Bulletin*, **109**(1): 14-45.

Mikula, R. J., K. L. Kasperski, R. D. Burns and M. D. MacKinnon (1996). Nature and fate of oil sands fine tailings. *Advances in Chemistry Series*,(251).

Mochida, I., K. Sakanishi, Y. Korai and H. Fujitsu (1986). Two-stage catalytic up-grading of vacuum residue of a Wandoan coal liquid. *Fuel*, **65**(8): 1090-1093.

Rico, M., G. Benito, A. Salgueiro, A. Díez-Herrero and H. Pereira (2008). Reported tailings dam failures: a review of the European incidents in the worldwide context. *Journal of Hazardous Materials*, **152**(2): 846-852.

Snars, K. and R. Gilkes (2009). Evaluation of bauxite residues (red muds) of different origins for environmental applications. *Applied Clay Science*, **46**(1): 13-20.

Somasundaran, P. and L. Zhang (2006). Adsorption of surfactants on minerals for wettability control in improved oil recovery processes. *Journal of petroleum science and engineering*, **52**(1): 198-212.

Sparks, B., L. Kotlyar, J. O'Carroll and K. Chung (2003). Athabasca oil sands: effect of organic coated solids on bitumen recovery and quality. *Journal of Petroleum Science and Engineering*, **39**(3): 417-430.

Sullivan, A. P. and P. K. Kilpatrick (2002). The effects of inorganic solid particles on water and crude oil emulsion stability. *Industrial & Engineering Chemistry Research*, **41**(14): 3389-3404.

Ukwuoma, O. and B. Ademodi (1999). The effects of temperature and shear rate on the apparent viscosity of Nigerian oil sand bitumen. *Fuel Processing Technology*, **60**(2): 95-101.

Wang, C., M. Geramian, Q. Liu, D. G. Ivey and T. H. Etsell (2015). Comparison of Different Methods To Determine the Surface Wettability of Fine Solids Isolated from Alberta Oil Sands. *Energy & Fuels*, **29**(6): 3556-3565.

Wang, J., J. Yin, L. Ge, J. Shao and J. Zheng (2010). Characterization of Oil Sludges from Two Oil Fields in China. *Energy & Fuels*, **24**(2): 973-978.

Yang, X., W. Tan and Y. Bu (2008). Demulsification of asphaltenes and resins stabilized emulsions via the freeze/thaw method. *Energy & fuels*, **23**(1): 481-486.

Leadership in Sustainable Infrastructure
*Leadership en Infrastructures Durables*

**Case Study**

Vancouver, Canada
May 31 – June 3, 2017/ *Mai 31 – Juin 3, 2017*

# EFFECT OF A NOVEL BIOCHAR SUPPORTED NANOSCALE IRON SULFIDE COMPOSITE ON *CORYNEBACTERIUM VARIABILE HRJ4* AND A CHEM-BIO HYBRID PROCESS FOR TRICHLOROETHYLENE DEGRADATION

Lyu, Honghong[1,2,3], Zhao, Hang[1], Mulligan, Catherine[4] and Tang, Jingchun[1,5]
[1] Key Laboratory of Pollution Processes and Environmental Criteria (Ministry of Education), Tianjin Engineering Center of Environmental Diagnosis and Contamination Remediation, College of Environmental Science and Engineering, Nankai University, Tianjin 300071, China
[2] Department of Agricultural and Biological Engineering, University of Florida, Gainesville, Florida 32611, United States
[3] Brook Byer Institute for Sustainable Systems and School of Civil and Environmental Engineering, Georgia Institute of Technology, Atlanta, Georgia 30332, United States
[4] Department of Building, Civil and Environmental Engineering, Concordia University, 1455 de Maisonneuve Blvd. W., Montreal, Quebec H3G1M8, Canada
[5] tangjch@nankai.edu.cn

## 1 Project Overview

Trichloroethylene (TCE) is one of the most common organic pollutants detected in both soil and groundwater (Bacik et al. 2012). Since TCE poses a number of risks, there has been an increasing concern over TCE usage, and growing attention on monitoring its environmental levels, fate, and remediation.

Several studies have been reported for degradation or transformation of TCE, including physico-chemical oxidation, UV/$TiO_2$ photocatalysis, and microbial degradation (Murugesan et al. 2011). However, microbial degradation of TCE is mainly hindered by the halogen substituents and their hyper-toxicity. Therefore, preceding dechlorination is essential for efficient microbial degradation to occur.

Traditional catalysts, such as nFe and nFe–Pd, were found to be toxic to microorganisms and the bactericidal activity (Lee et al. 2008), which is another important concern as they might affect the biodegradation process in the hybrid treatment. Therefore, an efficient, economic, and environmentally friendly multifunctional catalyst should be developed.

## 2 Innovation

In this project, a sequential chem-bio hybrid process with a novel biochar (BC) supported nanoscale iron sulfide (FeS) (CMC-FeS@BC) as chemical remover and *Corynebacterium variabile HRJ4* as biocatalyst for TCE degradation and the impacts of CMC-FeS@BC on TCE biodegradation by *HRJ4* were investigated.

## 3 Lessons Learned

Strain *HRJ4* grew well in the presence of CMC-FeS@BC up to 0.25 g/L in medium with TCE as carbon source, which is higher than the CMC-FeS@BC dosage for TCE chemical removal (0.18 g/L). The CMC-FeS@BC composite displayed an enhanced TCE removal capacity of 77.8 mg/g compared to 59.7 mg/g for CMC-FeS, 26.8 mg/g for bare FeS, and 56.5 mg/g for BC. At the equilibrium TCE concentration of 3.4 mg/L, 40.4% of TCE removal was due to reduction, and 59.6% was ascribed to sorption. The dechlorination pathway was proposed based on the results of GC-MS and XPS. 10 mg/L of TCE was dechlorinated to *cis*-1,2-dichloroethene (*cis*-DCE), vinyl chloride (VC), and acetylene within 12 h by 0.18 g/L CMC-FeS@BC. Addition of *HRJ4* strain into the reactor effectively degraded the *cis*-DCE and VC to ethylene and acetylene. XPS analysis suggested that sorption, reduction and biodegradation were dominant mechanisms for TCE removal. Results from this study forecast the potential of CMC-FeS@BC/*HRJ4* chem-bio hybrid treatment in TCE degradation

### Figures

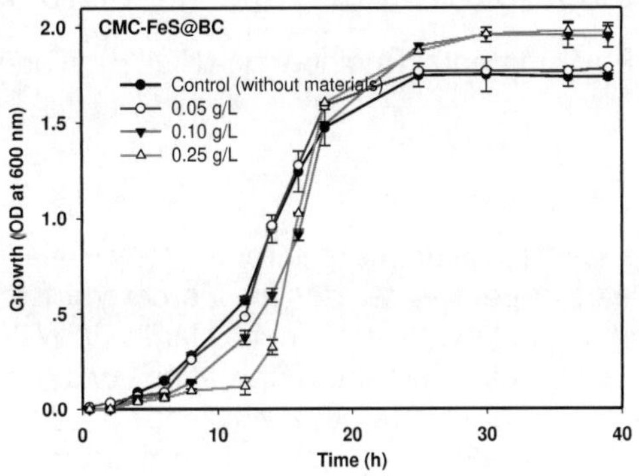


Figure 1: Example figure caption Effect of CMC-FeS@BC on the growth of *Corynebacterium variabile HRJ4* in MSM medium with DE as a sole carbon source.

### Acknowledgements

This work was supported by (1) National Natural Science Foundation of China (41473070, 31270544), (2) 863 Major Program (2013AA06A205), (3) 863 achievement transformation program in Tianjin (14RCHZSF00144). (4) Beautiful Tianjin Scientific Project (No. 14ZCDGSF00033), (5) The National Water Pollution Control and Treatment Science and Technology Major Project (2015ZX07203-011-06).

### References

Bacik, D.B., Zhang, M., Zhao, D., Roberts, C.B., Seehra, M.S., Singh, V. and Shah, N., 2012. Synthesis and characterization of supported polysugar-stabilized palladium nanoparticle catalysts for enhanced hydrodechlorination of trichloroethylene. *Nanotechnology,* **23**(29), 294004.

Murugesan, K., Bokare, V., Jeon, J.R., Kim, E.J., Kim, J.H. and Chang, Y.S., 2011. Effect of Fe-Pd bimetallic nanoparticles on Sphingomonas sp. PH-07 and a nano-bio hybrid process for triclosan degradation. *Bioresour. Technol.,* **102**(10), 6019-6025.

Lee, C., Kim, J.Y., Lee, W.I., Nelson, K.L., Yoon, J. and Sedlak, D.L., 2008. Bactericidal effect of zero-valent iron nanoparticles on Escherichia coli. *Environ. Sci. Technol.,* **42**(13), 4927-4933.

Leadership in Sustainable Infrastructure
*Leadership en Infrastructures Durables*

Vancouver, Canada
May 31 – June 3, 2017/ *Mai 31 – Juin 3, 2017*

# ENHANCED ELECTROCHEMICAL AND BIOLOGICAL PHOSPHORUS REMOVAL IN A SOLE REACTOR

Abdelmajeed, Adam.[1, 2], Maria, Elektorowicz [1] and Sharif, Ibeid[1].

[1] Department of Building, Civil & Environmental Engineering (BCEE), Concordia University, Canada
[2] ab_ada@encs.concordia.ca

**Abstract:** Excessive phosphorus levels in aquatic environments create toxic algal bloom resulting and eutrophication, which ultimately lead to poor water quality. To control this undesirable phenomenon, the phosphorus removal from municipal wastewater treatment plants (WWTPs) is essentially required. The removal of nutrients requires constructing various operation units to provide different conditions (i.e. anoxic and aerobic) intended to promote the removal efficiency of each single nutrient compound. Such designs have high energy consumption and a big footprint. In this work, the possibility of simultaneous electrochemical and biological phosphorus removal in a single reactor was investigated. Bench scale electro-bioreactor was fed with synthetic wastewater containing phosphorus in the range of 14−18 mg $PO_4 - P$/L. At bench scale steady state conditions, the removal efficiency of phosphorus (as $PO_4 - P$) reached over 99%. It was concluded that such high percentage of phosphorus removal was due to both electrocoagulation process and polyphosphate-accumulating organisms (PAOs) growth in the reactor since a single submerged membrane electro-bioreactor (SMEBR) was capable of creating anoxic and aerobic conditions suitable for PAO growth. Furthermore, the electrocoagulation generated $Al^{3+}$ due to the electrolytic dissolution of aluminum anode led to higher electrochemical phosphorus removal in the system.

Keywords: Advanced wastewater treatment processes, Phosphorus removal, Electrochemical treatment, Submerged Membrane Electro-Bioreactor (SMEBR), Polyphosphate-accumulating organisms (PAOs), Sustainable design.

## 1 INTRODUCTION

Over the recent years, the providing of a fresh water supply and safely treated effluent has become an urgent global issue. The discharge of insufficiently treated wastewater effluent can potentially create a significant squeeze to fresh water resources and environmental milieus. Among of these impurities that discharged effluent usually containing are nutrients, in particular nitrogen (N) and phosphorus (P), which are constantly challenging and expensive to remove (Bodini et al. 2015). The accumulation of these nutrients in aquatic environments such as lakes and artificial reservoirs can pose its deterioration. Discharged nutrients have numerous effects including toxic algal growth and eutrophication leading to oxygen depletion in surface waters. Phosphorus is considered as the key component of municipal sewage, as human waste from the toilet is a substantial source (Nedjah and Laskri 2015), and thus the removal of phosphorus during treating processes is of great importance.

Several techniques have been established for the removal of phosphorus (i.e. biological treatment (EBPR), chemical treatment (precipitation, sometimes supported with membrane filtration) (Nassef 2012). Each process does exhibit drawbacks when they operate separately. For instance, membrane filtration process (microfiltration and ultrafiltration) is not capable of an adequate removal of phosphorus from wastewater to an acceptable level. Reversible and irreversible fouling during membrane processes is also considered as a major obstacle facing this technique since it results in a flux decline and an increase in the overall resistance to the filtration process, and hence, increasing the energy requirement and decreasing the life-span of the membrane (Wilbey 2013, Basile et al. 2015). Likewise, biological means to enhance phosphate accumulation by biomass requires more than one biological reactor and pump systems to ensure recirculation of wastewater. Such processes can drop the reactive phosphorus concentration as low as 0.2 mg $PO_4 - P/L$, but it requires space and energy that not all wastewater treatment plants and municipalities can accommodate and handle. In the areas where population density is low and climate conditions are extreme, the above techniques would be difficult to apply. Chemical treatment, which uses coagulation agents to co-precipitate phosphorous, leads to the increase of the operation costs and generate huge sludge as by-products. Thus, these conventional processes have economic and sustainability limitations as a result of excess sludge disposal, high energy inputs, poor effluent quality, and limited phosphorus removal (Posadas et al. 2013). An alternative strategy is to use electrokinetic processes in combination with biological and membrane processes to eliminate phosphorous at low cost, while the requirement for space is minimal.

Electrocoagulation as a result of electrokinetic processes is a promising technology, which has been proved to be an efficient and inexpensive technique for wastewater treatment. Electrocoagulation (EC) for wastewater treatment provides a set of unique advantages over chemical coagulation processes. Electrocoagulation process as an example can operate with no chemical additives while, alkalinity is not consumed through the processes so that pH adjustment is not necessary (Al-Shannag et al. 2013). EC process does not require coagulation agents such as alum and ferric sulfate, and hence, reduces the cost of operation and produces less sludge as an end-product (Zhu et al. 2005). Moreover, EC has a potential in an efficient removal of metals, colloids, and soluble inorganic pollutants, resulting in higher quality of the effluent (Hasan et al. 2014).

The Submerged Membrane Electro-Bioreactor (SMEBR), presented in this paper, implements EC to combine it with conventional membrane bioreactor at a low direct current in a single vessel (Elektorowicz et al. 2009, Bani-Melhem and Elektorowicz 2010). The novelty of this technology bases on allowing the electrokinetic phenomena to create alternated reduction – oxidation conditions (e.g. between -150 mV to +150 mV) in the same vessel to provide an adequate environment for the growth of all diverse types of microorganisms including PAO (Ibeid et al. 2011). The process takes fully advantage of phenomena initiated by direct current applied intermittently into the vessel (Eq. 1– 4) (Ibeid et al. 2011). Such phenomena control physicochemical as well as biological processes in the reactor (Ibeid et al. 2013). The operation mode (DC current system ON and OFF) is a state-of-art and can be adjusted depending on treatment objectives (e.g. carbon removal, phosphorous or all impurities simultaneously, etc.) (Hasan et al. 2012, Ibeid 2011). Since physicochemical, and biological and filtration processes take place simultaneously in the same tank, the SMEBR leads to the possible minimizing of space requirements and footprint (Ibeid et al. 2016). Additionally, the system also requires less aeration supply and no additional of chemicals comparing to a conventional MBR system where intensive aeration and chemicals usage are inevitable. This system has been recognized as an efficient, operational cost-effective and environmentally friendly (Ibeid et al. 2016). Furthermore, SMEBR is capable of removing metallic impurities (Hasan et al. 2014). Major electrochemical reactions taking place around the anode and cathode area in SMBER system are presented in the equations (1– 4).

Anode area (oxidation):

[1] $\mathrm{Al} \rightarrow \mathrm{Al}^{+3} + 3\mathrm{e}^-$

[2] $2\mathrm{H_2O} \rightarrow \mathrm{O_2}\ (\mathrm{gas}) + 4\mathrm{H}^+(\mathrm{aq}) + 4\mathrm{e}^-$

Cathode area (reduction):

[3] $3\mathrm{H_2O} + 3\mathrm{e}^- \rightarrow \frac{3}{2}\mathrm{H_2}\ (\mathrm{gas}) + 3\mathrm{OH}^-$

[4] $0.5\ O_2 + 2e^- + H_2O \rightarrow 2OH^-$

This article displays an outstanding performance of integrated processes comprising electrokinetic, biological and membrane filtration in one vessel for treating wastewater with respect to phosphorous. Moreover, a potential of the simultaneous biological and electrochemical phosphorus removal in this single reactor was discussed.

## 2 MATERIALS AND METHODS

### 2.1 Material and feed media

The tested systems were supplied with a synthetic wastewater, where phosphorous was provided by potassium phosphate ($KH_2PO_4$) and potassium chloride (KCl). Such composition was selected to simulate real municipal wastewater in terms of organic compounds, nitrogen, phosphorus, and soluble salts content. Phosphorus in the form of potassium phosphate ($KH_2PO_4$) changed between 14 to 18 mg $PO_4 - P/L$ in feed wastewater over the experimental period. The organic portion of the feeding consisted of both glucose ($C_6H_{12}O_6$) and sodium acetate ($CH_3COONa.3H_2O$). Ammonium sulfate $(NH_4)_2SO_4$ was the major source of nitrogen. Other trace elements were also added to the synthetic wastewater. Table 1 shows the characteristics of influent ("feed wastewater").

Table 1: Initial physicochemical characteristics of synthetic wastewater

| Parameter | Abbreviate | Average Value/or Range |
|---|---|---|
| Chemical oxygen demand [mg/L] | COD | 280 |
| Ammonia-nitrogen [mg/L] | $NH_4^+ - N$ | 23.3 |
| Nitrate-nitrogen [mg/L] | $NO_3^- - N$ | 0.148 |
| Nitrite -nitrogen [mg/L] | $NO_2^- - N$ | ≤ 0.05 |
| Total nitrogen [mg/L] | T-N | 24.6 |
| Phosphorus [mg/L] | $PO_4 - P$ | 14−18 |
| Potential of hydrogen | pH | 6.62 |
| Conductivity [µS/cm] | EC, κ | 97.4 |
| Oxidation-reduction potential [mV] | ORP, $E_h$ | 6−8 |
| Total alkalinity (as CaCO3 mg/L) | TALK | 98.5 |
| Temperature [ºC] | T | 15.8−17.2 |

### 2.2 Experimental setup

Two bench scale continuous flow reactors: one submerged membrane electro-bioreactor– SMEBR (Figure 1) and one conventional MBR system without DC electrical field (reference reactor) were run in parallel to assess the phosphorus removal. Both reactors were continuously fed with the same influent composition and a constant flow rate of 7 L/d. The corresponding hydraulic residence time (HRT) and sludge residence time (SRT) were 24h and 20 days respectively in both reactors (Table 2). To obtain comparative conditions between the processes of SMBER and MBR, alum (as aluminum sulfate, $Al_2(SO_4)_3 \cdot 18H_2O$) was added to the supernatant of MBR system. The configuration of the electro-bioreactor was identical to patented SMEBR design (Ibeid et al. 2011, Elektorowicz et al. 2009). SMEBR consisted of a cylindrical polyethylene container with a total effective working volume of (7 L), a hollow fiber ultrafiltration membrane module (Zeeweed-1, GE, Canada), two perforated electrodes, and 8 fine bubble air diffusers. The air diffusers supplied oxygen for biological activities, achieving a proper mixing, and restraining the membrane fouling. The membrane, with a pore size of 0.04 µm and surface area of 0.047 $m^2$, was located centrically and surrounded by electrodes (Figure 1). A DC power supply connected to a timer was used to provide intermittent current (at a current density of 12.5 A/m$^2$). The system was inoculated with fresh activated sludge from a local WWTP. The activated sludge was diluted (0.5 L of activated sludge and 6.5 L water) to provide the initial mixed liquor suspended solid (MLSS) concentration of 3000 mg/L. About of 350 mL of sludge was wasted on a daily basis to maintain target SRT of 20 days. The temperature of 22±2 °C was kept at during the experiment while neutral pH was also preserved. Table 2 summarizes the major operational factors of SMEBR and MBR systems.

Table 2: Operation condition of SMEBR and MBR

| Parameter | SMEBR | MBR |
|---|---|---|
| DO (mg/L) | 0.5-3.5 | 5-6 |
| HRT (h) | 24 | 24 |
| SRT (d) | 20 | 20 |
| Wasted sludge (mL) | 350 | 350 |
| Effluent flowrate (L/d) | 7 | 7 |
| Effective working volume (L) | 7 | 7 |
| MLSS (mg/L) | 3000 | 3000 |
| Current density (A/m$^2$) | 12.5 | N/A* |

*N/A= not applicable

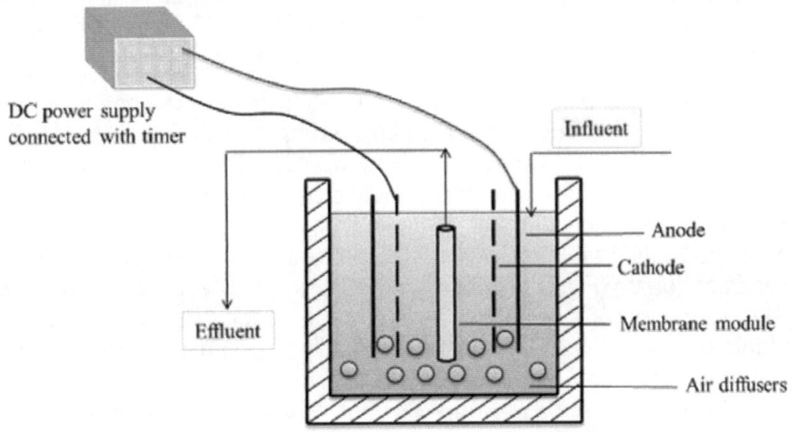


Figure 1: Schematic diagram of submerged membrane electro-bioreactor, SMEBR

## 2.3 Analytical methods

The concentrations of phosphorus in influent and effluent were determined as orthophosphate-phosphorus ($PO_4 - P$) using Hach procedure (Ascorbic Acid Method 10209, TNTplus™ 844, Hach, DR 2800, USA). The values of pH and temperature were measured by using a pH probe (HQ30d, multi- parameters meter, Hach, SA). Measurements of dissolved oxygen (DO) was conducted using a digital DO meter (YSI, Model 52, USA), while oxidation-reduction potential (ORP) was measured by a digital IntelliCAL™ MTC101 Standard Gel-Filled ORP electrode (Hach). These parameters are interrelated and are of importance in controlling the reactor performance. Besides, the measurement of ORP assures the accuracy of DO distribution. Calibration of DO and pH meters were performed on a weekly basis. The removal efficiency of phosphorus (% Removal) was calculated as per Eq. 5:

[5] $(\%\ \text{Removal}) = \frac{C_i - C_f}{C_i} \times 100$

Where; $C_f$ and $C_i$ are the final and initial phosphorus concentration in "mg/ L," respectively.

To assess the biomass activity, mixed liquor suspended solids (MLSS) and mixed liquor volatile suspended solids (MLVSS) were measured regularly according to a procedure described by the Standard Methods 2540 D and 2540 E, respectively (APHA 2012). Such techniques were mainly based on weight lost, at 105 °C and 550 °C for TSS and VSS, respectively. Calculations were done based on the given equations 6 and 7:

[6] mg TSS/L = ([103–105 °C] dry weight - weight of filter)*1000, mg/volume of sample, mL.

[7] mg VSS/L= (550 °C dry weight – [103–105 °C] dry weight)*1000, mg/volume of sample, mL.

Furthermore, the viability of bacteria in the electrical field were counted using the plate count agar (PCA) method. Such method is used to provide a rough estimation of the number of aerobic and facultatively anaerobic bacteria present in water sample. In this practice, dilution of tested samples "serial dilution technique" between $10^1$ to $10^5$ were performed as described in Standard Methods for Microbiological Examination of Dairy Products, (APHA Method, 7157) (Marshall 1993). Diluted samples that yielded a plate with an average of 30 ~ 300 colonies were chosen. 0.1 mL volume of the diluted specimen was utilized to inoculate or to quantify bacterial counts in the sample coming from the electro-bioreactor. Samples were incubated at 32±1°C and examined for growth after 48 hours. Duplicated plates were used to test bacteria count, and the value was recorded as an average. The number of bacteria in the representative samples were calculated using this formula: CFU/mL= CFU * dilution factor * 1/aliquot, where magnifying colony counter was used when needed.

Applying a DC electrical field in a biological reactor is most likely to impair the bacteria's metabolism, shape, movement, and membrane integrity. Thus, the available commercially LIVE/DEAD *Bac*Light bacterial viability kit (L13152) provided Molecular Probes, Inc. (Eugene, OR, USA) was applied to test the bacterial viability in view of assessing the impact of electrokinetics processes on microbial activity. The *Bac*Light viability stain package provides an indication of the fraction of physiologically active and dead "inactive" cells based on membrane integrity. In this protocol, cells are treated with two different nucleic acid-binding dyes: SYTO®9 and propidium iodide (PI). SYTO®9 is a small green-fluorescent nucleic acid that penetrates all bacterial membranes (both those with intact and damaged membranes), and hence it is used for assessing total cell counts (live and dead). In contrast, propidium iodide is a large red-fluorescent nucleic acid that penetrates only cells with a compromised membrane and consequently visualizing dead microbes. PI is, as a result, causing a reducing in the SYTO®9 stain fluorescence intensity when both dyes are present. The simultaneous use of SYTO9 and PI, therefore, enables differentiation and measurement of the relative ratio of active cells "green" with an intact membrane and dead cells "red" with a damaged cytoplasmic membrane.

Two fluorescent nucleic acid stain were either used alone or in combination: SYTO9 (Invitrogen AG, Basel, Switzerland) and propidium iodide (Invitrogen). The working solution of these dyes was prepared as recommended by the supplier. All stock solutions were stored at -20°C. Staining was done by mixing equal volumes of SYTO®9 and PI, dissolved in 5 mL deionized water. 1.5 mL of two staining mixture was added directly to 1.5 mL of the appropriate bacterial suspension. Incubation was accomplished in the dark at room temperature for 15 minutes. Calibration was done according to manufacturer's instructions except for *E. coli* suspensions which were replaced by the MBR activated sludge. Five samples were prepared to achieve effective concentrations ranging from 0 to 100% to create a standard curve (all mixtures were five specimens * 1.5 mL = 7.5 mL total). A standard curve was then prepared for each microplate. Stained bacterial suspensions were observed under a confocal laser-scanning microscopy (CLSM) with 1000X total magnification. The excitation/emission was 488/535 nm for viable cells, and 488/580 nm for dead cells. The cells were then visualized and quantitated using Image processing and analysis in Java.

## 3 RESULTS AND DISCUSSION

### 3.1 Phosphorus removal

The phosphorus concentration in effluent and removal efficiency by SMEBR and MBR systems are shown in Table 3 and Figure 2. The results have displayed that the ability of SMEBR system to remove phosphorus to an appreciable extent. Treatment by SMEBR process reduced orthophosphate with an average of 94.86% compared with 48.99% obtained by MBR. The maximum phosphorous removal in SMEBR was 99.28%, while in conventional MBR was only 64.44%. Yet, this removal percentage in MBR system was only achieved by adding of coagulant agents to help phosphorus removal. The same observations regarding to hinder the P-removal efficiency of the MBR system was also reported by (Hasan et al. 2012, Wei et al. 2012, Wei et al. 2009, Zuthi et al. 2013). Zuthi et al. (2013) indicated that the application of MBR treatment targets for P-removal and effluent quality requirements is feasible if the biological processes relevant to P-removal could be fully understood and linked to other process parameters.

Such results may suggest the necessity of an additional operation unit to MBR system to deal with the persisting phosphorus compounds. Contrary, the SMEBR system was enhanced with the electrokinetic process in the same operation unit which also is able to remove other impurities. For example, Hasan et al. (2014) reported that the introducing of electrical filed to the reactor contributed in electro-oxidation of organic impurities and making them possibly much more bioavailable due to the production of hydroxyl radicals ($OH^{\cdot}$) at the anode area. Hydroxyl radical is highly reactive aqueous and a strong oxidizing agent that may interact with organic impurities (Apaydin et al. 2009, Dohare and Sisodia 2014). Electrokinetic also controlled the reaction of dissolved ions with phosphorous and its fate within the reactor (e.g. deposition on electrodes) (Hasan et al. 2014). *In situ* generation of hydroxyl radicals for the treatment of non-biodegradable and refractory phosphorus fractions is displayed in the equation 8:

[8] $\mathrm{M} + \mathrm{H_2O_2} \rightarrow \mathrm{M^{n+}} + \mathrm{OH^-} + \cdot\mathrm{OH}$

Where; $(\mathrm{M})$ is the metal which is oxidized to its cation $(\mathrm{M^{n+}})$

*In situ* aluminum ions ($Al^{3+}$) generated due to the electrolytic dissolution of anode led to removal and precipitation of phosphorus (Eq. 9 and 10). Phosphorus was incorporated into a long chain of a complex compound that became a part of the suspended solids and removed due to the liquid-solid separation using membranes (Elektorowicz et al. 2014, Ibeid et al. 2011).

[9] $\mathrm{Al^{+3}} + 3\mathrm{H_2O} \rightarrow \mathrm{AL(OH)_3} + 3\mathrm{H^+}$

[10] $\mathrm{Al^{+3}} + \mathrm{H_nPO_4^{3-n}} \leftrightarrow \mathrm{ALPO_4} + \mathrm{nH^+}$

Table 3: Statistical parameters for the phosphorus concentration in effluent and its removal efficiency in SMEBR and MBR.

| Parameter | Conventional MBR | | SMEBR | |
|---|---|---|---|---|
| | Effluent (mg/L) | % Removal | Effluent (mg/L) | % Removal |
| Mean | 9.18 | 48.99 | 0.93 | 94.86 |
| Standard Deviation | 1.17 | 6.52 | 0.92 | 5.13 |
| Variance | 1.38 | 42.52 | 0.85 | 26.30 |
| Median | 9.60 | 46.67 | 0.73 | 95.96 |
| Mode | 10.10 | 43.89 | - | - |
| Minimum | 6.40 | 42.78 | 0.13 | 83.33 |
| Maximum | 10.30 | 64.44 | 3 | 99.28 |

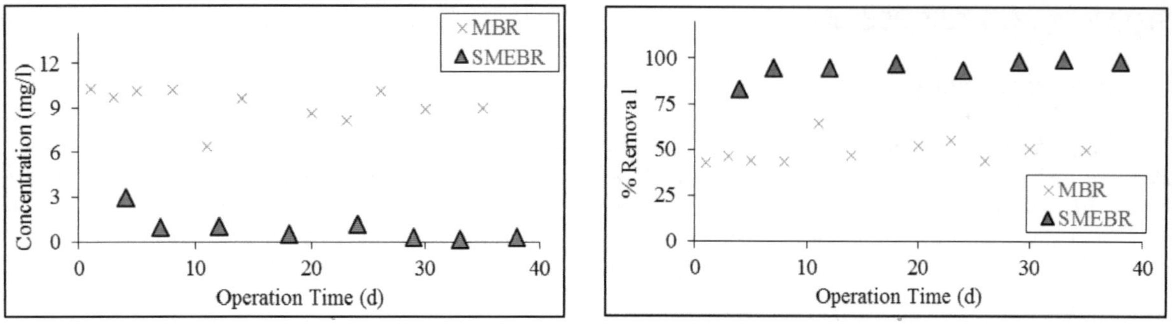


(a) (b)

Figure 2 : Performance of SMEBR and MBR processes with respect to phosphorus: (a) effluent concentration and (b) removal efficiency

It was concluded that excellent results in phosphorous removal by SMEBR system were due to both electrocoagulation and PAOs (bio-flocs) growth. It is speculated that the SMEBR system provided adequate (anoxic/aerobic) for the growth of polyphosphate-accumulating organisms due to the ability of the system to work in a wide range of redox potentials. In theory, biological phosphorus removal (Bio-P) requires the retention of biomass in sequential anaerobic and aerobic phases. During this process, P-release and organic matters uptake happen under anaerobic phase, while P-removal or uptake occurs under the following aerobic conditions (Sathasivan 2008). In SMEBR, the alternating anoxic/aerobic conditions were provided by periodic switching on/off the DC power supply. Thus, it is assumed that the electro-bioreactor was able to fluctuate oxidation-reduction potential (ORP), providing a suitable environmental for the activity of PAO organisms, and subsequently allowing to support the removal of large amounts of phosphorus. Such fluctuating ORP levels was ensured by the application of electrokinetic processes (Eq. 1– 4) and adequate air supply in the system where the measured DO in the reactor fluctuated between 0.5 and 3.5 mg/L. Ibeid et al. (2011) reported that the anaerobic condition was gradually supported when the electrical field was activated, while the aerobic condition was maintained by disconnecting the electrical current. In contrary, the MBR as all conventional systems preserved aerobic conditions without affecting phosphorous removal.

Previous studies have reported that an intermittent aeration membrane bioreactor (MBR) can achieve simultaneous carbon, nitrogen and phosphorous removal (Wang et al. 2015, Gnirss et al. 2003). However, while these studies have indicated the development of polyphosphate-accumulating organisms (PAOs) within the processes, the removal efficiency of phosphorous did not exceed 60%. The removal efficiency of phosphorus was also affected by the influent phosphorus concentration. It is speculated that in the present investigations, a high phosphorous removal was caused by both PAOs biomass growth and the electrolysis using a sacrificial aluminum anode.

### 3.2 Biomass activity assay

In order to support above conclusion with respect to biological removal of phosphorus, the biomass activity measurements for both the SMEBR and MBR systems were conducted. To assess the impact of electrokinetics processes on microbial activity, the LIVE/DEAD *Bac*Light bacterial viability kit was used to determine the ratio of viable bacteria in both reactors. The application of low current density with intermittent exposure regime to the bioreactor did not reduce significantly the percentage of live cells comparing to the control reactor. The average of live cells in both the SMEBR and MBR systems were 90% and 96%, respectively. Such test is based on the membrane integrity, while some bacteria having compromised membranes may be able to recover from the electrical field impact and resume their activity. According to Alshawabkeh et al. (2004), anaerobic cultures were able to recover their activities when the DC supply was switched off. The findings of our study suggest that intermittent electrical field of CD =12.5 A/m$^2$ applied into SMEBR were not negatively affecting the microbial biomass viability. These results were not in agreement with Wei et al. (2011) who reported 15% reduction in the live cell under CD =12.3 A/m$^2$. This slight discrepancy may be attributed to different electro-bioreactor configurations applied in both studies. In the current study, the continuous flow electro-bioreactor provided sufficient mixing and well-aerated condition that reduced the potential adverse effect of the DC on bacterial cells, while the batch electro-bioreactor mode used in the previous study may not provide a proper mixing and aeration conditions.

To further assess the activity of the microbial biomass, the bacterial plate count agar (PCA) procedure was undertaken to quantify the number of viable bacteria in both reactors. Remarkably, the applied current density of 12.5 A/m$^2$ to the bioreactor resulted in electro-stimulation of microbial cells and increasing in bacterial counts comparing to the reference reactor " without current." The highest bacterial counts were observed at the SMEBR system with an average of counted cells of 46 x 10$^4$ CFU/mL, whereas the MBR system had an average of bacterial counts of 31 x 10$^4$ CFU/mL. Such results support the hypothesis of that a low-level DC electric field enhances the bacterial growth, which was also confirmed in this study (Berg 1995, Fiedler et al. 1995, Alshawabkeh et al. 2004). Low current density stimulated biomass activity as a result of DNA changes, protein and biopolymer synthesis, enzyme reactions, cell membrane permeability and transport, and cell growth (Nakanishi et al. 1998, Berg 1995). Moreover, an intermittent current density below 25A/m$^2$ for the MLSS concentration of 6,000 mg/L was found in the literature to stimulate microbial biomass activity (Bayar and Karagunduz 2014). Wei et al. (2011) had studied the viability of the bacteria in the electrical system by which biological processes were not significantly affected by adequate interrupted electrical current density leading to great nutrient removal efficiencies. Furthermore, Yin et al. (2015) investigated the enhancement of the activity of anammox bacteria using electric current through which the total nitrogen removal rate (NRR) was significantly increased by 38.7 % over the control reactor. Chen et al. (2016) found that the electrical field could speed up the processes of hydrolysis and acidification. The elevated abundance of *Pseudomonas* (break down proteins) and *Methanoregula* (generate methane) were typically recorded at an electric voltage of 0.6 V/cm, with species richnesses of 19.1% and 53.3%, respectively. Such results support the fact of that the installation of electrodes on the SMEBR system enhanced the biomass activities. However, as the total number of polyphosphate-accumulating organisms was not directly investigated, it is uncertain to what extent the activity of PAOs have contributed in the removal of phosphorus.

Likewise, the investigation of mixed liquor suspended solids (MLSS) illustrated the high microbial growth and activity in the SMEBR comparing to conventional MBR, which was increased from 3000 mg/L to 4900 mg/L and hence played a key role removal efficiency of organics. Whereas, MLSS concentration in MBR system increased from 3000 mg/L to 3700 mg/L after 40 days' operation period. A significant increase in MLSS was also due to the dissolution of the aluminum anode at 12.5 A/m$^2$ experiment. MLVSS/MLSS ratio had shown high microorganism content of the suspended material. The investigation of MLVSS fraction illustrated the microbial growth and activity in the SMEBR, which was increased from 2520 mg/L to 3480 mg/L and hence played a key role in removal efficiency of phosphorous. Besides, the concentration of total phosphorus in activated sludge indicated that the phosphorous was likely stored it in biomass. The concentration of orthophosphate in the supernatant was higher than the concentration of phosphorous in the influent wastewater (date not shown). Sophisticated microbial and molecular techniques are continuing to be used in our upcoming research work. Even if we are not able to directly prove the presence of PAOs, it is evident from this research that the system had substantial phosphorous removal and electrode processes stimulated the biological activity in the system. The processes of electron transfer (ET) through microbial redox and their mechanisms have been previously reported by many studies (Watanabe et al. 2009, Summers et al. 2010, Ailijiang et al. 2016, and Liu et al. 2015). These studies have shown that some bacteria species are capable of performing anodic and cathodic half reactions via electrodes as electron donors and acceptors, respectively. Ailijiang et al. (2016) assumed that the conductive minerals would help electron transfer between microbial biomass and electrodes, and hence promote the processes of electrostimulation. Another study reported that electrically conductive magnetite nanoparticles permit the interspecies electron transfer (IET) through bacteria community such as *Geobacter Sulfurreducens* to *Thiobacillus Denitrificans*, leading to enhance simultaneous acetate oxidation and nitrate reduction (Zeyoudi et al. 2015). Such evidence support above conclusion with respect of POA stimulation by the electrical field and adequate redox conditions.

## 4 CONCLUSION

The potential of phosphorus removal from synthetic wastewater using a hybrid electro-bioreactor has been investigated. The system has apparently shown its capability to combine electrochemical and biological phosphorus removal in a single operation unit by providing an adequate direct current density (CD) and dissolved oxygen (DO) concertation. A substantial phosphorus removal of nearly 99% was attained in the SMEBR process, whereas the average of observed removal in conventional MBR was 47.84%. The SMEBR system distinguished by a compact hybrid unit, where electrokinetic, biological (biodegradation) and membrane processes (ultrafiltration) take place simultaneously. It was also assumed that due to alternative anoxic/aerobic conditions, PAOs are also participating in phosphorous removal. Such promising technique for wastewater treatment can improve treatment efficiency, reduce phosphorus to an acceptable level and decrease fouling rate while minimizing footprint and overall energy costs. The developed system may have greatest potential application to decentralized wastewater treatment facilities.

## 5 ACKNOWLEDGEMENT

The authors would like to thank for financial support of NSERC Discovery Grant (156161) awarded to Dr. Elektorowicz; they are also thankful to Mrs. Seyedehfatemeh Hosseini for assistance/suggestions with the laboratory experiment.

## 6 REFERENCES

Al-Shannag, M., Bani-Melhem, K., Al-Anber, Z., and Al-Qodah, Z. 2013. Enhancement of COD-Nutrients Removals and Filterability of Secondary Clarifier Municipal Wastewater Influent Using Electrocoagulation Technique. *Separation Science and Technology,* **48** (4):673-680.

Alshawabkeh, A. N., Shen, Y., and Maillacheruvu, K. Y. 2004. Effect of DC Electric Fields on COD in Aerobic Mixed Sludge Processes. *Environmental Engineering Science,* **21** (3):321-329.

Apaydin, Ö., Kurt, U., and Gonullu, M. 2009. An Investigation on The Treatment of Tannery Wastewater by Electrocoagulation. *Global NEST Journal,* **11** (4):546-555.

Bani-Melhem, K., and Elektorowicz, M. 2010. Development of a Novel Submerged Membrane Electro-bioreactor (SMEBR): Performance for Fouling Reduction. *Environmental Science & Technology,* **44** (9):3298-3304.

Basile, A., Cassano, A., and Rastogi, N. K. 2015. *Advances in Membrane Technologies for Water Treatment: Materials, Processes and Applications*: Elsevier Science.

Bayar, S., and Karagunduz, A. 2014. Influence of Electrical Field on COD Removal and Filterability of Activated Sludge. *Desalination and Water Treatment,* **52** (7-9):1316-1323.

Berg, H. 1995. Possibilities and Problems of Low Frequency Weak Electromagnetic Fields in Cell Biology. *Bioelectrochemistry and Bioenergetics,* **38** (1):153-159.

Bodini, S., Sanfilippo, L., Savino, E., and Moscetta, P. 2015. *Automated Micro Loop Flow Reactor Technology to Measure Nutrients in Coastal Water: State of The Art and Field Application.*OCEANS 2015-Genova.

Chen, Y., Yu, B., Yin, C., Zhang, C., Dai, X., Yuan, H., and Zhu, N. 2016. Biostimulation by Direct Voltage to Enhance Anaerobic Digestion of Waste Activated Sludge. *RSC Advances,* **6** (2):1581-1588.

Dohare, E. D., and Sisodia, T. 2014. Applications of Electrocoagulation in Treatment of Industrial Wastewater: A Review. *International Journal of Engineering Sciences & Research Technology,* **3**: 379-386.

Elektorowicz , M., Arian, Z., and Ibeid, S. 2014. Submerged Membrane Electro-bioreactor for Water Recovery. *Monographs of the Environmental Engineering Committee,* **119**:93-98.

Elektorowicz, M., Bani Melhem, K., and Oleszkiewicz, J. 2009. Submerged Membrane Electro-bioreactor-SMEBR. *US patent,* **12** (553-680).

Fiedler, U., Gröbner, U., and Berg, H. 1995. Electrostimulation of Yeast Proliferation. *Bioelectrochemistry and Bioenergetics,* **38** (2):423-425.

Gnirss, R., Lesjean, B., Adam, C., and Buisson, H. 2003. Cost Effective and Advanced Phosphorus Removal in Membrane Bioreactors for a Decentralised Wastewater Technology. *Water Science and Technology,* **47** (12):133-139.

Hasan, S. W., Elektorowicz, M., and Oleszkiewicz, J. A. 2012. Correlations between trans-membrane pressure (TMP) and sludge properties in submerged membrane electro-bioreactor (SMEBR) and conventional membrane bioreactor (MBR). *Bioresource technology* **120**:199-205.

Hasan, S. W., Elektorowicz, M., and Oleszkiewicz, J. A. 2014. Start-up Period Investigation of Pilot-scale Submerged Membrane Electro-bioreactor (SMEBR) Treating Raw Municipal Wastewater. *Chemosphere,* **97**:71-77.

Ibeid, S. 2011. "Enhancement of the submerged membrane electro-bioreactor (SMEBR) for nutrient removal and membrane fouling control." Citeseer.

Ibeid, S., Elektorowicz, M., and Oleszkiewicz, J. 2011. Processes and Apparatuses for Removal of Carbon, Phosphorus and Nitrogen. Google Patents.

Ibeid, S., Elektorowicz, M., and Oleszkiewicz, J. 2013. Modification of activated sludge properties caused by application of continuous and intermittent current. *Water research* **47** (2):903-910.

Ibeid, S., Elektorowicz, M., and Oleszkiewicz, J. A. 2016. Impact of electrocoagulation of soluble microbial products (SMP) on membrane fouling at different volatile suspended solids (VSS) concentrations. *Environmental technology* (just-accepted):1-15.

Marshall, R. T. 1993. *Standard Methods for The examination of Dairy Products.* Washington, DC, USA: American Public Health Association (APHA).

Nakanishi, K., Tokuda, H., Soga, T., Yoshinaga, T., and Takeda, M. 1998. Effect of Electric Current on Growth and Alcohol Production by Yeast Cells. *Journal of Fermentation and Bioengineering,* **85** (2):250-253.

Nassef, E. 2012. Removal of Phosphorous Compounds by Electrochemical Technique. *Engineering Science and Technology: An International Journal (ESTIJ), ISSN: 2250-3498* **2** (3).

Nedjah, N., and Laskri, N. 2015. Phosphorus Removal from Urban Wastewater via Chemical and Combined Treatment Against Eutrophication of Receiving Environments. *International Journal of U-and E-Service, Science and Technology,* **8** (7):303-312.

Posadas, E., García-Encina, P.-A., Soltau, A., Domínguez, A., Díaz, I., and Muñoz, R. 2013. Carbon and Nutrient Removal from Centrates and Domestic Wastewater Using Algal–bacterial Biofilm Bioreactors. *Bioresource technology,* **139**:50-58.

Sathasivan, A. 2008. Biological phosphorus removal processes for wastewater treatment. *Water and Wastewater Treatment Technologies-Encyclopedia of Life Support Systems (EOLSS), Online at: www.desware.net/sample-chapters/d13/e6-144-10.pdf*.

Wang, Y.-K., Pan, X.-R., Geng, Y.-K., and Sheng, G.-P. 2015. Simultaneous Effective Carbon and Nitrogen Removals and Phosphorus Recovery in An intermittently Aerated Membrane Bioreactor Integrated System. *Scientific reports,* **5**.

Wei, V., Elektorowicz, M., and Oleszkiewicz, J. 2011. Influence of Electric Current on Bacterial Viability in Wastewater Treatment. *Water Research,* **45** (16):5058-5062.

Wei, V., Elektorowicz, M., and Oleszkiewicz, J. 2012. Electrically enhanced MBR system for total nutrient removal in remote northern applications. *Water Science & Technology* **65** (4).

Wei, V., Oleszkiewicz, J., and Elektorowicz, M. 2009. Nutrient removal in an electrically enhanced membrane bioreactor. *Water Science & Technology* **60** (12).

Wilbey, A. 2013. Membrane Processing: Dairy and Beverage Applications (2013), edited by AY Tamime, John Wiley & Sons Ltd, The Atrium, Southern Gate, Chichester, West Sussex PO19 8SQ, UK. ISBN: 978-0-4706-5584-9. *International Journal of Dairy Technology,* **66** (3):452-453.

Yin, X., Qiao, S., and Zhou, J. 2015. Using Electric Field to Enhance The activity of Anammox Bacteria. *Applied Microbiology and Biotechnology,* **99** (16):6921-6930.

Zhu, B., Clifford, D. A., and Chellam, S. 2005. Comparison of Electrocoagulation and Chemical Coagulation Pretreatment for Enhanced Virus Removal Using Microfiltration Membranes. *Water Research,* **39** (13):3098-3108.

Zuthi, M., Guo, W., Ngo, H., Nghiem, L., and Hai, F. 2013. Enhanced biological phosphorus removal and its modeling for the activated sludge and membrane bioreactor processes. *Bioresource technology* **139**:363-374.

Leadership in Sustainable Infrastructure
*Leadership en Infrastructures Durables*
**Case Study**
Vancouver, Canada
May 31 – June 3, 2017/ *Mai 31 – Juin 3, 2017*

# ENHANCEMENT OF PRIMARY TREATMENT OF COMBINED SEWER OVERFLOW (CSO) BY USING FERRATE (VI)

Alumairi, Abdulrahim[1], Mohamed, Abdul[2] and Gamal El-Din, Mohamed[1,3]
[1] University of Alberta, Canada
[2] EPCOR Water Services, Canada
[3] mgg1@ualberta.ca

## 1 Project Overview

The main objective of this project is to conduct a comprehensive study in the area of enhancement of primary treatment of combined sewer overflow (CSO) by using different chemicals. During periods of high flow, exceeding the secondary treatment capacity of the plant, some flow receives preliminary and primary treatment, which has a potential impact on the ecological life of the receiving water body. In this paper, we highlight the use of ferrate, a well-known green and effective coagulant and oxidant (Jiang, 2013; Gombos et al., 2013). Numerous studies have shown the high removal capacity of ferrate for different types of contaminants such as diclofenac, carbamazepine, estrone COD, TSS, *E. coli*, bisphenol A (Lee et al.,2009, Yang et al.,2012, Zhou and Jiang, 2015; Han et al., 2015; Sun et al., 2016; Wang et al., Yates et al., 2014, Sharma, 2011; Dobosy et al., 2016).The main goal of the project is to reduce regulated parameter such as total suspended solids (TSS), turbidity and chemical oxygen demand (COD) to the regulated limits as per the Alberta guidelines. Laboratory scale experiments using a jar test apparatus were designed to optimize both the doses of the used chemical for the targeted parameters and the mixing conditions.

## 2 Innovation

Samples were collected from the wastewater treatment plant in four sampling events during November 2016, December 2016, and February 2017. Factorial design was designed to assess the jar test performance, aiming to optimize the ferrate dose levels and mixing conditions. Both high and low ferrate dose levels and the mixing condition for rapid and slow mixing were tested at high and low levels as shown in Table 1. Two different doses of ferrate were used 0.5 and 8 mg/L with and without cationic polymer of 1.25 mg/L. A jar treated with polymer only and blank jar were used as a control.

Table 1. Factorial design for four experiments

| Experiment | Rapid mixing (rpm/min) | Slow mixing (rpm/min) | Symbol |
|---|---|---|---|
| 1 | 300/3 | 30/20 | (1/1) |
| 2 | 300/3 | 15/10 | (1/-1) |
| 3 | 150/1 | 30/20 | (-1/1) |
| 4 | 150/1 | 15/10 | (-1/-1) |

## 3 Results and Discussion

As shown in Table 2 and Figure 1, 0.5 mg/L of ferrate with 1.25 mg/L of polymer showed the best removal efficiencies for the tested parameters in terms of mg of contaminant removed per mg of ferrate added. High dose of ferrate (8 mg/L) didn't show good results and that might be attributed to overdosing and reverse effect for the tested parameters. Moreover, the addition of polymer improved the removal efficiency of ferrate for 0.5 mg/l of Ferrate. The removal mechanism of TSS and COD might attribute to good coagulation and oxidation function of ferrate for the tested parameters. The pH of the samples was around 7 without any adjustment during the course of the entire experiments.

Table 2. The obtained values of mg removed /mg of coaglant.

| Sample ID | mg TSS removed/mg Fe | | | | mg COD removed/mg Fe | | | |
|---|---|---|---|---|---|---|---|---|
| | (1/1) | (1/-1) | (-1/1) | (-1/-1) | (1/1) | (1/-1) | (-1/1) | (-1/-1) |
| Ferrate 0.5 mg/L | 241.0 | 216.0 | 210.0 | 271.0 | 450.6 | 266.4 | 301.9 | 219.0 |
| Ferrate 0.5 mg/L + Polymer | *259.0* | *248.0* | *244.0* | *257.0* | *472.2* | *369.0* | *399.5* | *282.6* |
| Ferrate 8 mg/L | 15.1 | 12.8 | 13.3 | 16.3 | 27.2 | 19.7 | 20.2 | 19.3 |
| 8mg/L Ferrate + Polymer | 15.9 | 14.5 | 15.3 | 15.4 | 25.6 | 21.7 | 22.4 | 17.6 |
| FeCl3 -8 mg/l Fe+ Polymer | 18.3 | 16.6 | 20.3 | 17.9 | 23.1 | 21.0 | 26.8 | 19.1 |

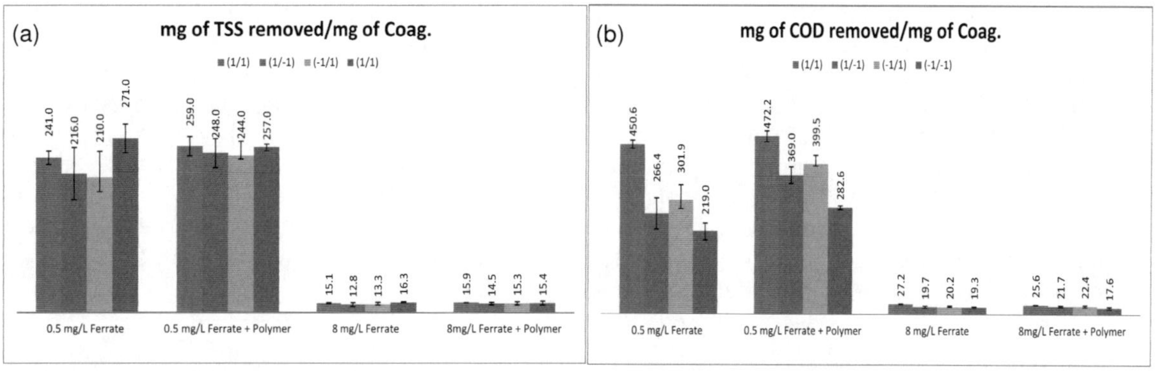


Figure 1. (a) mg of TSS removed per mg of Fe+ added; (b) mg of COD removed per mg of Fe+ added.

## 4 Lessons Learned and Future Work

1) The results showed that mixing condition is not a major factor affecting the ferrate effectiveness on removing target contaminants. . The results also indicated that 0.5 mg/L of Fe+ with polymer (-1/-1) was the optimum condition for the tested parameters.
2) The disinfection capacity of ferrate will require more investigations and might be more effective if ferrate is used for post treatment.
3) Different water matrices will be used to check the removal capacity of ferrate with different model micropollutants (MPs). Moreover, kinetics study for selected MPs will be conducted.

## Acknowledgements

We wish to acknowledge the financial support from the Natural Sciences and Engineering Research Council of Canada (NSERC) Collaborative Research and Development (CRD) grant and EPCOR Water

Services. Thanks also go to Dr. Pamela Chelme-Ayala and Dr. Selamawit Ashagre for their critical and useful comments.

**References**

Dobosy Péter, Éva Cseperke Vizsolyi, Imre Varga, József Varga, Győző G. Láng, Gyula Záray. 2016. Trichloroethylene removal from water by ferrate treatment. *Microchemical Journal* **127**: 74-78.

Gombos Erzsébet, Katalin Barkács, Tamás Felföldi , Csaba Vértes, Magdolna Makó, György Palkó and Gyula Záray. 2013. Removal of organic matters in wastewater treatment by ferrate (VI)-technology. *Microchemical Journal* **107**: 115-120.

Han Qi, Hongjie Wang, Wenyi Dong, Tongzhou Liu, Yulei Yin, Hongkai Fan. 2015. Degradation of bisphenol A by ferrate (VI) oxidation: Kinetics, products and toxicity assessment .*Chemical Engineering Journal* **262**: 34-40.

Yang Bin, Guang-Guo Ying, Jian-Liang Zhao, Shan Liu, Li-Jun Zhou, Feng Chen.2012.Removal of selected endocrine disrupting chemicals (EDCs) and pharmaceuticals and personal care products (PPCPs) during ferrate (VI) treatment of secondary wastewater effluents. *Water research* **4 6**: 2194-2204.

Yates Brian J., Radek Zboril and Virender K. Sharma. 2014. Engineering aspects of ferrate in water and wastewater treatment – a review. *Journal of Environmental Science and Health*, **Part A: 49**: 603-1614.

Jiang Jia-Qian. 2013. The role of ferrate (VI) in the remediation of emerging micro pollutants. *Procedia Environmental Sciences* **18**: 418-426.

Sharma Virender K.. 2011. Oxidation of inorganic contaminants by ferrates (VI, V, and IV) kinetics and mechanisms: A review. *Journal of Environmental Management* **92:**1051-1073.

Sun Xuhui, Qi Zhang, He Liang, Li Ying, Meng Xiangxu, Virender K. Sharma. 2016. Ferrate (VI) as a greener oxidant: Electrochemical generation and treatment of phenol. *Journal of Hazardous Materials* **319**: 130-136.

Lee Yunho,Saskia Gisela Zimmermann, Anh Trung Kieu, and Urs Von Gunten. 2009. Ferrate (Fe(VI)) Application for Municipal Wastewater Treatment: A Novel Process for Simultaneous Micropollutant Oxidation and Phosphate Removal.*Environ. Sci. Technol.* **43**: 3831-3838.

Wang Yingling, Haijin Liu, Guoguang Liu, Youhai Xie and Shuyan Gao. 2015. Oxidation of diclofenac by potassium ferrate (VI): Reaction kinetics and toxicity evaluation. *Science of the Total Environment* **506–507**: 252-258.

Zhou Zhengwei, Jia-Qian Jiang . 2015. Reaction kinetics and oxidation products formation in the degradation of ciprofloxacin and ibuprofen by ferrate (VI).*Chemosphere* **119**: S95-S100.

Leadership in Sustainable Infrastructure

*Leadership en Infrastructures Durables*

Vancouver, Canada

May 31 – June 3, 2017/ *Mai 31 – Juin 3, 2017*

# EVALUATION OF ELECTROCOAGULATION AS A METHOD OF PRE-TREATMENT FOR HIGHLY CONCENTRATED BREWERY WASTEWATER

Sultana, Sharmin[1], Anwar, Nawrin[1], Okoro, Oluchi[1] and Rahaman, M. Saifur[1,2]

[1] Building Civil and Environmental Engineering, Concordia University, Montreal, QC, Canada

[2] saifur.rahaman@encs.concordia.ca

**Abstract:** This study investigated the effectiveness of electrocoagulation as a pre-treatment for brewery wastewater. The performance of a batch electrocoagulation system with four electrode plates was explored with respect to turbidity and chemical oxygen demand (COD) removal, electrical energy and electrode consumption, and sludge production. A parallel connection mode was used to connect the electrodes of the sacrificial materials, Fe and Al. All experiments were conducted at a current density of 150 A/m$^2$ and initial pH of 5 and 7 for the Fe and Al electrodes, respectively. Different combinations of the two electrode materials (Fe/Fe, Al/Al, Fe/Al and Al/Fe (anode/cathode configuration)) were investigated with regards to the electrode materials and operating time. At the current density applied and initial pH used for the respective electrode materials, the optimum operating time for electrocoagulation was found to be 10 minutes and the corresponding turbidity removal was 99%, 98.5%, 97%, and 99% for Fe/Fe, Al/Al, Fe/Al and Al/Fe (anode/cathode configuration) electrodes, respectively. Both electrode materials exhibited somewhat similar characteristics in turbidity removal and, therefore, other parameters were considered to base the selection of an optimum electrode configuration. Aluminum electrodes consumed lower electrode materials yet required higher electrical potential. Experiments with the Fe/Al configuration produced the lowest sludge volume and required the lowest potential as compared to others, and therefore were chosen as the optimum electrode configuration. The COD removal efficiency was examined using this optimum configuration and showed an around 60% removal rate. The effluent from the optimized electrocoagulation reactor would substantially reduce the wastewater load for a post-treatment setup such as anodic oxidation or electro-Fenton.

## 1 Introduction

The rapid growth of industrialization and urbanization have stimulated the need for water resources and reclamation, along with the increment in the levels of environmental pollution, simultaneously from urban and industrial waste while industrial effluent is mainly responsible to increase the environmental disproportion due to its physical and chemical composition (Hilario Garcia, Matzenbacher et al. 2017). Brewing industry has high economic value and beer is considered as one of the most consumed beverage globally (Sultana 2016, Maintinguer, Lazaro et al. 2017). Around the globe, the production of beer has increased about 37.26 million kiloliters from 2004 to 2014 and among all countries China ranked top in beer production volume, followed by Brazil (about 5.42 million kiloliters) and Vietnam (Maintinguer, Lazaro et al. 2017). To produce this alcoholic beverage, the brewery consumes large volume of water and generates highly polluted effluent, which accounts 70% of consumed water. Generally, it produces 3-10 L of waste effluent for per liter of beer production (Subramaniyam, Subashchandrabose et al. 2016, Maintinguer, Lazaro et al. 2017). Primarily, two factors control the amount of wastewater generation: (i) production, (ii) specific water usage. Typically, this type of wastewater has a high organic matter concentration such as sugars, soluble starch, ethanol, volatile fatty acids, and very low concentrations of heavy metals. Therefore, if not treated properly before discharge it may cause the potential threat to the environment, aquatic life and human beings (Choi 2015, Subramaniyam, Subashchandrabose et al. 2016). In addition, regulations regarding the management of environmental issues has become more stringent than before and therefore, it is mandatory for the brewing industry to comply with environmental laws and ensure the treatment and safe disposal of brewery wastewater (Simate, Cluett et al. 2011, Choi 2015).

As a treatment method of treating brewery wastewater, conventionally physical, chemical and biological processes are used (Maintinguer, Lazaro et al. 2017). Although various treatment methods have been reported in the literature for treating brewery wastewater (Choi 2015, Eyvaz 2016), the application of electrocoagulation (EC) has the potential for pre-and post treatment of wastewater due to it's simple, reliable and cost effective process. However, there is limited information on the treatment of brewery wastewater using electrocoagulation. To the best of our knowledge, only one research study reported the use of electrocoagulation for treating brewery wastewater. Therefore, the objective of this study is to investigate the potential of electrocoagulation process as a method of pre-treatment for highly concentrated brewery wastewater.

## 1.1 Overview of Electrocoagulation

Electrocoagulation (EC) is an electrochemical treatment method which is associated with the anodic dissolution of a sacrificial metal electrode (typically aluminum, iron, or stainless steel (SS), because these metals are cheap, readily available, proven effective, and non-toxic). In this electrolytic system, hydroxy ions and hydrogen gas forms simultaneously, the hydroxy ions remove the suspended solids and colloidal particles from the electrochemical system by coagulation and flocculation (Terrazas, Vazquez et al. 2010, Butler, Hung et al. 2011, Kuokkanen, Kuokkanen et al. 2013). A summary of some chemical reactions which occur on electrodes and in the bulk wastewater in EC process, is shown in Table 1 (Inan, Dimoglo et al. 2004).

Table 1: Reactions at the electrodes and in the bulk wastewater

| Anode | Cathode (Simate et al. 2011) |
|---|---|
| $4OH^- = 2H_2O + O_2(g) + 4e^-$<br>$2H_2O = O_2(g) + 4H^+ + 4e^-$<br>$2Cl^- = Cl_2(g) + 2e^-$<br>$Cl_2(g) + H_2O = HOCl + H^+ + Cl^-$ | $2H_3O^+ + 2e^- = H_2(g) + 2H_2O$<br>(in acid solutions)<br>$2H_2O + 2e^- = H_2(g) + 2OH^-$<br>(in alkaline solutions)<br>$O_2 + 2H_2O + 4e^- = 4OH^-$ |
| Al-anode,<br>$Al(s) = Al(aq)^{3+} + 3e^-$<br>$Al(aq)^{3+} + 3H_2O = Al(OH)_3 + 3H^+$ | $Al(s) + 4OH^- = [Al(OH)_4]^- + 3e^-$<br>(at very high pH) |
| Fe-anode,<br>$Fe(s) = Fe(aq)^{2+} + 2e^-$<br>$Fe(aq)^{2+} + 2H_2O = Fe(OH)_2 + 2H_2O$<br>$Fe(OH)_2 + HOCl = Fe(OH)_3 + Cl^-$<br><br>$Fe^{2+} = Fe^{3+} + e^-$<br>$Fe^{3+} + 3H_2O = Fe(OH)_3 + 3H^+$ | $Fe(OH)_3 + OH^- = [Fe(OH)_4]^-$<br>$[Fe(OH)_4]^- + 2OH^- = [Fe(OH)_6]^{3-}$<br>(at very high pH) |

## 2 Materials and Methods

### 2.1 Physical-chemical Characteristics of Brewery Wastewater

Brewery wastewater was collected after fermentation from the Brewhouse, Labatt Brasserie, LaSalle, Quebec every two months and send directly to us. The sample was stored at 2-8°C and before running each experiment the required amount allowed to reach room temperature. Analyses were performed to characterize the Labatt brewery wastewater by measuring some parameters which are listed in the Table 2:

Table 2 Labatt brewery wastewater characteristics

| Parameters | Labatt Brewery Sample |
|---|---|
| **pH** | 6.91±0.38 |
| **Conductivity, μS** | 552 ± 59 |
| **TS, mg/L** | 913 ± 5 |
| **TSS, mg/L** | 148 ± 15 |
| **TDS, mg/L** | 765 ± 20 |
| **VTSS, mg/L** | 135 ± 9 |
| **FTSS, mg/L** | 13 ± 7 |
| **COD, mg/L** | 1860 ± 70 |
| **TOC, mg/L** | 574 ± 32 |
| **TN, mg/L** | 74 ± 10 |
| **TP, mg/L** | 6.71 ± 3.5 |
| **TFe, ppb** | 66 ± 12 |

### 2.2 Experimental Procedure

A glass cylindrical EC cell with 850 ml of homogenously brewery wastewater sample was employed in this study. Iron and Aluminum was used as electrode material with an inter electrode gap of 2 cm and a monopolar arrangement with four electrode plates (two anodes and two cathodes) in parallel connection mode (Figure 1). The EC cell used electrodes with 38.7 $cm^2$ of effective area with a current density of 150 $A/m^2$. A DC power supply (Agilent Technologies) was used to provide the electrical connection. To provide mixing in the system, a magnetic bar with 250 rpm was employed at the bottom of the cell reactor. For all experiments with Fe electrode an initial pH of 7 and for Al an initial pH of 5 was employed. All experiments were carried out at room temperature. The effluent from EC cell was filtered through a filter paper (Whatman 42 ashless-NJ, USA) after each run and then analyzed the filtrate for turbidity. In the study of sludge production, the sludge was filtered through a filter paper (Whatman 110 mm diameter) and then the dry weight of the solid residue on the paper was taken to calculate the sludge generation rate. Before each run, electrolytic cell (including the electrodes) was cleaned with 1M hydrochloric acid (HCl) solution for at least 15 min and then rubbed with a sponge and rinsed with tap water and at the end of the run, the electrodes were washed thoroughly with water to remove any solid residues on the surfaces, dried and re-weighted (Kobya, Can et al. 2003, Kobya, Demirbas et al. 2010).

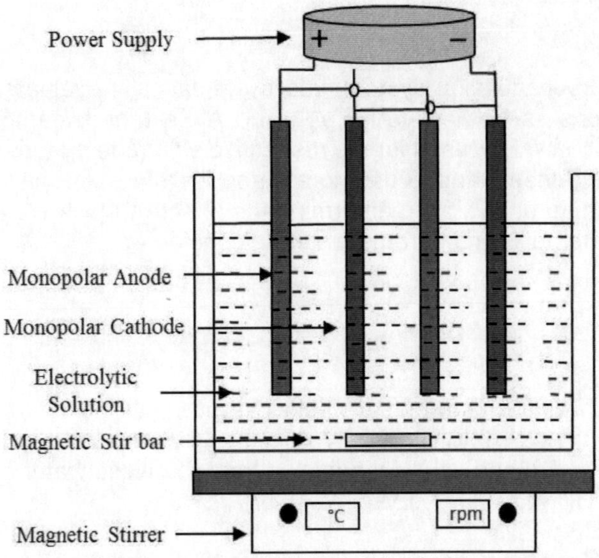


Figure 1: Electrocoagulation Experimental Setup

## 2.3 Evaluation techniques

For pH, turbidity and conductivity measurement, a calibrated pH meter (Oakton Instruments, 310 series, Vernon Hills, IL, USA), turbidity meter (Orion, Thermo scientific) and conductivity meter (Oakton Instruments, Vernon Hills, IL, USA) were respectively used. COD, TN, and TP were measured by reactor digestion method using Hach kits. In brief, the measurement of parameters by Hach kits involves the addition of sample solution into a reagent containing vial, which is heated for a specified period of time, then cooled to room temperature. The concentration of the parameter sought is then measured using a spectrophotometer (DR2800, Hach). The range of the COD vials was 20-1500 mg/L and a proper dilution factor was used to measure concentrations higher than this range.

The removal efficiencies (R%) with respect to turbidity and COD have been calculated using the following equation (Kuokkanen, Kuokkanen et al. 2013):

[1] $$R\ (\%) = \frac{[C_0 - C_t]}{C_0} \times 100$$

Where, $C_0$ and $C_t$ are pollutant concentration before and after EC treatment, respectively.

Theoretically, the dissolution of the anode material can be determined based on Faraday's law as Kg/m$^3$ (Kobya, Ulu et al. 2011, Fajardo, Rodrigues et al. 2015),

[2] $$m_{metal} = \frac{M_w It}{FzV}$$

Where I is the current intensity (A), t corresponds to time of electrocoagulation process, M to the molar mas of the metal (g/mol), z to the valence electrons of the metal (2 for Fe and 3 for Al) and F to the Faraday's constant (96485 C/mol).

## 3 Results and Discussion

The performance of the electrocoagulation system was systematically evaluated in this study with four different electrode combinations as Fe/Fe, Al/Al, Fe/Al and Al/Fe (anode/cathode) configuration. At a current density of 150 A/m$^2$ and initial pH used for the respective electrode materials, the system performed well for the first 10 minutes and then starts to deteriorate and therefore, 10 minutes was selected as the optimum operating time for further analysis. To determine the effective electrode configuration for the pre-treatment of brewery wastewater following parameters were considered,

### 3.1 Turbidity removal

All four combinations used in this study showed satisfactory results in terms of turbidity removal. For Fe/Fe, Al/Al, Fe/Al and Al/Fe removal was 99%, 98.5%, 97%, and 99%, respectively (Figure 2). The possible mechanism of removing suspended and colloidal particles from the wastewater can be stated as (Kobya, Demirbas et al. 2010, Kobya, Ulu et al. 2011, Eyvaz 2016),

(i) Chemical precipitation: the formation of the insoluble compounds, with iron electrodes precipitates such as $Fe(OH)_2$, $Fe(OH)_3$ or $FePO_4$ and with aluminum electrodes insoluble compounds such as $Al(OH)_3$, $AlPO_4$,
(ii) Co-precipitation
(iii) Adsorption – charge neutralization.

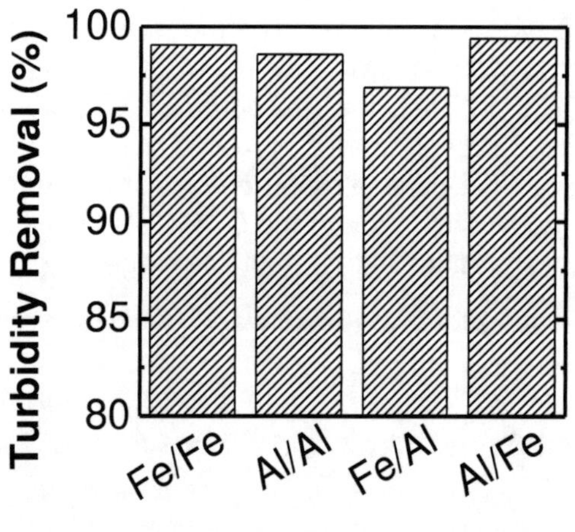


Figure 2: Turbidity removal in different electrode configurated electrocoagulation cell [Experimental condition: reaction time = 10 minutes, initial pH 7 (for Fe) and 5 (for Al)]

### 3.2 Electric potential requirement and electrode consumption

The requirement of electric potential and theoretical electrode consumption were investigated to determine the most effective electrode configuration to treat brewery wastewater. Figure 3 displays the results of this study, Aluminum electrode as anode dissociates lower electrode materials yet requires higher electrical potential as well as higher electrical energy consumption which increases the system cost significantly (Enric Brillas 2009).

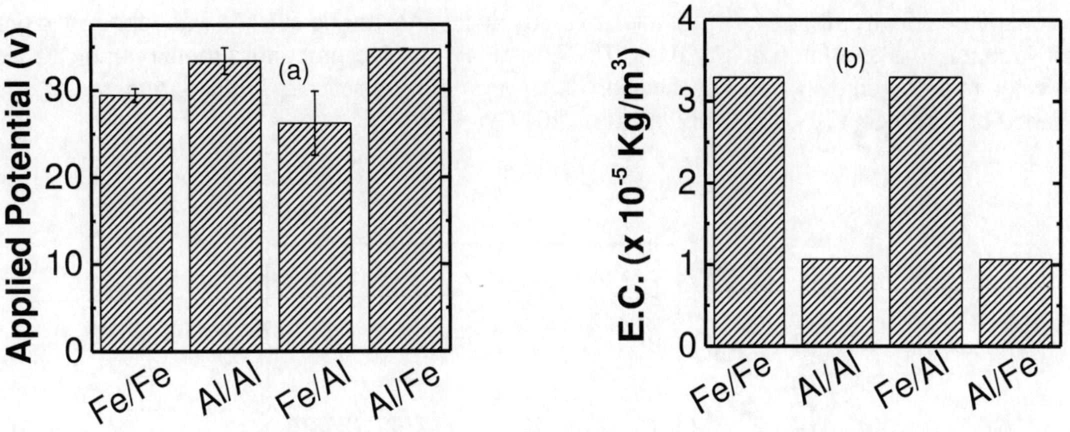


Figure 3: Electrical potential requirement and theoretical electrode consumption for different electrode configurated electrocoagulation cell [Experimental condition: reaction time = 10 minutes, initial pH 7 (for Fe) and 5 (for Al)]

### 3.3 Sludge production

The study of sludge production is an important parameter in electrocoagulation process as it is associated with the problems of solid waste generation and disposal. Most of the contaminants in the electrocoagulation process are present in the form of sludge either by isolation or flotation on the surface of the wastewater (Kobya, Ulu et al. 2011). The study of sludge generation showed lower quantity of sludge production with Fe/Al electrode configuration as seen in Figure 4, therefore, this combination was selected as the optimum electrode configuration.

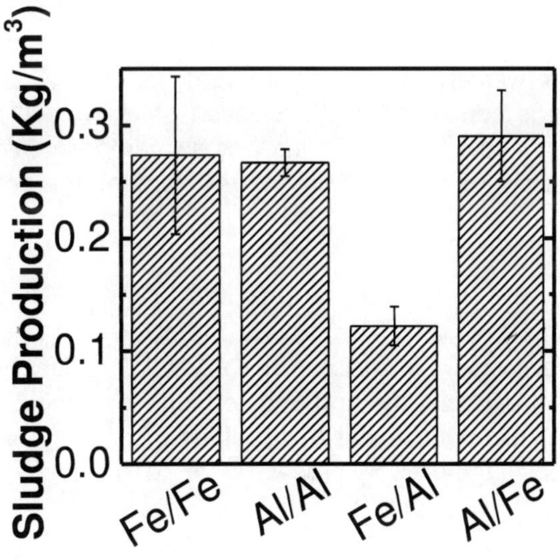


Figure 4: Sludge generation in electrocoagulation process [Experimental condition: reaction time = 10 minutes, initial pH 7 (for Fe) and 5 (for Al)]

### 3.4 COD removal Efficiency

To observe the COD removal efficiency using Fe/Al configuration a set of experiments was employed and it showed 60% of COD removal. The major limitation associated with electrocoagulation is electrode passivation which affects the process performance in long run. Deposition of calcium carbonate and magnesium hydroxide at cathode and formation of an oxide layer at anode surface was reported in previous studies (Kabdaşlı, Arslan-Alaton et al. 2012). The chemical cleaning procedure mentioned in section 2.2 plays a vital role in removing the passivation layer from the electrode surface and regenerate the performance of the system (Eyvaz, Gürbulak et al. 2014).

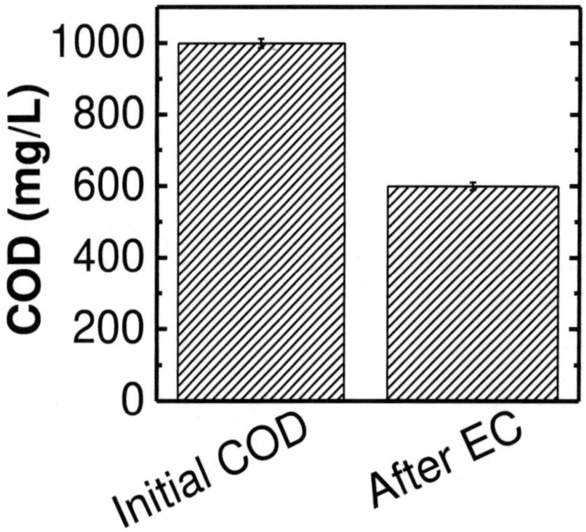


Figure 5: COD removal in electrocoagulation cell using Fe/Al configuration [Experimental condition: reaction time = 10 minutes, initial pH 7]

## 4 Conclusion

Electrocoagulation seemed to be an effective pre-treatment method for removal of turbidity and a significant portion of COD from brewery wastewater. A turbidity removal of 97% and COD removal of 60% was achieved using Fe/Al electrode configuration. Passivation of electrodes can be removed by chemical cleaning. For a post treatment method, the incoming wastewater load could be significantly minimized by using electrocoagulation as a pre-treatment which in turn increase the operating life span of the overall system.


## Acknowledgements

The authors gratefully acknowledge the Ministre de l'Agriculture, des Pêcheries et de l'Alimentation, Québec (MAPAQ) and the Natural Sciences and Engineering Research Council (NSERC) of Canada for providing funding supports for this project. We thank Claude Danis and Julien Beaulieu Sabourin (from Labatt) for collecting and shipping the brewery process water for this project.


## References


Butler, E., Y.-T. Hung, R. Y.-L. Yeh and M. Suleiman Al Ahmad 2011. "Electrocoagulation in Wastewater Treatment." *Water* **3**(4): 495-525.

Choi, H.-J. 2015. "Effect of Mg-sericite flocculant for treatment of brewery wastewater." *Applied Clay Science* **115**: 145-149.

Enric Brillas, I. S. a. M. A. O. 2009. "Electro-Fenton Process and Related Electrochemical Technologies Based on Fenton's Reaction Chemistry." *Chemical Reviews* **109**(12): 6570-6631.

Eyvaz, M. 2016. "Treatment of Brewery Wastewater with Electrocoagulation: Improving the Process Performance by Using Alternating Pulse Current." *International Journal of Electrochemical Science*: 4988-5008.

Eyvaz, M., E. Gürbulak, S. Kara and E. Yüksel (2014). *Preventing of Cathode Passivation/Deposition in Electrochemical Treatment Methods – A Case Study on Winery Wastewater with Electrocoagulation.*

Fajardo, A. S., R. F. Rodrigues, R. C. Martins, L. M. Castro and R. M. Quinta-Ferreira 2015. "Phenolic wastewaters treatment by electrocoagulation process using Zn anode." *Chemical Engineering Journal* **275**: 331-341.

Hilario Garcia, A. L., C. A. Matzenbacher, M. S. Santos, L. Prado, J. N. Picada, S. M. Premoli, D. S. Corrêa, L. Niekraszewicz, J. F. Dias, I. Grivicich and J. da Silva 2017. "Genotoxicity induced by water and sediment samples from a river under the influence of brewery effluent." *Chemosphere* **169**: 239-248.

Inan, H., A. Dimoglo, H. Şimşek and M. Karpuzcu 2004. "Olive oil mill wastewater treatment by means of electro-coagulation." *Separation and Purification Technology* **36**(1): 23-31.

Kabdaşlı, I., I. Arslan-Alaton, T. Ölmez-Hancı and O. Tünay 2012. "Electrocoagulation applications for industrial wastewaters: a critical review." *Environmental Technology Reviews* **1**(1): 2-45.

Kobya, M., O. T. Can and M. Bayramoglu 2003. "Treatment of textile wastewaters by electrocoagulation using iron and aluminum electrodes." *Journal of Hazardous Materials* **100**(1-3): 163-178.

Kobya, M., E. Demirbas, A. Dedeli and M. T. Sensoy 2010. "Treatment of rinse water from zinc phosphate coating by batch and continuous electrocoagulation processes." *Journal of Hazardous Materials* **173**(1-3): 326-334.

Kobya, M., F. Ulu, U. Gebologlu, E. Demirbas and M. S. Oncel 2011. "Treatment of potable water containing low concentration of arsenic with electrocoagulation: Different connection modes and Fe–Al electrodes." *Separation and Purification Technology* **77**(3): 283-293.

Kuokkanen, V., T. Kuokkanen, J. Rämö and U. Lassi 2013. "Recent Applications of Electrocoagulation in Treatment of Water and Wastewater—A Review." *Green and Sustainable Chemistry* **03**(02): 89-121.

Maintinguer, S. I., C. Z. Lazaro, R. Pachiega, M. B. A. Varesche, R. Sequinel and J. E. de Oliveira 2017. "Hydrogen bioproduction with Enterobacter sp. isolated from brewery wastewater." *International Journal of Hydrogen Energy* **42**(1): 152-160.

Simate, G. S., J. Cluett, S. E. Iyuke, E. T. Musapatika, S. Ndlovu, L. F. Walubita and A. E. Alvarez 2011. "The treatment of brewery wastewater for reuse: State of the art." *Desalination* **273**(2–3): 235-247.

Subramaniyam, V., S. R. Subashchandrabose, V. Ganeshkumar, P. Thavamani, Z. Chen, R. Naidu and M. Megharaj 2016. "Cultivation of Chlorella on brewery wastewater and nano-particle biosynthesis by its biomass." *Bioresource Technology* **211**: 698-703.

Sultana, S., Rahaman, M.S. and Baltodano, A. 2016. *Highly effective electro-fenton oxidation treatment for concentrated brewery wastewater* CSCE Annual Conference, London, Ontario.

Terrazas, E., A. Vazquez, R. Briones, I. Lazaro and I. Rodriguez 2010. "EC treatment for reuse of tissue paper wastewater: aspects that affect energy consumption." *Journal of Hazardous Materials* **181**(1-3): 809-816.

Leadership in Sustainable Infrastructure
*Leadership en Infrastructures Durables*

Vancouver, Canada
May 31 – June 3, 2017/ *Mai 31 – Juin 3, 2017*

# EVAPOTRANSPIRATION LANDFILL BIOCOVER TO CONTROL ATMOSPHERIC RELEASE OF LANDFILL METHANE

E.M. Bartholameuz[1, 6], J.P.A. Hettiaratchi[2], M. A. Steele[3], T.T. More[4], S.S. Gunasekera[5]
[1, 2, 3, 4, 5] Centre for Environmental Engineering Research and Education (CEERE). Schulich School of Engineering, University of Calgary, Calgary, Alberta, Canada T2N 1N4
[6] embartho@ucalgary.ca

**Abstract:** There is considerable interest in quantifying and deploying cost-effective methods to control landfill emissions of methane ($CH_4$), which is a key greenhouse gas (GHG). Recent studies by the University of Calgary and others indicate one of the most promising options is biological oxidation of $CH_4$ in a passive system known as a Landfill Biocover (LBC). The LBC utilizes a permeable granular medium placed over a landfill as a final cover, which allows the landfill gas (LFGs) to "breathe out". The LBC also supports the growth of methanotrophs, which can convert the $CH_4$ into carbon dioxide ($CO_2$). Although, successful LBC projects have been developed, there is a risk that the permeable LBC will allow water percolation into the waste matrix, increasing landfill leachate production and potentially contaminating groundwater. A novel approach could be used that integrates components of the Evapotranspiration (ET) cover into the LBC design. The ET-LBC is a novel, hybrid landfill cover design that utilizes a granular medium to promote the growth of methanotrophs and to store water to facilitate evapotranspiration by plants.
A number of issues prevent ET-LBCs adoption in Canada. One major barrier is the inability to accurately assess emissions reductions and carbon credit value associated with ET-LBC projects. Research currently being conducted in Leduc, Alberta is being used as a model to develop protocols to assess these emission reductions. Other ongoing research in this technology includes: determination of the most suitable medium and thickness for maximum $CH_4$ oxidation; identification and adaptation of a compatible ET configuration for LBC and minimization of water percolation; and assessment and identification of robust and cost-effective method(s) to determine emissions from landfills and the $CH_4$ oxidation capacity of the ET-LBC system throughout the year. Accurate assessment of emission credits will be possible with the help of increased understanding of the functioning of ET-LBC system.

Key words: Landfill, Biocover, Evapotranspiration

## 1 Introduction

Landfill covers are constructed to contain waste emissions and prevent infiltration of water to the waste layer. Reducing water infiltration reduces percolation through the waste, decreasing leachate generation from the waste mass and the risk of groundwater contamination. Standard conventional landfill covers have accomplished this goal by using low permeability, resistive materials to prevent water flow into the landfill cover and the waste below. These covers have typically employed some combination of a clay cap layer, geomembrane, gas collection layer, and drainage layer. Conventional covers have been shown to function well, generally limiting percolation to < 1.5 mm/year or 0.2% of precipitation in arid and semi-arid regions. However, due to the high cost and failure rate of conventional cover systems, the use of alternative landfill covers has been increasingly pursued. Most proposed alternative covers are water balance type covers designed as either monolithic, capillary barrier, or anisotropic cover systems, collectively known as Evapotranspirative (ET) covers.
The decomposition of organic solid waste in landfills generates landfill gas (LFG) consisting mainly of methane ($CH_4$) and carbon dioxide ($CO_2$). The emission of $CH_4$ into the atmosphere is a serious concern because its global warming potential (GWP) is approximately 34 times that of $CO_2$ (Myhre et al., 2013). If volumes are high, LFG can be recovered to produce energy (Haubrichs and Widmann, 2006). However, volumes are too small for the clear majority of landfills and LFG is allowed to escape into the atmosphere. Landfills in Canada are estimated to release 27 million tonnes of $CO_2$ equivalent (t $CO_2$eq) of $CH_4$ per year (Climate Change, 2012) and their emissions constitute approximately 25% of Canadian anthropogenic $CH_4$

emissions. The estimated annual $CH_4$ emissions from landfills in Alberta total 4.2 million t $CO_{2eq}$. Considering these facts, it is imperative that alternative methods are found to control fugitive $CH_4$ emissions from landfills.

Recent studies by our team and others indicate that one of the most promising options is the biological oxidation of $CH_4$ in a passive system known as a landfill biocover (LBC) (Huber-Humer et al., 2008). The LBC utilizes a permeable granular medium placed over a landfill as a final cover, which allows the LFG to "breathe out". The LBC also supports the growth of methanotrophs, which are capable of utilizing $CH_4$ as an energy source, thereby converting the $CH_4$ into $CO_2$ without creating toxic byproducts (Bogner et al., 1997). While it is estimated that a traditional cover oxidizes only 10% of the total methane passing through it, LBC systems have been proven to potentially oxidize over 90% of methane emissions from landfills. However, there is a risk that the permeable LBC may encourage water percolation into the waste matrix, resulting in increased landfill leachate production and potentially creating groundwater contamination.

To counter this serious environmental problem, a novel approach may be used that integrates components of the ET cover into the LBC design. The ET cover is compatible with the LBC because it minimizes rainwater percolation by utilizing the water storage capacity of the granular medium and the transpiration capacity of plants, without depending on a low permeable compacted clay or geo-membrane barrier (Lamb et al., 2014). As such, an ET-LBC would allow unhindered LFG migration to promote $CH_4$ oxidation, but the ET function of the hybrid design would minimize percolation of precipitation into waste.

Since Alberta has a semi-arid climate, the ET-LBC should succeed; however, experiences with these systems in cold climates is extremely limited. The following discussion, therefore, focuses on the current knowledge of methanotrophy and ET-LBC systems to identify knowledge gaps that need to be filled in order to implement this technology in the future.

## 2 Evapotranspiration (ET) Technology

Typically, ET covers employ a monolithic granular layer to store water and support vegetation (Hauser VL, 2009); however, ET covers can also be constructed to utilize a capillary effect and increase water retention (Abdolahzadeh, 2011, Lacroix Vachon, 2015 and Parent & Cabral, 2005). The granular medium captures and stores precipitation until it is either transpired through vegetation or evaporated from the cover surface. These systems can be aesthetically pleasing because they employ naturalized vegetation, require less maintenance once the plants are established, and may require fewer repairs than a conventional cover (Rock et al., 2012).

There are several different ET cover designs; however, every ET cover should fulfill at least three main criteria: i) The soil should store water to prevent/reduce the downward water movement below cover; ii) the soil should provide the support for excellent root growth of the plants; iii) The plants/vegetation should be well adapted to the soil and the site (preferably native plant species) and meet the requirements of the site. The most common ET cover types are the Monolithic covers, capillary barriers, and anisotropic barriers. A monolithic cover consists of a single layer of vegetated media designed to store precipitation until it is removed by evapotranspiration. A capillary barrier consists of a layer of vegetated fine grained media set on top of a much coarser drainage layer to create a capillary barrier between these layers and increase moisture retention in the finer, top layer. An anisotropic barrier is a capillary barrier system that contains a vegetated soil layer on top of a sand drainage layer, which is set on top of a coarser drainage layer. This creates two consecutive capillary barriers to promote lateral drainage and prevent downward flow.

Each of these types of alternative covers have been shown to be effective at limiting percolation rates to those of a conventional cover, and they are collectively referred to as ET covers. ET covers (or water balance covers or store-and-release covers) are vegetated landfill covers that control water percolation into the waste zone through water balance mechanisms instead of the resistive mechanism employed by conventional landfill caps (Abdolahzadeh et al., 2011; Parent & Cabral, 2005). ET systems are widely acknowledged as an alternative to conventional landfill covers, at least in arid and semi-arid climates (Albright & Benson, 2005), albeit in jurisdictions other than Canada (Albright et al., 2004). The U.S. Environmental Protection Agency's (USEPA) Alternative Cover Assessment Program (ACAP) tested different cover designs in a variety of arid, semi-arid, and humid regions at 11 locations in the US (Albright, et al., 2004). It was found that monolithic ET covers were unable to control percolation in humid climates with high precipitation rates, although others have found ET covers performed better than conventional covers even under humid conditions (Abichou et al., 2006c); however, in arid and semi-arid locations, they achieved comparable or better performance than conventional covers.

A variety of granular media can be effectively used for construction of the ET cover, which allows for ease of construction and cost savings through the use of local soils (Zornberg et al., 2003), provided the chosen medium encourages robust plant growth (Hauser et al., 2001). ET cover thickness is a function of site-specific precipitation and selected granular material (Jacobson et al., 2005). The selection of appropriate vegetation is a critical part of the design. Native vegetation is generally preferable, as it is adapted to local climate (Rock, 2003) and non-native vegetation tends to naturally be supplanted by better-adapted native vegetation over time.

Most of the field research studies reported in literature are from systems located in the US (Hauser et al., 2009, Abichou et al., 2004 and ITRC, 2003). There is very little information available on ET system design and performance in cold climatic conditions. The amount, distribution, and form of precipitation define the effectiveness of an ET cover at a given site (Hauser et al., 2001). For example, during a sudden spring thaw or chinook in Southern Alberta, a large amount of snow melts, but the vegetation is still dormant. The cover may not have sufficient water storage capacity, and percolation may occur (Hauser, 2001; EPA, 2000). These facts point to the necessity of conducting site-specific experiments to generate design data, such as the most appropriate granular material and thickness, and types of plants.

For design and performance evaluation purposes, it is necessary to use two models, i.e. a water balance model and a numerical model (Khire et al 1997). Although several models are available, the unsaturated soil water and heat flow (UNSAT-H) and the HYDRUS-2D/3D are two numerical models that have been used frequently in the design of ET covers (Hauser, 2009).

According to the economic analysis by Hauser et al. (1999, 2001) of the construction of conventional landfill covers and ET covers for similar landfills, estimated construction costs of ET covers were less than half as much as conventional covers. This was because of the fact that ET covers don't need barrier and drainage layers. Moreover, due to the self-renewing nature of ET covers, the maintenance costs are comparatively small. Repairing conventional covers is more difficult and expensive than ET covers. In the case of ET covers, soil filling and re-seeding vegetation can solve most common issues; however, climate, soil, plant cover, and the site requirements are unique and therefore, demand a site-specific design. Unlike conventional covers, there is a scarcity of the information about design parameters for ET landfill covers. Extensive investigations to increase the methods and innovative strategies to overcome these limitations are critical for the wider acceptance and implementation of this technology.

## 3 Landfill Biocover (LBC)Technology

Although information on field LBCs is limited, especially in Canada, there exists a large body of fundamental and applied information generated by researchers worldwide. LBCs are constructed with the most suitable granular medium to provide an ideal environment for the growth of methanotrophs. Potential media include organic-rich soil, compost, native soil amended with biosolids, and wood chips (Scheutz et al., 2009). Experiments with soil (Stein and Hettiaratchi, 2001) and compost (Wilshusen et al., 2004) determined that the $CH_4$ oxidation rate of compost was about three times that of soil; however, the oxidation rate of compost varied due to differences in their moisture holding capacities and nutrient contents. Other media tested include sawdust, lava rock, soil/compost, and sludge/soil (Goya and Hettiaratchi, 2015), but field data are not available for these media. The other main parameters that impact oxidation efficiency are temperature, moisture content, and nutrient availability. The optimal temperature for methanotrophy is about 30°C (Stein and Hettiaratchi, 2001; Whalen et al., 1990), but research has shown that methanotrophic activity continues even at temperatures as low as 4°C (Kettunen et al., 2006). Yet, the current knowledge on the behaviour of methanotrophs in field LBCs under cold climatic conditions is limited. Since the field moisture content changes over time, the design of an LBC entails using a hydrologic model to select a granular media that would maintain optimal moisture content (McCartney and Zornberg, 2006). Even though compost material is the choice of researchers testing top cover layers, there are a number of issues related to the use of this material that should be highlighted. Occurrence of competitive inhibition of methanotrophic activity due to the presence of additional carbon sources for heterotrophic activity in the organic materials makes them potentially problematic and ineffective for landfill gas mitigation. Maturity of the compost is one of the critical design considerations for an effective bio-based system (Scheutz et al. 2009; 2011), as a more mature compost will have less bioavailable carbon sources for heterotrophic competition.

To determine the LBC performance, mathematical models can be used to predict time-dependent changes to the LBC moisture profile (Bohnhoff et al., 2009), but these models should be field-validated under various operating conditions. Since macro- and micro-nutrients, including nitrogen, are needed for methanotroph growth, some researchers have treated media with digested sludge [de Visscher et al., 1999] and nutrient-rich landfill leachate (Watzinger et al., 2005; Chiemchaisri et al., 2010). However, considering the contradictory results obtained and the variability of the nutrient content of these additives, further site-specific research is needed before nutrient addition is adopted in Canada.

Methanotrophs produce exo-polysaccharides (EPS) as a by-product, and the accumulation of EPS causes $CH_4$ oxidation to decline over time (Hilger et al., 1999; Wilshushen et al., 2004). EPS coats the base biofilm, thereby limiting the diffusion of gases into cells. It also reduces the air-filled porosity of the granular medium, thereby limiting the depth of oxygen ($O_2$) penetration, which is a critical factor in selecting the type and thickness of granular media (Wilshushen et al., 2004; Rachor et al., 2011). In narrow columns, we can note a narrow zone within which most oxidation occurs (Rachor et al., 2011) and EPS forms. However, the time-

dependent changes in the LBC performance due to EPS and the importance of EPS in large surface field LBCs are not well understood.
Researchers have developed reactive-transport models that, when given $CH_4$ source strength and physical and biological parameters as input, can predict $CH_4$ oxidation rates (Stein et al., 2001; Hettiarachchi et al., 2007; Perera et al., 2002). Using these models in conjunction with a hydrologic model, the optimal media type for LBCs can be determined; however, relationships between the required biological kinetic parameters and soil properties have not been developed for use in these models in a prescriptive manner. The key to developing an LBC design procedure, therefore, lies in establishing relationships between kinetic parameters and media type, gas concentrations and nutrient status.

## 4 Combined ET-LBC Technology

The top layers of the ET-LBC, with widely distributed rooted zones and high organic content, can provide ideal conditions for the growth of methanotrophs. Vegetation can indirectly contribute to improved $CH_4$ oxidation in soils by altering important soil properties, including moisture content (Reichenauer et al., 2011; Ndanga et al., 2015), porosity (Bohn et al., 2011), and inhibition of ammonia toxicity (Hilger et al., 1999; de Visscher and Van Cleemput, 2003; Reichenauer et al., 2011) Additionally, up to 45% of a plant's annual net fixed carbon from the atmosphere can be exuded from the roots as a variety of organic acids, sugars and amino acids (Reichenauer and Germida, 2008). Wang et al. (2008) compared soil samples subjected to $CH_4$ exposure, with and without vegetation, and determined that the presence of vegetation supports methanotrophy. Similarly, the composition of bacterial species within the rhizosphere can be influenced by the plant exudate, with Type 1 methanotrophs becoming more abundant (Stralis-Pavese et al., 2006). In addition, the aerenchyma of vascular plants are able to transport oxygen from the atmosphere to the rhizosphere surrounding their roots, which can stimulate methanotrophy by creating aerobic conditions within an otherwise anoxic environment (Wu et al., 2009). Conversely, root systems may also compete with methanotrophs for oxygen and nutrients (Wang et al., 2008). Considering the competing processes observed in the presence of plants, further work is required to understand the influence of native Alberta plants on soil methanotrophy and ET potential. Leachate production from landfills presents operational and environmental challenges, since the leachate must be collected and properly treated. Leachate irrigation is one simple and inexpensive treatment option to reduce the volume and improve the quality of landfill leachate on site. Some work has been performed showing that leachate may be used to irrigate an ET-LBC cover without negatively impacting $CH_4$ oxidation (Chiemchaisri et al., 2010). The maximum rate of leachate irrigation depends on the amount of natural precipitation and type of vegetation on the landfill cover; excess soil moisture may reduce methane oxidation in the landfill cover and high nitrogen levels in the leachate may increase nitrous oxide emissions from the landfill surface, while vegetation may be detrimentally affected by high levels of leachate (Watzinger et al., 2005).

## 5 Ongoing research/Future work

Research has been conducted in a number of areas related to LBCs, ET covers, and ET-LBCs. There are several barriers to fully implement ET-LBCs as a viable alternative landfill cover. The main barrier is the inability to accurately assess emission reductions and carbon credit value associated with ET-LBC projects. The ability to claim emission credits from LBC projects is a major incentive for industry to implement these projects.
The emission credits largely depend on the ability to accurately measure emissions from a landfill. Both initial (baseline) and project completion emissions have to be measured to accurately measure the emissions reductions achieved by the project. In Alberta, the Specified Gas Emitters Regulation under the Climate Change and Emission Management Act defines a series of protocols for clear implementation of technology and measurement of gas emissions. Currently, LBC technology does not have a developed protocol, hence there is particular interest to develop an acceptable method of quantifying landfill gas emissions to develop an appropriate protocol.
Several technologies are available to estimate the landfill surface emissions. These include theoretical calculations using mathematical models, in-situ techniques, and ex-situ techniques.
Our current research is being conducted in collaboration with industry and the Leduc and District Regional Waste Management Authority in Alberta to demonstrate the different measurement techniques in ET-LBCs and determine the most feasible technique for accurate emissions measurements.
When analyzing a LBC, one of the main problems faced is the large surface area. Many methods have been developed to obtain instantaneous field measurements of landfill methane surface emissions. These include dynamic or static flux chambers, micrometeorological methods, or path integrated optical remote sensing (PI-ORS) techniques. These methods all provide estimates of $CH_4$ emission levels under the

prevailing conditions at the time of measurement. The main method of determining the surface emission rates is using a closed flux chamber method (Perera, 2002; Abichou et al., 2006b; Héroux et al., 2010). The closed flux chamber method is a slow process that can take up to 1 hour to record a single reading, depending on the flux rates; hence, the amount of information that can be obtained under constant environmental conditions is minimal. Several attempts have been made by researchers to address this issue (Héroux et al., 2010; Abichou et al., 2006b; US-EPA, 2005, Huber-Humer et al., 2009).
Some of these available techniques are being tested in Leduc, Alberta, as detailed below. For some of these techniques a limited number of initial results are available.

### Optical remote sensing-radial plume mapping (ORS-RPM)

USEPA (2005) proposed the use of equipment called "open-path Fourier transform infrared (OP-FTIR) spectrometer". The study involved a technique developed through research funded by the USEPA's National Risk Management Research Laboratory (NRMRL) that uses optical remote sensing-radial plume mapping (ORS-RPM). ORS-RPM uses a horizontal radial plume mapping (HRPM) to map surface concentrations, and a vertical radial plume mapping (VRPM) method to measure total emissions fluxes downwind of the site. Once either the concentration values and flux values are determined the total emission from the landfill could be estimated.

### ISM and flux chamber method

Most methods to estimate landfill surface emission follow the approach where the concentration is mapped and corresponding flux values are measured to find a co-relationship between flux and concentration (Héroux et al., 2010). Field measurements of landfill methane emissions show natural variability. The determination of an average emission rate for a given field site requires sampling designs and statistical techniques, which consider spatial and temporal variability. But such an approach necessitates many sampling points using an enclosure method (flux chamber). To mitigate this problem, correlations between methane concentrations on the ground, in general by the instantaneous surface monitoring (ISM) method, and the surface methane emission measured with a flux chamber have to be developed in order to minimize the number of samplings with flux chamber (Fécil et al., 2003). In the ISM method, a portable FID is used to instantaneously measure the concentration of total organic compounds (TOCs measured as methane) at the landfill surface divided into grids.
An initial investigation was conducted at the Leduc and District Regional Waste Management Facility (The Landfill) to compare methane emissions measurements using both a portable FID device (Thermo TVA 100B FID) and closed, static flux readings using a University of Calgary built flux chamber (Plexiglass; cylinder shape; Dimensions: H=0.19 m, D=0.19 m). The concentrations of methane and carbon dioxide in the chamber were measured with either a portable Rki Eagle 2 or a GEM5000 gas analyzer. Comparison data was recorded for three discrete existing "biowindow" features in a closed section of The Landfill: North Biowindow, Central Biowindow, and South Biowindow. Locations identified as emissions "hotspots" generally showed moderate to good spatial agreement between the FID and flux chamber methods, although no direct and consistent relationship between the surface concentration and measured flux rates could be discerned at this stage (representative results shown in Figures 1 and 2).

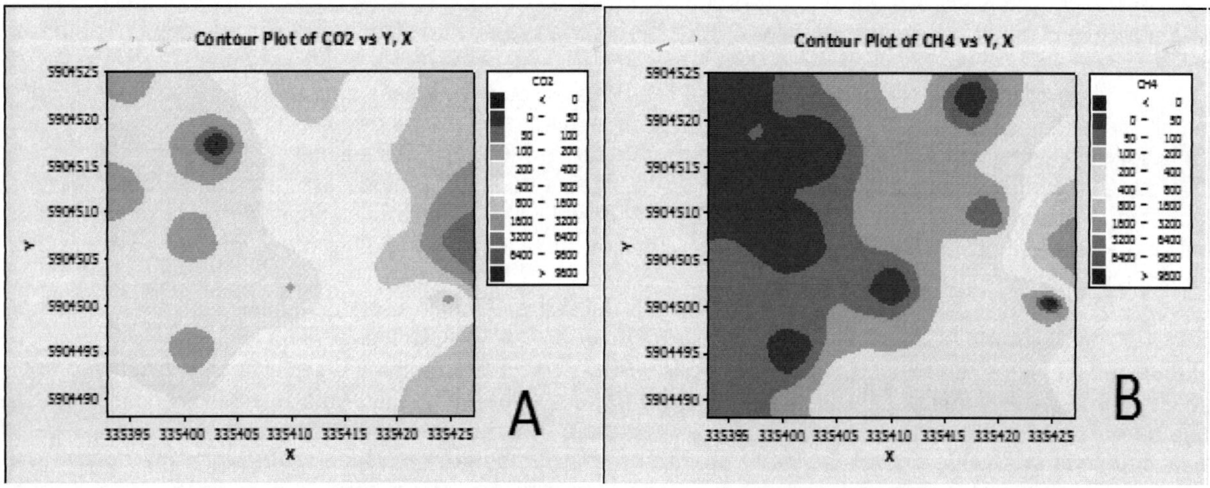

Figure 1: Flux emissions heat maps of the North biowindow (estimated dimensions 35 m by 35 m). Emissions shown as g/m²/day. Distance shown in metres and represents Northing and Easting of the measured emissions. A) $CO_2$ emissions; B) $CH_4$ emissions.

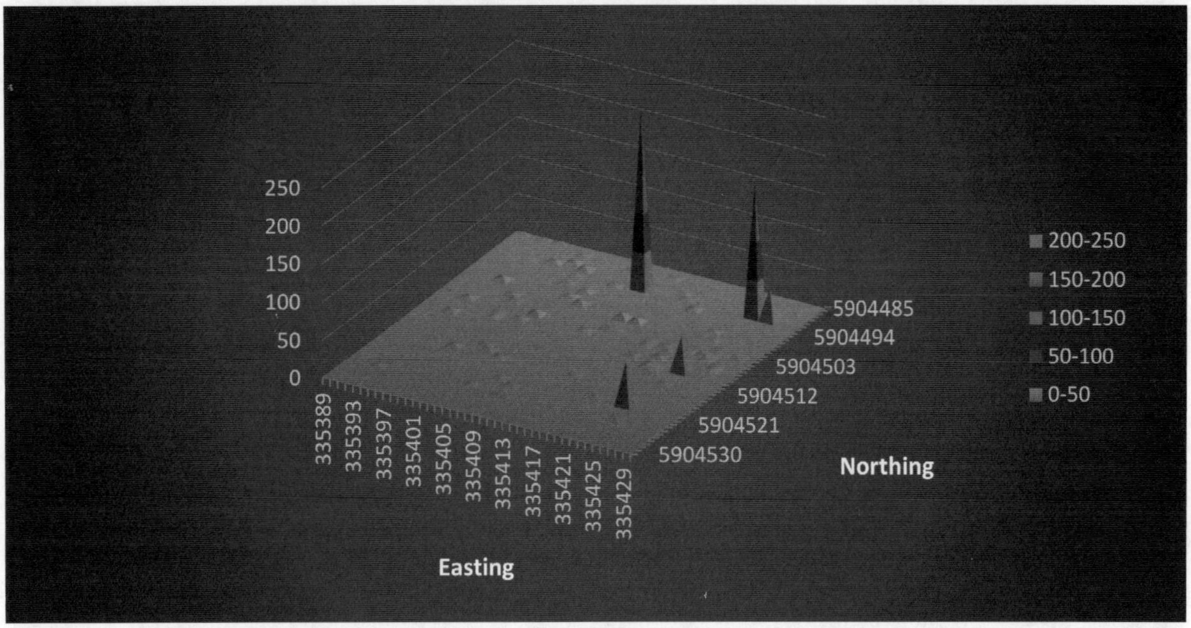


**Figure 2: Flux emissions heat map of the North biowindow (estimated dimensions 35 m by 35 m). FID measurements denote total VOCs in ppm. Distance shown in metres and represents Northing and Easting of the measured emissions.**

## Flux chambers with interpolation

In another similar approach, Abichou et al., (2006a) has attempted to develop surface emission maps without using the ISM method. Here, interpolation methods between different flux values were used. Two commonly used interpolation methods are kriging and inverse distance weighing (IDW). The method used by Abichou et al., (2006a) was the IDW model.

In IDW, the interpolation contours are calculated by weighing neighboring data using the inverse of the separation distance to a power. IDW uses weighted averaging techniques to fill the elevation matrix. The interpolated value of a cell is determined from values of nearby data points considering the distance of the cell from those input points. In kriging, a model of the overall spatial measured variance structure is used to generate the interpolated contours. The measured variance structure is shown as a variogram with half the variance on the y-axis and sample separation distance on the x-axis. Key variables for a variogram are the nugget (unexplained or error variance), sill (total model variance, equal to nugget plus "scale"), and range (distance where the variance reaches the sill) (Abichou et al., 2006a; Abichou et al., 2006b). In addition, Abichou et al., (2006b) has found that IDW method and kriging method do not produce largely different results.

Part of our ongoing research includes analyzing the flux data from the ISM-Flux chamber method using the proposed interpolation in this method. However, the flux data to date has not produced the required resolution to be analyzed. Further data collection and analysis is planned to evaluate this method.

## Other continuing research

Compost is identified as an ideal media for LBCs, yet, it is not well known how compost would behave in an ET-LBC. Laboratory research is currently being conducted to identify combined properties needed for several different types of material.

Potential cost savings are a key driver for the shift from conventional to ET cover systems. Although ET covers typically have higher design and regulatory costs, they have typically had much lower construction and operations costs. Increased design and regulatory costs stem from the site specific water balance requirements of each ET cover, as opposed to a design hydraulic conductivity for a conventional clay cap. The savings on construction costs are possible both due to the potential to use locally available soil materials for construction, rather than compacted clay, and through the use of simple construction methods, including thicker lifts and decreased requirements for compacting to optimal moisture. ET covers also reduce maintenance costs associated with typical causes of failure to maintain the designed saturated hydraulic conductivity in a conventional clay cap, including desiccation, free-thaw cycles, settlement, or improper compaction during construction (Salt et al., 2011). These cost savings averaged roughly $23,500 per acre for alternative covers built as part of the Alternative Cover Assessment Program run by the USEPA (Albright and Benson, 2005)

Since the main cost savings of an ET cover arise from using locally available soils, the availability and cost of materials was the main consideration in selecting potential materials for this ongoing project. Soils available at The Landfill and possible compost amendments available near the site were tested for their ability to sustain methanotrophic populations, with the goal of identifying the most inexpensive materials that could fulfill the minimum functional requirements of this ET cover. As initial laboratory screening must allow for simple and low cost screening of materials, the primary tests conducted for screening were total organic content (measured as LOI) and pH (Figures 3 and 4). Materials with acceptable measures on these initial tests were screened for C-H-N ratio (Figure 5) and directly tested for their maximum methane oxidation potentials. The topsoil available at The Landfill and a residual compost waste product from a nearby waste management facility were identified as materials with no direct costs, and both of these materials were capable of supporting robust methanotrophic populations at a level capable of oxidizing the expected emissions of 100 – 150 g/m$^2$day$^{-1}$. Determination of the hydraulic characteristics of these selected materials and a mix of the two components is ongoing.

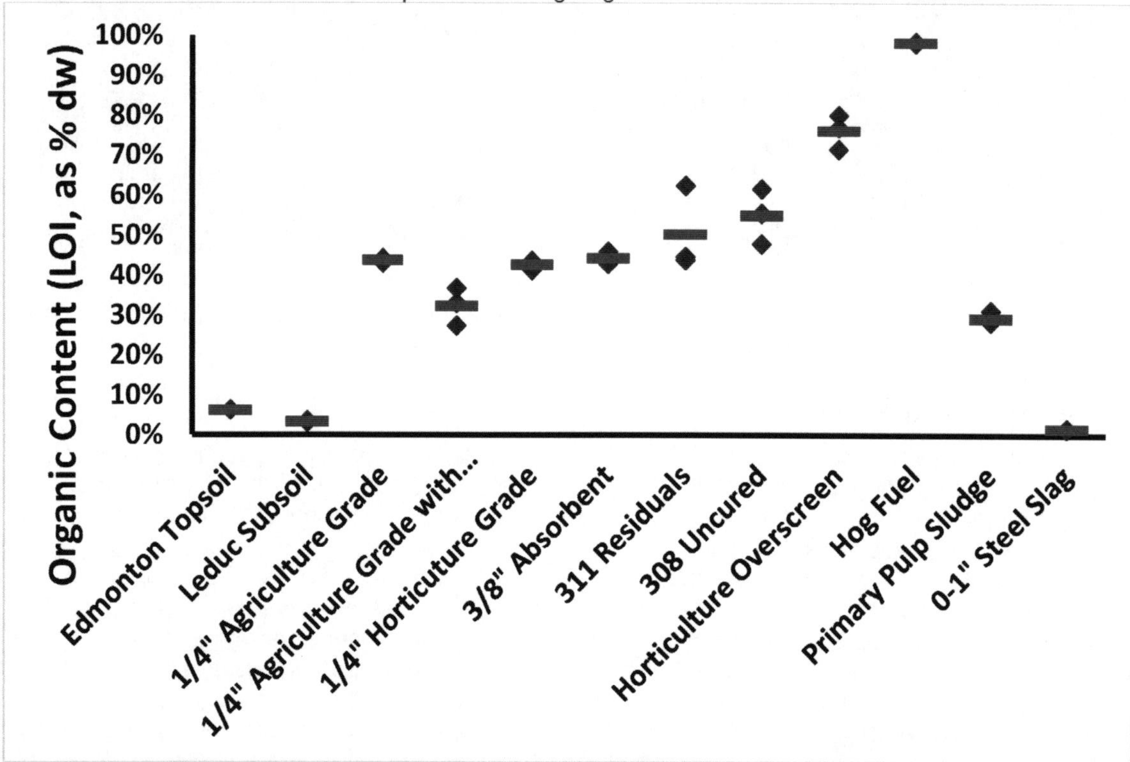


**Figure 3: Organic content of materials considered for construction of Leduc ET-LBC. Organic content shown as loss on ignition at 440°C in muffle furnace for 8 hours. Individual replicates shown as point, with means as short line. N=3.**

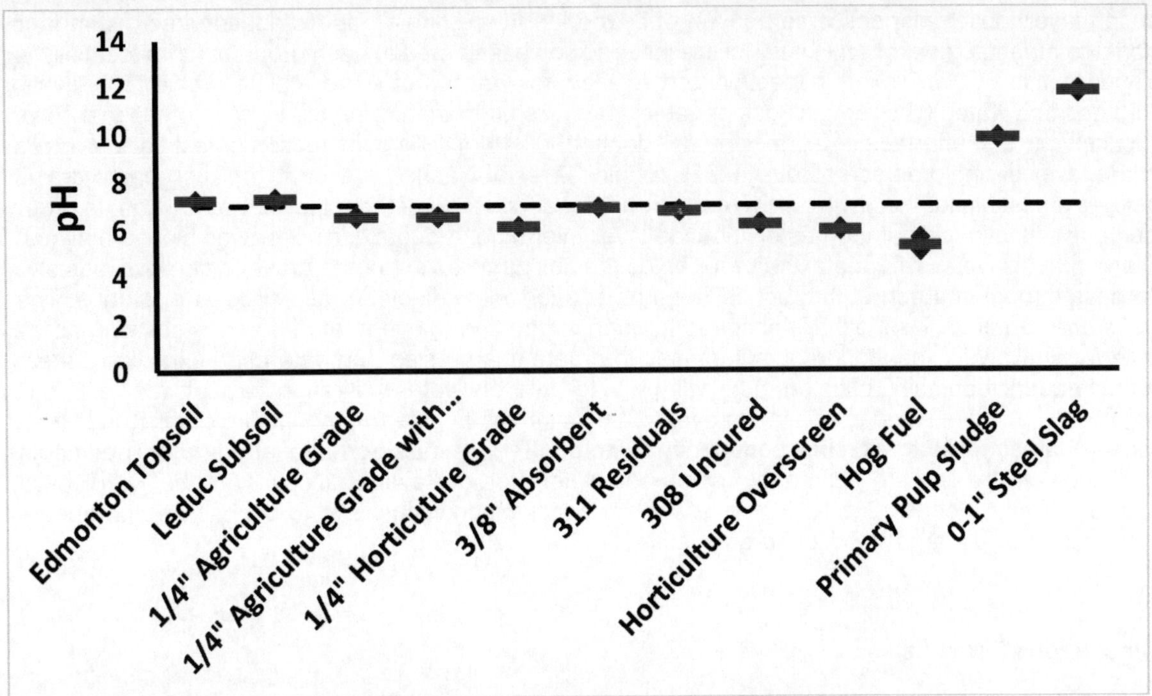


**Figure 4: pH of materials considered for construction of Leduc ET-LBC. Individual replicates shown as point, means as short line, and neutral pH (ideal) shown as dotted line. N=3.**

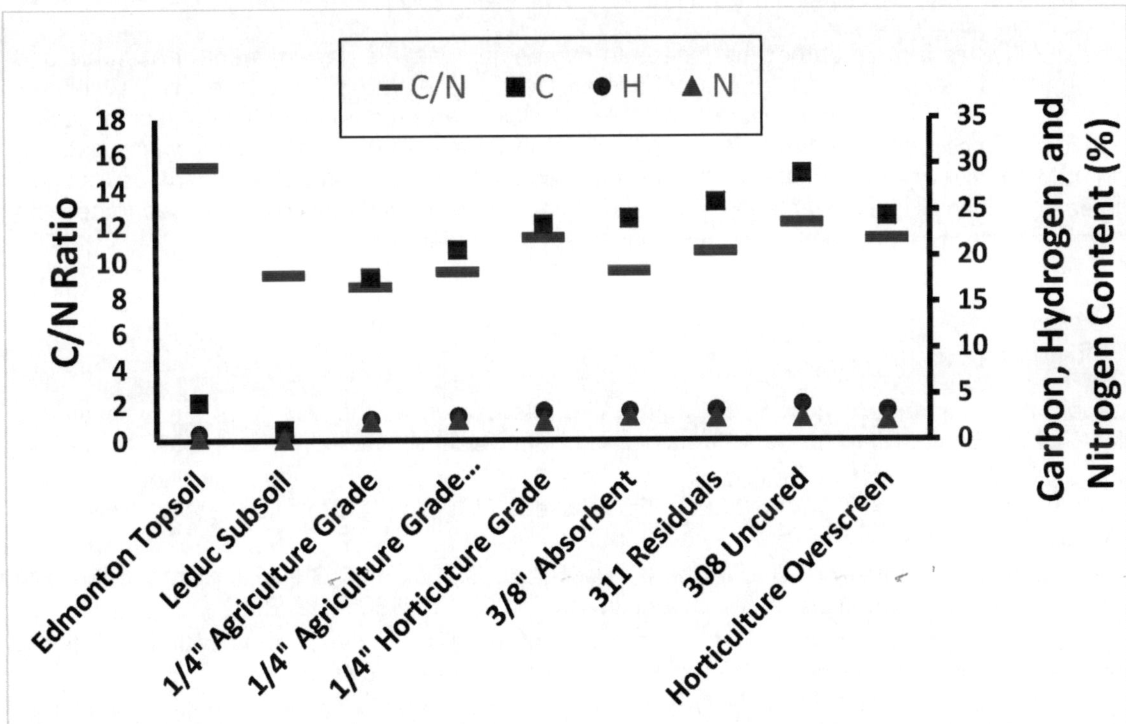


**Figure 5: Elemental analysis of materials considered for construction of Leduc ET-LBC. N=1.**

ET-LBC vegetation is another interesting area of research that is currently conducted. Considering the competing processes observed in the presence of plants, further work is required to understand the influence of native Alberta plants on soil methanotrophy and ET potential.

There is an ongoing scale-up study in Okotoks, Alberta to demonstrate the effects of vegetation on the water storage and ET capabilities, as well as $CH_4$ oxidation, of cover media. This project involves 8 test cells specifically designed to demonstrate ET-LBC systems at field scale. The intent of the test cells construction is to simulate the landfill biocover at a controllable scale by growing vegetation over the soil layer and supplying controlled $CH_4$ flow through distribution piping within an underlying coarse gravel layer (distribution layer). Each test cell is comprised of a wooden frame, liner, gas distribution layer, sump for collection and measurement of any water infiltration, and soil based media; each group of fours test cells is built around a 1 m by 1 m Lexan inspection port to allow for viewing of the complete profile. The liners, which are gas and water tight, prevent loss of gas or water, other than through the surface of the test cells. Each test cell has a soil surface of 3.3 $m^2$ and soil depth of 0.915 m. The construction of test cells involves placing the layer of fine-grained soil over 0.279 m thick layer of washed gravel layer. The coarse gravel layer acts as a gas distribution layer, which creates homogenous load of the supplied natural gas. In order to preserve the functionality of the gas distribution layer over time, a geotextile cover was placed between the soil and gravel layers. The geotextile cover holds the soil particles out of the gravel, yet allows gas and liquid to pass through it. Each of the test cells will be seeded with selected grass species or shrubs. The soil layer stores water until it is either transpired through vegetation or evaporated from the soil surface. The system will allow for monitoring of soil gasses to create gas profiles and determine oxidation zones. Moreover, frequency-domain reflectometers will provide data on water infiltration through the test cells following natural and simulated precipitation events. Soil samples at different monitoring depths and times will used to characterize the microbial community using qPCR. The data from the laboratory experiments and field study will be used to accurately model soil water storage, evapotranspiration, for hydrological modeling to design an effective ET-LBC cover for site specific conditions, and to choose an appropriate mixed media to provide target media characteristics.

## 6 Conclusions/Remarks

Interest has grown recently in the area of methane reduction and elimination. Conversion of $CH_4$ to $CO_2$ emitted from landfilled surface is an ideal solution to reduce the GHG effect. ET-LBC technology is an emerging technology that can be used for GHG emission reduction while maintaining the conventional requirements of landfill covers.

ET-LBC technology has a main barrier that needs to overcome, which is development of accurate and acceptable emission measurement methods to incorporate these methods into developing regulatory protocols for industry to be able to measure carbon offsets. Our current demonstration project in Alberta attempts to identify acceptable methods for landfill emission measurement. The technologies used are ORS-RPM, ISM and flux chamber method, and Flux chambers with interpolation. Preliminary results show that, while each method has its advantages and disadvantages, these methods could be used to develop standard emission measurement protocols for carbon offset monitoring.

## References


Abdolahzadeh, A.M., Lacroix Vachon, B. & Cabral, A.R. Assessment of the Design of an Experimental Cover with Capillary Barrier Effect Using 4 Years of Field Data. Geotech. Geol. Eng. 29, 783–802 (2011).

Abichou, T., Liu X.& Tawfiq K. Design of cost effective lysimeters for alternative landfill cover demonstrations projects, Report #04-0232007. Tallahassee (FL): Florida State University, College of Engineering (2004).

Abichou, T., Chanton, J., Powelson, D., Fleiger, J., Escoriaza, S., Lei, Y. & Stern, J. Methane flux and oxidation at two types of intermediate landfill covers. Waste Manag. 26, 1305– 12 (2006).

Abichou, T., Chanton, J. & Powelson, D. Field performance of biocells, biocovers, and biofilters to mitigate greenhouse gas emissions from landfills. Waste Management, 14: 47–60. (2006)

Abushammala, M.F.M., Basri, N.E.A., Irwan, D. & Younes, M. K. Methane Oxidation in Landfill Cover Soils: A Review. Asian Journal of Atmospheric Environment 8, 1–14 (2014).

Albright, W.H., Benson, C.H., Gee, G.W., Roesler, A.C., Abichou, T., Apiwantragoon, P., Lyles, B. F. & Rock, S.A. Field Water Balance of Landfill Final Covers. J. Environ. Qual. 33, 2317 (2004).

Albright, W. & Benson, C. Alternative cover assessment program: report to Office of Research and Development National Risk Management Research Lab Land Remediation and Pollution Control Division [Phase II Report—Monitoring]. Reno (NV): Desert Research Institute and University of Wisconsin (2005).

Bohn, S., Brunke, P., Gebert, J. & Jager, J. Improving the aeration of critical fine-grained landfill top cover material by vegetation to increase the microbial methane oxidation efficiency. Waste Management 31, 854-863 (2011).

Bogner, J., Meadows, M. & Czepiel, P. Fluxes of methane between landfills and the atmosphere: natural and engineered controls. Soil Use and Management 13,268-277 (1997).

Bohnhoff, G.L., Ogorzalek, A.S., Benson, C.H., Shackelford, C.D. & Apiwantragoon, P. Field Data and Water-Balance Predictions for a Monolithic Cover in a Semiarid Climate. J. Geotech. Geoenvironmental Eng. 135, 333–348 (2009).

Chiemchaisri C., Chiemchaisri, W., Chittanukul, K., Wiwattana, S. & Nathiya, T. Effect of leachate irrigation on methane oxidation in tropical landfill cover soil. J Mater Cycles Waste Manag. 12, 161–168 (2010).

Czepiel, P.M., Mosher, B., Crill, P.M. & Hariss, R.C. Quantifying the effect of oxidation on landfill methane emissions. Journal of Geophysical Research 101, 16721-16729 (1996).

de Visscher, A.,Thomas, D., Boeckx, P. & van Cleemput, O. Methane Oxidation in Simulated Landfill Cover Soil Environments. Environmental Science and Technology 33, 1854- 1859 (1999).

de Visscher, A. & Van Cleemput, O. Induction of enhanced CH4 oxidation in soils: NH4+ inhibition patterns. Soil Biology and Biochemistry 35, 907-913 (2003).

EPA. Introduction to phytoremediation. Office of Research and Development. EPA/600/R-99/107. Washington (DC): US EPA (2000).

Fécil, B., Héroux, M., Guy, C. Development of a Method for the Measurement of Net Methane Emissions From MSW Landfills. *9th Int. Waste Manag. Landfill Symp.* (2003).

Goya, J., Hettiaratchi, J.P.A. Development of an alternative media to sustain methanotrophs in landfill biofilters, 65th Canadian Chemical Engineering Conference, Calgary, Canada (2015).

Haubrichs, R. & Widmann, R. Evaluation of aerated biofilter systems for microbial methane oxidation of poor landfill gas. Waste Manag. 26, 408–16 (2006).

Hauser, V.L., Weand B.L. & Gill, M.D. Natural covers for landfills and buried waste. J Environ Eng. 127: 768–775 (2001).

Hauser, V.L. Evapotranspiration covers for landfills and waste sites. ISBN: 13: 978-1-4200-8651-5. Boca Raton (FL): CRC Press (2009).

Héroux, M., Guy, C. & Millette, D., 2010. A statistical model for landfill surface emissions. Journal of the Air & Waste Management Association (1995), 60(2), pp.219–228.

Hettiaratchi, J.P.A. & Pokhrel, D. A new approach to quantify methane oxidation in a landfill bio-cover: Experience with a pilot scale landfill test cell. SARDINIA 2003, 9th International Waste Management and Landfill Symposium, Cagliari, Italy, 6-10 October (2003).

Hettiarachchi, V.C., Hettiaratchi, J.P.A. & Mehrotra, A. Comprehensive One-Dimensional Mathematical Model for Heat, Gas and Moisture Transport in Methane Biofilters, ASCE Practice Periodical of Hazardous, Toxic, and Radioactive Waste Management Vol. 11(4), 225-234 (2007).

Hilger, H.A., Liehr, S.K. & Barlaz, M.A. Exopolysaccharide Control of Methane Oxidation in Landfill Cover Soil. J. Environ. Eng. 125, 1113–1123 (1999).

Huber-Humer M., Gebert, J. & Hilger, H. Biotic systems to mitigate landfill methane emissions. Waste Management and Research 26(1), 33-46 (2008).

Huber-Humer, M., Röder, S. & Lechner, P., Approaches to assess biocover performance on andfills. Waste Management, 29(7), pp.2092–2104 (2009)

Interstate Technology & Regulatory Council (ITRC). Technical and regulatory guidance for design, installation, and monitoring of alternative final landfill covers. Washington (DC): ITRC (2003).

Jacobson, J.J., Heydt, H., Piet, S., Sehlke, G., Soto, R. & Visser, J. Dynamic Modeling of an Evapotranspiration Cover. Pract. Period. Hazard. Toxic Radioact. Waste Manag. 9, 223– 236 (2005).

Kettunen, R.H., Einola, J.K.M. & Rintala, J.A. Landfill Methane Oxidation in Engineered Soil Columns at Low Temperature. Water. Air. Soil Pollut. 177, 313–334 (2006).

Khire, M.V., Benson, C.H. & Bosscher, P.J. Water balance modeling of earthen final covers. J Geotech Geoenviron Eng. 123(8), 744–754 (1997).

Kutzbach, L., Wagner, D. & Pfeiffer, E. Effect of microrelief and vegetation on methane emission from wet polygonal tundra, Lena Delta, Northern Siberia. Biogeochemistry 69, 341–362 (2004).

Lacroix Vachon, B., Abdolahzadeh, A.M. &Cabral, A.R. Predicting the diversion length of capillary barriers using steady state and transient state numerical modeling: case study of the Saint-Tite-des-Caps landfill final cover. Can. Geotech. J. 52, 2141–2148 (2015).

McCartney, J.S., Zornberg, J.G., Decision Analysis for Design of Evapotranspirative Landfill Covers. In: Proceedings UNSAT '06. April 2-6, 2006; Carefree, AZ. ASCE (2006).

Myhre, G., Shindell, D., Bréon, F.M., Collins, W., Fuglestvedt, J., Huang, J., Koch, D., Lamarque, J.F., Lee, D., Mendoza, B., Nakajima, T., Robock, A., Stephens, G., Takemura T. & Zhang, H. Anthropogenic and Natural Radiative Forcing. In: Climate Change 2013: The Physical Science Basis. Contribution of Working Group I to the 5th Assessment Report of the intergovernmental Panel on Climate Change [Stocker, T.F., et al., (eds.)]. Cambridge University Press, Cambridge, United Kingdom and New York, NY, USA (2013).

Ndanga, É.M., Bradley, R.L. & Cabral, A.R. Does vegetation affect the methane oxidation efficiency of passive biosystems? Waste Manag. 38, 240–249 (2015).

Perera, L.A.K., Achari, G. & Hettiaratchi, J.P.A., Determination of Source Strength of Landfill Gas: A Numerical Modeling Approach. J. Environ. Eng. 128, 461–471 (2002).

Parent, S.É. & Cabral, A.R. Material selection for the design of inclined covers with capillary barrier effect. In Geo-Frontiers 2005. Austin, TX. 24-26 January, CD-Rom (2005)

Rachor, I., Gebert, J., Gröngröft, A. & Pfeiffer, E. M. Assessment of the methane oxidation capacity of compacted soils intended for use as landfill cover materials. Waste Manag. 31, 833–842 (2011).

Reichenauer, T.G. & Germida, J.J. Phytoremediation of organic contaminants in soil and groundwater. ChemSusChem 1, 708–717 (2008).

Reichenauer, T.G., Watzinger, A., Riesing, J. & Gerzabek, M.H. Impact of different plants on the gas profile of a landfill cover. Waste Manag. 31, 843–853 (2011).

Rinnan, R., Impiö, M., Silvola, J., Holopainen, T. & Martikainen, P.J. Carbon dioxide and methane fluxes in boreal peatland microcosms with different vegetation cover--effects of ozone or ultraviolet-B exposure. Oecologia 137, 475–483 (2003).

Rock, S. A. Vegetative covers for waste containment. Adv. Biochem. Eng. And Biotechnol. 78, 157–170. (2003).

Rock, S., Myers, B. & Fiedler, L. Evapotranspiration (ET) Covers. International Journal of Phytoremediation. 14, 1-25 (2012).

Salt, M., Lightbody, P., Stuart, R., Albright, W. H. & Yeates, R. *Guidelines for the Assessment, Design, Construction and Maintenance of Phytocaps as Final Covers for Landfills. October* (2011).

Scheutz C., Kjeldsen P., Bogner J.E., De Visscher A., Gebert J., Hilger H.A., Huber-Humer M. & Spokas, K. Microbial methane oxidation processes and technologies for mitigation of landfill gas emissions. Waste Management & Research. 27. 409–455 (2009).

Stein, V.B. & Hettiaratchi, J.P.A. Methane oxidation in three Alberta soils: influence of soil parameters and methane flux rates. Journal of Environmental Technology 22:, 1-11 (2001).

Stralis-Pavese, N., Bodrossy, L., Reichenauer, T. G., Weilharter, A. & Sessitsch, A. 16S rRNA based T-RFLP analysis of methane oxidising bacteria—Assessment, critical evaluation of methodology performance and application for landfill site cover soils. Appl. Soil Ecol. 31, 251–266 (2006).

USEPA, Measurement of Fugitive Emissions at a Landfill Practicing Leachate Recir- culation and Air Injection; EPA-600/R-05/088; U.S. Environmental Pro- tection Agency; Office of Research and Development: Washington, DC, (2005)

Wang, Y., Wu, W., Ding, Y., Liu, W., Perera, A. & Chen Y. Methane oxidation activity and bacterial community composition in a simulated landfill cover soil is influenced by the growth of Chenopodium album L. Soil Biol. Biochem. 40, 2452–2459 (2008).

Whalen, S., Reeburgh, W. & Sandbeck, K. Rapid CH4 Oxidation in a Landfill Cover soil. Applied and Environmental Microbiology 56, 3405-3411 (1990).

Watzinger, A., Reichnauer, T.G., Blum, W.E.H., Gerzabek, M.H. & Zechmeister-Boltenstern, S. The effect of landfill leachate irrigation on soil gas position: methane oxidation and nitrous oxide formation. Water Air Soil Pollut. 164, 295–313 (2005).

Wilshusen, J.H., Hettiaratchi, J.P.A., De Visscher, A. & Saint-Fort, R. Methane oxidation and formation of EPS in compost: effect of oxygen concentration. Environ. Pollut. 129, 305– 14 (2004).

Wu, L., Ma, K. & Lu, Y. Rice roots select for type I methanotrophs in rice field soil. Syst. Appl. Microbiol. 32, 421–8 (2009).

Zornberg, J.G., LaFountain, L. & Caldwell, J.A. Analysis and Design of Evapotranspirative Covers for Hazardous Waste Landfill. Journal of Geotechnical and Geoenvironmental Engineering 129, 427-438 (2003).

Leadership in Sustainable Infrastructure
*Leadership en Infrastructures Durables*

Vancouver, Canada
May 31 – June 3, 2017/ *Mai 31 – Juin 3, 2017*

# EXPLORATION OF PARAMETERS EFFECTING ENERGY PERFORMANCE OF BUILDING

Tabesh, Tofigh [1, 3], Ghorbanalavi, Farrin [2]
[1] Sama Technical and Vocational Training College, Islamic Azad University, Urmia Branch, Urmia, Iran
[2] Department of Environmental Control and Building Technology, Faculty of Architecture, Istanbul Technical University, Turkey
[4] ttabesh@itu.edu.tr

**Abstract:** The reduction of the effects of global warming has become a priority all over the world. This resulted in the need for reducing the greenhouse gas emissions. As the building sector has the major amount, precautions must be taken in the existing and new building. This paper based on literature review tries to highlight significant factors which effect building energy consumption and to find a proper way in order to increase energy efficiency of the building, reduce running costs and environmental effects. It is considerably achieved that high levels of energy-saving usually can be achieved by optimal combination of several measures, such as exterior heating, ventilation and air condition system, walls, windows, roofs, thermal insulation, and lighting system. The most effective strategies contributing in energy-saving of building design, however, are those applied before construction. As a result, it can be concluded that these strategies can be useful for both the refurbishment of existing buildings and the decision making during the preliminary design stage of new buildings. Thus, it is expected that outcomes of this study would help decision makers and building designer important parameters which effects building energy conservation and environmental concern specifically in early design stages.

***Keywords***— Building energy modeling; Energy efficient building; Envelope design; Retrofit building.

## 1 INTRODUCTION

By retrospect to 70's decade oil crisis, energy becomes one of the major topics in agendas of developed and developing countries, which provide their energy needs from others.

Moreover, rapidly growing world energy use has already raised concerns over supply difficulties, exhaustion of energy resources and heavy environmental impacts (ozone layer depletion, global warming, climate change, etc.). According to Oluklulu (2012), during the last two decades, primary energy has grown by 49% and $CO_2$ emissions by 43%, with an average annual increase of 2% and 1.8% respectively. Certainly, reducing emissions of greenhouse gases and other pollutants, decreasing energy consumption and providing security of energy supply are main global targets against environmental problems. Beside environmental problems, world economy also requires energy savings since economic load of energy use is one of the major actors of the global economy. Energy Efficiency Directive of European Commission (2012), proposed that due to ever-increasing demand for fossil fuels brings ever increasing energy prices, world is forced to use less. All over the industrialized world, the building sector accounts for more than 1/3 of the global energy consumption. Focus on the European situation it is responsible for about 40% of the total energy use for 36% of $CO_2$ emissions. In that respect, importance of buildings energy consumption became controversial issues all over the world. These data could not be neglected because they directly involve the share of global greenhouse emissions with all its negative impacts on the environment.

## 2 Literature survey

A probe conducted by Olklulu (2012), revealed that global surface temperature increased 0.4 ~ 0.8°C between the start and the end of the 20th century and the temperature will rise between 1.8 ~ 4.0°C at the end of the 21st century with an increase in extreme events. Climate change is strongly affecting diverse aspects of human society and the natural world. Based on Energy Efficiency Directive of European Commission (2012), the significant warming climate of the whole world due to the global climate change is considered to have strong effects on a building's energy requirement or usage as their heating and cooling needs are related to temperature conditions and weather variations. Many previous studies have concerned the climate impact on building energy usage in recent years. Revealing the impact of climate change on building energy consumption is beneficial for not only making efficient energy saving measures but also reducing pollutant or greenhouse gas emission. In the past decades, buildings designed according to climatic condition which is neglected in our present time. In this study, the climate impacts on extreme energy consumption of different types of buildings are determined and how the climate impact on energy consumption in building was analyzed. Maximum and minimum temperature, solar radiation, type of material, isolation of walls and roofs, climate condition and other parameters are among the factors that affected by our building design process.

### 2.1 Hot-humid and Mediterranean

A life cycle model was presented to a case study located in hot humid (Antalya) of Turkey by Mangan and Oral (2015). This study focused on implementing effective energy retrofit strategies, thermal insulation and Photovoltaic (PV) systems in terms of improving building energy performance. Then the effect of each measure on life cycle energy consumption and $CO_2$ emission was determined by using the "Life Cycle Energy (LCE)", and "Life Cycle $CO_2$ ($LCCO_2$)" analysis. In total, 12 evaluated alternatives were divided into two groups that have been shown in Table 1. The alternative with an optimum performance for Antalya, among the described alternative groups, was A4.

Table 1: The alternatives related to the energy retrofit strategies.

| Alt No. | Description | Uwall1 ($W/m^2K$) | Uwall 2 ($W/m^2K$) |
|---|---|---|---|
| A1 | No thermal insulation layer in the exterior wall components | 0.79 | 3.25 |
| A2 | Thermal insulation value = base case | 0.37 | 0.58 |
| A3 | | 0.34 | 0.49 |
| A4 | | 0.31 | 0.43 |
| A5 | | 0.28 | 0.39 |
| A6 | | 0.26 | 0.35 |
| A7 | Thermal insulation value ≥ U value in | 0.24 | 0.32 |
| A8 | TS 825 | 0.20 | 0.25 |
| A9 | | 0.18 | 0.22 |
| A10 | | 0.16 | 0.18 |
| A11 | | 0.14 | 0.17 |

- A12 Mono crystalline silicon module for terrace roof (190Wp- PV surface area: 148.36 $Wp/m^2$
- A13 Amorphous silicon module for opaque areas of south façade (340 Wp) PV surface area: 55.30 $Wp/m^2$
- Turkish Standard -TS 825

Comparing A4 (alternative for Antalya) with A1(no thermal insulation layer in the exterior wall components) showed an increase in embodied energy and embodied carbon values respectively with the ratio of 3% and 1% respectively. By applying PV system, A12 has become the alternative with an optimum performance.

An analysis of the current situation of the public school in Rome was carried out by De Santoli et al. (2014) to identify strategies for energy saving through investigating dimensional, technological and architectural features. In this study, several standard interventions were defined according to the historical and architectural features. By comparing the costs the analysis was performed and parameterized for standard retrofit interventions of the existing envelopes. Benefits achievable by the interventions in terms of energy and money saving were evaluated through a simple payback time analysis (PBT) useful to identify priorities for action. The result showed a total cost save about 0.09 €/kWh. The emerged outcome on investigation of type and quality of the refurbishment and its payback time was led to innovate the problem-solving activities and opportunities. These solutions aimed to gain appropriate standards speedily and consciously, and also avoiding the waste of time and money in the same time[4].

A calculation conducted by Buonomano et al. (2014), focused on energy and economic savings of four hospital buildings. Four strategies were analyzed to improve energy performance (Table 2).

Table 2: Suggested energy efficiency alternatives

| Alt. No | Description |
|---|---|
| CASE A: | Roofs thermal insulation by 3.0 and 8.0 cm of polyurethane. |
| CASE B: | Installation of a 3-way valve in the buildings substation, suitable for varying the supply water temperature as a function of the external temperature. |
| CASE C: | Installation of thermostatic valves for each radiator. |
| CASE D: | Installation of a time-programmable regulation for classroom air heating unit (AHU). |

Case D was possibly the most profitable action among the proposed energy saving scenarios determining major savings both during winter and summer operations with a very low capital cost. According to the strict economic point of view, a centralized heating regulation in case B was more profitable than a room in case C.

An investigation conducted by Ascione et al. (2013), focused on varying heating, ventilation, air-conditioning and cooling systems (HVAC) for energy saving by refurbishing thermal- physical features of the building envelope in Mediterranean climates. Impact of envelope renovation on energy demand of three kinds of HVAC system (Table 3) of two buildings envelope was analyzed.

Table 3: The alternatives related to the energy retrofit strategies

| No. | Description | Energy Reduction |
|---|---|---|
| A1 | Fan coil without outside air- HVAC system | 2458 GJ (i.e., -50%) |
| A2 | Fan coil with outdoor air- HVAC system | 2887 GJ (i.e., -31%) |
| A3 | Air–water HVAC system, with fan coils and dedicated outdoor air system | 2945 GJ (i.e., -16%) |

Overall, the renovation of the building envelope showed improvement in indoor thermal conditions with regards to all HVAC systems being considered. However, the improvements are more evident for HVAC systems Nos. A 1 and A 2.

In order to evaluate energy performance and the dynamic thermal behavior of the green roof house-holiday building, Gagliano et al. (2014), conducted a survey in Italy using DesignBuilder. The available Energy savings through the introduction of the green roof showed the significant reduction of the total cooling and heating about 80%, 34%, respectively. Furthermore, compared to the traditional roof the green roof provided lower fluctuation of its superficial temperatures, diminishing the average daily temperature fluctuations from 12 °C to 6 °C.

Dogan Sahin et al. (2015), adopting a historical building approach, focused on energy efficient retrofitting (EER) interventions in Izmir-Turkey, with the aim of energy consumption reduction. The alternative which have been considered were weather stripping, Indoor air temperature control, changing the heating system and/or fuel type, insulation of the attic floor, additional insulation to the roof, interior insulation of the walls and additional insulation to the ground floor. Appropriate retrofits were gathered into three packages to evaluate their effects on the energy consumption as shown in Table 4.

Table 4: Energy efficient retrofit packages for the case building.

| Alt. No | Description |
|---|---|
| Package 1: | Without altering the building appearance and damaging the heritage values |
| Package 2: | Interventions in terms of improving the thermal performance |
| Package 3: | Some inappropriate retrofits to better understand methodology. |

The results show that energy savings of more than 34% can be obtained without damaging the heritage values [8].

## 2.2 Cold and Continental

Sekki et al. (2015), investigated heating and electricity consumption to find out differences by comparing primary energy consumption. The studied buildings were day care centers, schools and university buildings located in southern Finland and classified based on the construction year. The assessed educational buildings' characteristics were evaluated in terms of volume, heat consumption, energy class, gross area, electricity consumption, heat production, ventilation. These evaluations were based on the actual energy consumption measured to find out both heating and electricity consumption.

Based on the results, differences in energy consumption and primary energy consumption of studied buildings were high and newer buildings consumed less heating energy. For instance, differences between the primary heating and electricity consumption in day care centers, schools and university buildings were 83%, 84% and 76% respectively. However, there was no obvious correlation for electricity consumption.

In order to improve building energy performance, a study conducted by Mangan et Koçlar Oral (2015) in cold climate (Erzurum) of Turkey. Some effective energy retrofit strategies, thermal insulation and PV systems were carried out using life cycle model. "Life Cycle Energy (LCE)", and "Life Cycle CO2 (LCCO2)" analysis-based on life cycle assessment (LCA) method, was implemented to evaluate the effect of each strategy. In total, 12 evaluated alternatives were divided into two groups (Table 1).The alternative with an optimum performance for Erzurum, among the described alternative groups, was A11 (Table 1). As to the PV system application, the alternative with an optimum performance for Erzurum was A12.

In Korea, an on-site survey was piloted by Chunga and Rhee (2014), to find out potential opportunities using the software eQuest for energy conservation in educational buildings. Three groups of building were evaluated. Chosen Energy saving strategies were based on their technical and practical detailed in Table5:

Table 5: Energy saving strategies

| Energy classification | Construction year | Parameters | Reduction (%) |
|---|---|---|---|
| New and energy guzzling | 2005-2010 | Space usage, Age | |
| Old and low-energy | 1961-1999 | Orientation, Lighting | 6%–30%. |
| Old and energy guzzling | 1974- 1991- 1969 | Window area, Shading-device, Heating and cooling equipment | |

Strategies such as replacement of windows, reduction of power for lightings, adjustment of indoor air temperature, change in U-value of roofs and walls, automatic standby power cutoff switches were established for buildings. The energy analyses of the surveyed buildings determined the potential for energy conservation in the range 6%–29%. Energy consumption can be reduced in old buildings by10–22%, if windows are replaced and insulation are applied to the building envelope are added. By improving the thermal performance of the envelope, it is possible to reduce energy usage by 18–29 %, while changing the behavior of its occupants.

## 2.3 Temperate, Maritime Temperate, Temperate - humid

Raji et al. (2015) suggested an envelope design to decrease energy consumption of high-rise office buildings in mild climates. Four significant façade parameters such as window to wall ratio (WWR), glazing, shading, and roof parameters were defined. The effects of applied parameters and their combination were quantified as the percentage of reduction or increase in the total heating, cooling, lighting compared to the reference model. The results revealed that the integration of high-performance design solutions provides a considerable savings in energy by 42%, 64%, and 34% for total energy, heating energy, and electric lighting energy use, respectively.

A comparative study conducted by Han et al. (2015) focused on the accuracy of building energy simulations associated. The methodology focused on the wind pressure coefficients (Cp) measured with computational fluid dynamics (CFD) simulations and building leakage areas. A commercial office building located in Saginaw, Michigan, US was deployed as the framework of study in a case study. In this study four simulation scenarios of infiltration rate have been conserved which are shown in Table 6:

Table 6: Simulation Scenarios

| Scenario No. | |
|---|---|
| Scenari 1 | Default setting of infiltration rates in DesignBuilder considering three envelopes tightness levels (leaky, medium, and tight) |
| Scenario 2 | Simulates building infiltration rates using CONTAM (multi zone building airflow and contaminant transport computer program often used for ventilation and indoor air quality analysis) in which data sources of wind pressure loads (simulated by CFD) is Air Infiltration and Ventilation Centre (AIVC) database |
| Scenario 3 | Simulated by CONTAM, input yearly-averaged infiltration rates |
| Scenario 4 | Simulated by CONTAM, input monthly-averaged infiltration rates |

The simulation-predicted data of scenarios were compared with the actual utility data and evaluated according to ASHRAE guideline 14-2002. The result revealed that the approximately 12% of the total annual energy consumption (standard deviation: 3.3%) was taken by energy consumption due to air infiltration. The simulations using monthly and yearly inputted infiltration rates, compared to the simulation results associated with default setting and database, had more accurate results [12].The study conducted by Tahsildoost and Zomorodian (2015) examined techniques of energy retrofit and the necessity of appropriate solutions for two school buildings.

Table 7: Building description

| Classification | Construction year | Construction year |
|---|---|---|
| CASE A | Before 2000 | Low quality construction, Non-isolated exterior envelope, Single layer glazing's with metal frame windows |
| CASE B | After 2000 | Only with insulated roof |

The procedure contains three steps: first preliminary building performance assessment and presenting retrofit proposals (Table 7), second optimizing and prioritizing scenarios based on energy simulation and payback time (PBT), and third assessment of the implemented techniques. Based on evaluation, 29.87%

and 38.29% primary energy reduction were observed after retrofit implementations in the old and new cases, respectively.

Gul and Patidar (2015) focused on the electrical consumption of an academic building of Heriot-Watt University, Edinburgh, Scotland, adopting a case study approach, to determine the influence of room activities and occupancy in terms of lectures, seminars and meetings on the energy consumption of the building from January 2013 to May 2013. Therefore, daily average electricity consumption, total daily occupancy and daily room activities were considered. An online questionnaire was distributed to staff and students as well as interviews to have a clear view of relationship between electricity and occupancy. The results of the questionnaire indicated that 92% of the building users are visitors. 67% of the respondents were found to use a laptop, 14% use a PC and 37% use a tablet or an IPad. About 6% said that they use an electric kettle and table lamp, 4% use microwave ovens and only 2% use portable heaters.

## 2.4 Very hot and Hot

To detect operating problems and find out the opportunities for energy conservations (ECOs), Alajmi (2012) focused on energy audit (EA). Sampled building was an educational building in a hot summer climate located in State of Kuwait- Kuwait. A list of issues was identified which were violating the energy conservation principles, such as those outlined in the ASHRAE Standard 100-2006. To measure how much energy could be saved, different recommended effects (HVAC system, Lighting, Plug-in equipment) on building's thermal performance had been simulated using DesignBuilder (DB). In order to assess the feasibility of utilizing the ECOs, the auditor had to calculate the payback period. It was indicated that if all the recommendations had been implemented, 52% of the total energy could be saved. Furthermore, it was found that the $CO_2$ emission due to electricity use in the building would be reduced by 648 tons per year.

To achieve energy efficient building design, Kocagil and Oral (2015) were set to evaluate design parameters of traditional Diyarbakır houses on heating and cooling loads in Turkey by Design Builder. They evaluated the thermal performance of building forms, generated from traditional architecture, within the defined settlement textures. Each building form, alternatively derived from plan type and A/V ratio combinations. Four commonly seen reference standardized plan type: L-type plan, U-type plan, Inner courtyard plan and central courtyard plan were considered for evaluation.

In this study, the impact of form factor is investigated by six different A/V ratios to determine building form alternatives; 0.50, 0.60, 0.70, 0.80, 0.90 and 1.00. In order to evaluate the impact of the street width and urban blocks on energy loads, four alternative settlement textures (ST1, ST2, ST3, ST4) were defined;

- ST1 and ST2; settlement texture with respectively 6m and 3m street width,
- ST3 and ST4; settlement texture consisted of blocks respectively with four buildings and six buildings

Consequently, building form and settlement texture was found to be influential in heating and cooling loads, and there was a clear correlation between A/V ratio and the heating and cooling loads. By an increase in A/V ratio, the heating and cooling loads increase too. Inner courtyard plan type, among different building form alternatives, was found to need the lowest heating and cooling loads while L type plan need was the highest. On the other hand, ST4, the buildings within the settlement texture alternative, consume the lowest cooling energy, and the reference buildings consume the lowest heating energy.

# 3 Discussion

For the past 50 years, a wide variety of building energy simulation programs have been developed, enhanced and are in use. An overview of articles provides features and capabilities of major building energy simulation programs. Concerning all kinds of climate condition, building envelope design, energy effective retrofit methods, thermal energy storage (TES) technology and heating, ventilation and air-condition systems (HVAC) are considered the most typical energy-saving techniques. Numerous factors have been affected energy consumption of building. Regarding the literature, the most energy savings are gained by changing heating, cooling, air conditioning and ventilation (HVAC), glazing type window to wall ratio and roof system.

Exploring changes of building energy consumption and its relationships with climate can provide basis for energy-saving and carbon emission reduction. In this study, findings of the previous papers and practical implementations on the total energy consumption are summarized. The most important conclusions are:

- HVAC system can offer great potential for energy conservation when proper control and operation strategies are employed. Based on different climate and building type specific HVAC system were applied in building. For example, countries in harsh hot climate condition like Saudi Arabia and Kuwait, given the heavy reliance on HVAC system to achieve thermal comfort, energy savings of up to 30% can be obtained. However, in humid climates like California evaporative coolers are not suitable and thus are not considered in the optimization analysis for the HVAC system selection. In terms of building type, heating, cooling, ventilation and air conditioning (HVAC) systems are the major part of electrical energy consumption especially at the hospitals.

- In comparison with common roof construction and Green roofs, simulations confirm the effectiveness of the green roofs in reducing the energy needs for both cooling and heating respectively of about 80 and 34%.

- The effect of window size on building energy consumption is investigated for different values of window-to-wall ratio (WWR). As the window-to-wall ratio increased, it resulted in a tendency to increase the variance of heating demand in winter and cooling in summer. For a high-performance envelope (low U-values for opaque parts and for glazing) in a heating dominant climate, energy-saving for different window-to-wall ratio (30%, 50%, 80% and 100%) is highest when the WWR is around 50%, which is the right balance for reducing heat transmission and increasing solar heat gains in winter. Based on the simulation results resulted from studies showed that for lower values of WWR, a higher percentage of energy-saving is achieved for both cooling and heating; meanwhile, energy use for electric lighting increased slightly compared to a fully glazed façade

- Selecting a glazing for window system is crucial which offers a balance between visual and energy aspects. Window glazing improved optical and thermal properties. Thickness, coating, tinted and spacer between panes are important parameters for determining thermal and daylight aspects of the glazing. Results that outcome from studies showed that replacement of windows, by adopting low-emissive glass, photo-chromic or electro-chromic glass and lower U-value would doubtlessly reduce the heat losses. For example, electrochromic glazing has one of the best overall energy performance among the glazing type and shading strategies with considerable energy-savings for lighting (16.9%) and cooling (11.3%) and a total energy-saving of around 1%. Static glazing offered fixed thermal and optical properties while dynamic glazing was considered more advanced due to its switchable thermal and optical properties. Out of all dynamic glazing, electrochromic glazing is the most popular, and its performance was widely investigated by researchers. Thermotropic windows may be less popular, due to its visual provision, but their thermal performance have been confirmed by the researchers.

- Energy reduction and cost reduction in building is more dependent on the amount of glass used in the facilities. Based on studies approximately well-made double and triple glazing can reduce cooling and heating costs by as much as 5–10% and 7.31% respectively.

- Based on literature review, generally, as the orientation was changed from north (N) to south (S), the heating demand decreased and the cooling demand increased.

## 4 Conclusions

The increase of energy consumption and $CO_2$ emissions in the built environment has made energy efficiency and savings strategies a priority objective for energy policies in most countries. Scientist across the world is working on energy modeling and control in order to develop strategies that would result in an overall reduction of a building's energy consumption. In that respect, this paper analyses available information concerning energy consumption in buildings related to heating, cooling and lighting.

Numerous studies for various building types located in different climatic regions have been reviewed. It is concluded that the most effective strategies contributing in energy-saving of building design, however, are those applied before construction. Moreover, based on literature review, it seems that retrofitting of building is a reasonable act for increasing energy efficiency of building.

The development and improvement of the HVAC system section of the building consider the most important factors for efficiency requirements presented in literature. In this case, simulation tools for energy use estimation seem to be the best choice. With the demand for thermal comfort, HVAC systems (and its associated energy consumption) have become an unavoidable asset, accounting for almost 50% the energy consumed in buildings.

In addition to great emphasis of HVAC system, building envelope plays an increasingly crucial role in energy reduction. Among all features of the building envelope, effectiveness and efficiency of glazing types is remarkable. One of the strategies of energy efficient building design is the inclusion of daylighting and/or the natural lighting into the building design.

Using simulation programs to solve the complex relationships between climate, occupancy requirements, mechanical and electrical systems, energy-efficiency issues and design characteristics is a strong way of coping with these problems. Designbuider and EnergyPlus are used most-widely among other simulation tools.

References

Alajmi, A., *Energy audit of an educational building in a hot summer climate.* Energy and Buildings, 2012. **47**: p. 122-130.

Ascione, F., Biancoa. F., De Masi. R.F. and Vanoli, G.P, *Rehabilitation of the building envelope of hospitals: Achievable energy savings and microclimatic control on varying the HVAC systems in Mediterranean climates.* Energy and Buildings, 2013. **60**: p. 125-138.

Buonomano, A., Calise, F., Ferruzzi, G. and Palombo, A., *Dynamic energy performance analysis: Case study for energy efficiency retrofits of hospital buildings.* Energy, 2014. **78**: p. 555-572.

Chung, M.H. and Rhee, E.K., *Potential opportunities for energy conservation in existing buildings on university campus: A field survey in Korea.* Energy and Buildings, 2014. **78**: p. 176-182.

De Santoli, L., Fraticelli, F., Fornari, F. and Calice C., *Energy performance assessment and a retrofit strategies in public school buildings in Rome.* Energy and Buildings, 2014. **68**: p. 196-202.

Energy Efficiency Directive of European Commission ; Available from: <http://ec.europa.eu/energy/efficiency/buildings/buildings_en.htm>, date retrieved 01.12.2012.

Gagliano, A., Detommaso, M., Nocera, F. and Patania, F., *The retrofit of existing buildings through the exploitation of the green roofs–a simulation study.* Energy Procedia, 2014. **62**: p. 52-61.

Gul, M.S. and Patidar, S., *Understanding the energy consumption and occupancy of a multi-purpose academic building.* Energy and Buildings, 2015. **87**: p. 155-165.

Han, G., Srebric, J. and Enache-Pommer, E., *Different modeling strategies of infiltration rates for an office building to improve accuracy of building energy simulations.* Energy and Buildings, 2015. **86**: p. 288-295.

Kocagil, I.E. and Oral, G.K., *The effect of building form and settlement texture on energy efficiency for hot dry climate zone in turkey.* Energy Procedia, 2015. **78**: p. 1835-1840.

Mangan, S.D. and Oral, G.K. *A Study on Life Cycle Assessment of Energy Retrofit Strategies for Residential Buildings in Turkey.* Energy Procedia, 2015. **78**: p. 842-847.

Oluklulu, Ç., *A research on the photovoltaic modules that are being used actively in utilizing solar energy, sizing of the modules and architectural using means of the modules.* MSc, Gazi University, Ankara, Turkey, 2001.

Raji, B., Tenpierik, M.J. and Van Den Dobbelsteen, A., *An assessment of energy-saving solutions for the envelope design of high-rise buildings in temperate climates: A case study in the Netherlands.* Energy and Buildings, 2016. **124**: p. 210-221.

Sahin, C.D., Arsan, D.Z., Tuncoku. C.C., Broström. T. and Akkurt. G.G., *A transdisciplinary approach on the energy efficient retrofitting of a historic building in the Aegean Region of Turkey.* Energy and Buildings, 2015. **96**: p. 128-139.

Sekki, T., Airaksinen, M. and Saari, A., *Measured energy consumption of educational buildings in a Finnish city.* Energy and Buildings, 2015. **87**: p. 105-115.

Tahsildoost, M. and Zomorodian, Z.S., *Energy retrofit techniques: An experimental study of two typical school buildings in Tehran.* Energy and Buildings, 2015. **104**: p. 65-72.

Leadership in Sustainable Infrastructure
*Leadership en Infrastructures Durables*

**Case Study**

Vancouver, Canada
May 31 – June 3, 2017/ *Mai 31 – Juin 3, 2017*

# FIELD EVALUATION OF SULFOLANE DEGRADATION IN GROUNDWATER USING UV/$H_2O_2$

Yu, Linlong[1,6], Iranmanesh, Sobhan[2], Achari, Gopal[3], Langford, Cooper[4] and Keir, Ian[5]
[1, 2, 3, 4, 5] University of Calgary, Canada
[6] linyu@ucalgary.ca

## 1. Background

Sulfolane ($C_4H_8SO_2$) is a highly polar, water miscible and colorless organic solvent. It is used extensively in natural gas processing plants to remove $H_2S$ and other polar compounds from natural gas and petroleum refining operations for aromatics extraction. Leachates from disposal areas and leaks during processes and maintenance activities have led to extensive sulfolane contaminated soil and groundwater around gas processing plants and refineries. Sulfolane has been identified as an emerging contaminant with a potential hazard to human health and environmental receptors. A recognition of this problem by industry, coupled with more attention from regulators, has pushed sulfolane contamination issues to forefront in some regions of North America. Recent studies have indicated that sulfolane can be degraded using advanced oxidation processes (AOPs) (Yu et al, 2016; Mehrabani et al. 2016). Yu et al. (2016) found that a combination of UV and $H_2O_2$ was a promising method to degrade sulfolane in aqueous media; however, the study was conducted at laboratory scale and prior to application at commercial scale, field pilot scale studies are needed. In this project, we evaluated a pilot scale AOP system on sulfolane degradation at a sulfolane contaminated site using sulfolane contaminated groundwater. The impact of different hydrogen peroxide dosages on sulfolane degradation was investigated.

## 2. Contaminated site and pilot UV/$H_2O_2$ unit

A sulfolane contaminated site in Alberta, Canada was selected for this field pilot test. The field test was conducted from May to August, 2016. A sour gas plant has been operating on the site since 1978. The plant processed sour gas and produced sweet natural gas and liquid condensate until 2015 when the sour processing units were decommissioned. Recently, sulfolane contamination has been observed in both groundwater and soil on the site. In this field test, a pilot scale UV/$H_2O_2$ system was evaluated for treatment of sulfolane contaminated ground waters. The UV/$H_2O_2$ pilot testing system was setup on site. The testing system consisted of one reservoir, several water pumps, one $H_2O_2$ peristaltic pump, one $H_2O_2$ storage tank and one UV Unit (UV LOGIC 8AL20, Trojan Technologies). The reservoir, UV unit and water pumps were integrated through flexible PVC pipes. The power consumption of the UV unit is approximately 2140 W, and is equipped with eight germicidal lamps. The reservoir is made of high density polyethylene with a volume of 5.6 $m^3$. The water in the reservoir was pumped into the UV unit by using a variable controllable valve. Hydrogen peroxide was introduced through a peristaltic pump. Water was mixed with $H_2O_2$ by a motionless mixer and then passed through the UV unit.

## 3. Testing procedure

For each test, three cubic meters of groundwater was pumped into the reservoir. The UV Trojan unit was turned on and warmed up for 30 minutes before each test. The $H_2O_2$ dosage was adjusted based on the ratio of $H_2O_2$ delivered to the water flow rate. There are three water sampling ports: (#1) before adding

$H_2O_2$; (#2) after adding $H_2O_2$ but before UV chamber and (#3) after UV chamber. Samples were collected at varying time intervals and immediately placed in ice. Sulfolane in the water samples was then extracted with dichloromethane and analyzed with gas chromatography-flame ionization detector.

### 4. Findings

Our field testing results (Figure 1. a) show that sulfolane in groundwater can be successfully treated using UV/$H_2O_2$.With 8 hours of retention time, sulfolane decreased from 12.6 ppm to less than 0.5 ppm. Electrical energy per order reduction (EEO) of sulfolane was used to evaluate the energy efficiency of the system. As shown in Figure 1.b, higher $H_2O_2$ dosages resulted in a smaller EEO values. EEO decreased from 12 KWH to 2KWH as $H_2O_2$ increased from 10 ppm to 240 ppm;.however, the chemical expense is increased with $H_2O_2$ dosage. The trade-off between energy consumption and chemical expense led to an optimal $H_2O_2$ concentration, which was determined to be 40 ppm in this project. The required EEO for treating sulfolane in water is 2.3 KWH. The one order sulfolane reduction operational cost including electricity consumption, chemical expense and UV lamps replacement cost was estimated to be 0.35 \$/m$^3$. The success of this field demonstration has provided useful and practical information for system scale-up and full application of the technology.

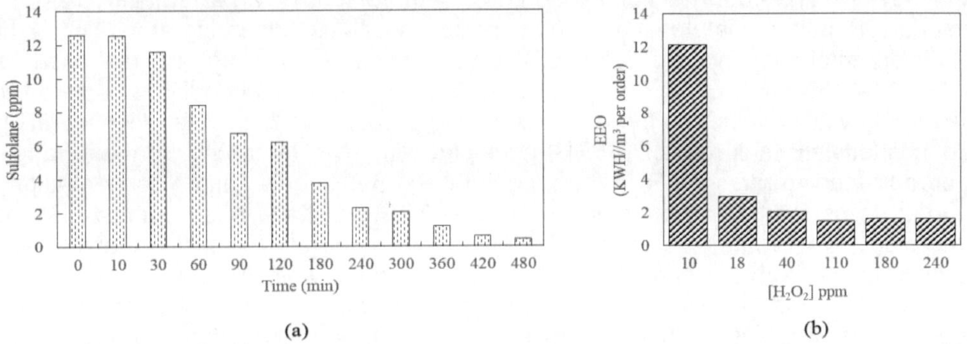


Figure 1. (a) Degradation of sulfolane in two different groundwater with 30 ppm of $H_2O_2$; (b) The impact of $H_2O_2$ dosage on EEO.

### 5. Challenges

It was observed that some groundwater turned yellow and became turbid after it was brought to the surface. This might be due to the oxidative environment of ambient condition that led to iron precipitation. The precipitates can decrease UV transmittance and cause fouling problem on the protection sleeves of UV lamps. In order to achieve a higher treatment efficiency, a pre-treatment step might be required for full-scale application.


### Acknowledgement

The authors acknowledge the financial support provided by Bonavista Energy Corporation and Natural Science and Engineering Council of Canada to conduct this research.


### References


Mehrabani-Zeinabad, M., Yu, L., Achari, G. and Langford, C. H. 2016. Mineralisation of sulfolane by UV/O3/H2O2 in a tubular reactor. Journal of Environmental Engineering and Science, 11(2): 44-51.

Yu, L., Mehrabani-Zeinabad, M., Achari, G. and Langford, C.H., 2016. Application of UV based advanced oxidation to treat sulfolane in an aqueous medium. Chemosphere 160:155-161.

Leadership in Sustainable Infrastructure
*Leadership en Infrastructures Durables*

Vancouver, Canada
May 31 – June 3, 2017/ *Mai 31 – Juin 3, 2017*

# FOAMING IN AN INNOVATIVE ANAEROBIC MEMBRANE BIOREACTOR TREATING HIGH SOLIDS WASTEWATER

Joshua J. Snowdon [1, 3] and Kripa S. Singh [2]
[1,2] Department of Civil Engineering, University of New Brunswick, Canada
[2] h557d@unb.ca

**Abstract:** An innovative anaerobic membrane bioreactor (AnMBR) using an external tubular nanofiltration membrane module is being operated under mesophilic conditions for the treatment of a high-solids wastewater (influent total suspended solids (TSS) of 5658 mg/L (± 1771 mg/L) and a chemical oxygen demand (COD) of 84,156 mg/L (± 4570 mg/L)) to determine operational conditions required to achieve good system stability while minimizing the reactor foaming propensity. The system configuration consists of a 1000 L continuous stir tank reactor (CSTR) equipped with an external tubular membrane. Thus far, operation at organic loading rates (OLR) in the range of 0.8 – 8.2 kg/m$^3$•d, and a concentration of mixed-liquor suspended solids (MLSS) in the range of 13.6 – 33.9 g/L, have been studied. Throughout the duration of the study five events of excessive foaming were recorded. Excessive foaming events were defined as events where both the primary and back-up gas traps were filled with foam, and additional foam poured down the sides of the reactor through gaps between the CSTR cover and wall created during these instances of excessive foaming. Increases (shock) in OLR, temperature, reactor ammonia nitrogen and total alkalinity concentrations, and the accumulation of colloidal materials within the reactor have all been known to impact reactor foaming propensity. This research project found shock increases in OLR, and its associated accumulation of colloidal materials, to be the most influential factor affecting reactor foaming propensity. Following an analysis of the causation of each excessive foaming event, a mitigation plan was developed to prevent future foaming events.

**Keywords:** Anaerobic membrane bioreactor, High-solids wastewaters, Foaming, Shock loading, Temperature.

## 1 INTRODUCTION

Anaerobic bioreactor technologies for the treatment of wastewaters offer advantages over traditional aerobic wastewater treatment systems such as high rate suspended growth systems, high removal efficiencies of biochemical oxygen demand (BOD) and chemical oxygen demand (COD), resource recovery (methane in biogas and value added chemicals), low sludge yield, and low carbon footprint, (Kinnunen et al. 2015, Ozgun et al. 2013, Pilli et al. 2016, Meng et al. 2009, and Skouteris et al. 2012). Additionally, the integration of membrane technologies within anaerobic bioreactors offer advantages over non-membrane integrated anaerobic bioreactor wastewater treatment systems such as high biomass retention (independent of sludge settle-ability through the membrane's complete solid-liquid separation capabilities), compact design, and most importantly the ability to produce a superior quality effluent for possible recycle or high quality discharge.

An operational issue that has been common to anaerobic membrane bioreactors (AnMBRs) since their first applications is bioreactor foaming. Foaming in AnMBRs results from a complex three-phase phenomenon (liquid-solid-gas) contributed from surface-active materials and water in sludge, as well as biogas generated through the anaerobic digestion pathway (Subramanian and Pagilla 2015). Reactor foaming causes inefficient gas recovery from the digester, reduction in the active digester volume, mechanical blockages, fouling of gas pipes, and the loss of biomass from spill carry-over (Ganidi et al. 2009, Kale and Singh 2016, Subramanian and Pagilla 2015, and van Niekerk et al. 1987).

Although foaming has been studied extensively, and factors having a tendency to affect its propensity have been identified, further studies into which of these factors contribute the most significantly towards this operational issue would help develop control mechanisms to prevent it. The goal of this study is to evaluate instances of excessive bioreactor foaming exhibited during the operation of a pilot-scale AnMBR system integrated with an external nanofiltration tubular membrane for the treatment of a high-solids wastewater. Instances of excessive bioreactor foaming will be evaluated to gain a better understanding into what causes this operational issue, and to develop a mitigation strategy to prevent this issue in future operations and applications.

# 2 MATERIALS & METHODS

## 2.1 Pilot-scale study

The pilot reactor configuration used for the duration of the experiment consists of a 1000 L stainless steel complete stir tank anaerobic reactor (CSTR), equipped with an external tubular nanofiltration membrane unit capable of backwashing processes. The reactor contents are maintained at 36°C, providing mesophilic conditions. The schematic flow diagram of the experimental set-up is shown in Figure 1.

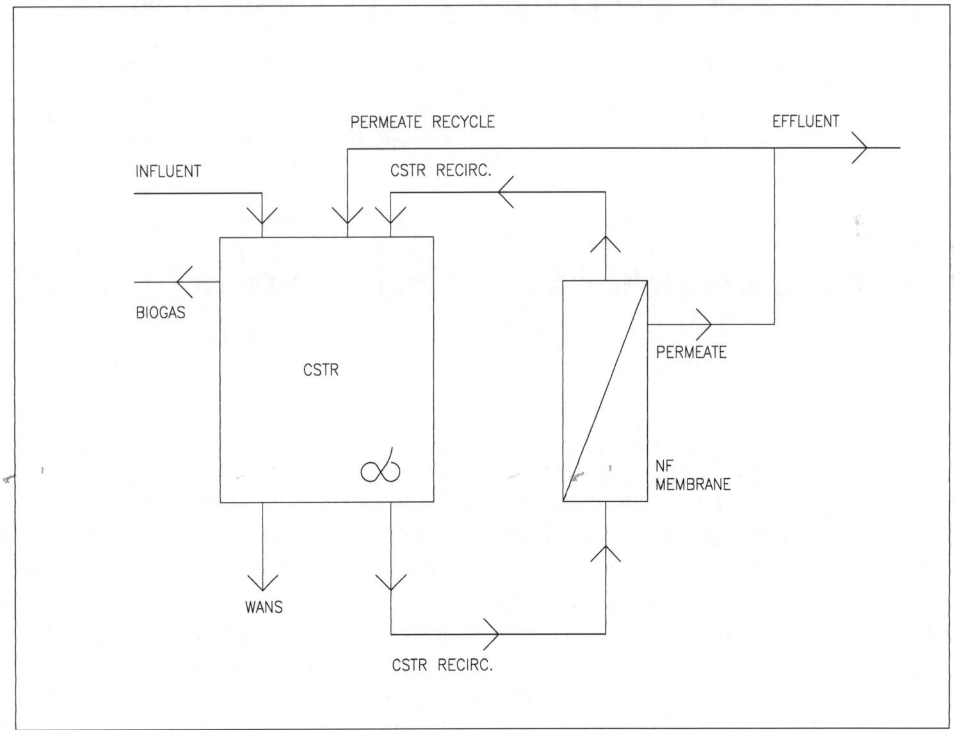


Figure 1: Schematic flow diagram of experimental set-up.

The CSTR contents are circulated through the external tubular membrane unit where a portion flows through the membrane unit, being removed as permeate, and the rest is recycled back into the CSTR.

Permeate is then wasted as effluent or backwashed through the membrane to remove reversible membrane fouling.

### 2.2 Seed sludge & Synthetic wastewater

The CSTR was seeded with a granular sludge obtained from an anaerobic digester treating wastewater generated during fruit-juice production. The seed sludge was run through a coarse metal screen prior to seeding to remove larger granules, and other suspended matter. The seed sludge was then circulated in the CSTR prior to membrane attachment, to allow it to develop into its own culture without granules. Reactor mixed liquor suspended solids (MLSS) has varied from a minimum value of 13.6 g/L to a maximum value of 33.9 g/L, with an average of 23.8 g/L (± 5.0 g/L).

The CSTR is fed a synthetic wastewater composed of both particulate (supplied through potato starch) and soluble (supplied through granulated sugar) chemical oxygen demand (COD). Macronutrients are supplied continuously in the influent synthetic wastewater, and micronutrients are bulk-dosed based on the COD loading fed to the system on a weekly basis. A partial characterization of the synthetic influent wastewater may be viewed in Table 1.

Table 1: Partial characterization of the synthetic wastewater.

| Parameter | Unit | Value | |
|---|---|---|---|
| | | (Avg.) | (S.D.) |
| Chemical Oxygen Demand (COD) | mg/L | 84,156 | ± 4570 |
| Particulate COD | mg/L | 4900 | ± 2073 |
| Filtered COD | mg/L | 79,256 | ± 4278 |
| Soluble COD | mg/L | 79,058 | ± 4108 |
| Total Suspended Solids (TSS) | mg/L | 5658 | ± 1771 |
| Volatile Suspended Solids (VSS) | mg/L | 5632 | ± 1868 |

### 2.3 Analysis

CSTR operational flows and pressures are collected daily and displayed through a PLC system. Influent, effluent, and CSTR contents have their water quality parameters analysed according to HACH Water Analysis Handbook and Standard Methods for the Examination of Water & Wastewater (Hach Company 2002, and Eaton et al. 2005). Biogas production rate is monitored using an AALBORG Mass Flow Meter (GFM) calibrated for a gas possessing a composition of 70% $CH_4$ and 30% $CO_2$. To gauge the actual methane content of the biogas being produced, gas samples are collected weekly and their composition is determined using a combination of gas reaction tubes and a Varian CP-3800 gas chromatography machine. When foaming was present, it was collected through buckets located beneath the CSTR if possible.

## 3 RESULTS & DISCUSSION

### 3.1 Reactor foaming

There are a number of factors that have been identified to contribute towards foaming in anaerobic reactor systems. Amongst these factors are rapid (shock) increases in organic loading rate (OLR), temperature changes, and the accumulation of alkalinity, ammonia nitrogen, and colloidal material (Ganidi et al. 2009, Kale and Singh 2016, Subramanian and Pagilla 2015, and van Niekerk et al. 1987). Although exact volumes of foam were not able to be collected, the foaming events that occurred throughout the operation of the pilot-scale AnMBR may be categorized into two primary categories. The first category is small foaming events, where only the primary gas trap was filled with foam. These foaming events resulted in minimal biogas losses, and were remediated quite quickly. The second category is excessive foaming events, where both the primary and back-up gas traps were filled with foam, and additional foam poured down the sides of the reactor through gaps between the CSTR cover and wall created during these instances of excessive foaming. In a full-scale system, these excessive foaming events would likely be sufficient to force either system shutdown, or result in large energy losses through the biogas collection issues associated with them and loss of reactor biomass. There were five events of excessive foaming noted, and they occurred on days 17, 28, 42, 100, and 116, with day 100 being an event large enough to force the pilot-scale system shutdown. Operational data from February 22nd – August 8th, 2016, will be examined with a specific focus on reactor temperature, OLR, alkalinity, ammonia nitrogen, and the potential for the accumulation of colloidal material leading up to each excessive foaming event. The goal will be to establish a primary causation for each excessive foaming event to develop mitigation strategies to prevent future events.

### 3.1.1 Organic loading rate (OLR)

A number of researchers have remarked that a shock loading in the feed supplied to an anaerobic digester may result in foaming due to an excess of compounds not being readily degraded by the bacteria, which then causes an abrupt accumulation of surface active by-products (surfactants) (Ganidi et al. 2009). As may be viewed in Figure 2 there were a number of instances where foaming was noticed after a shock loading was applied.

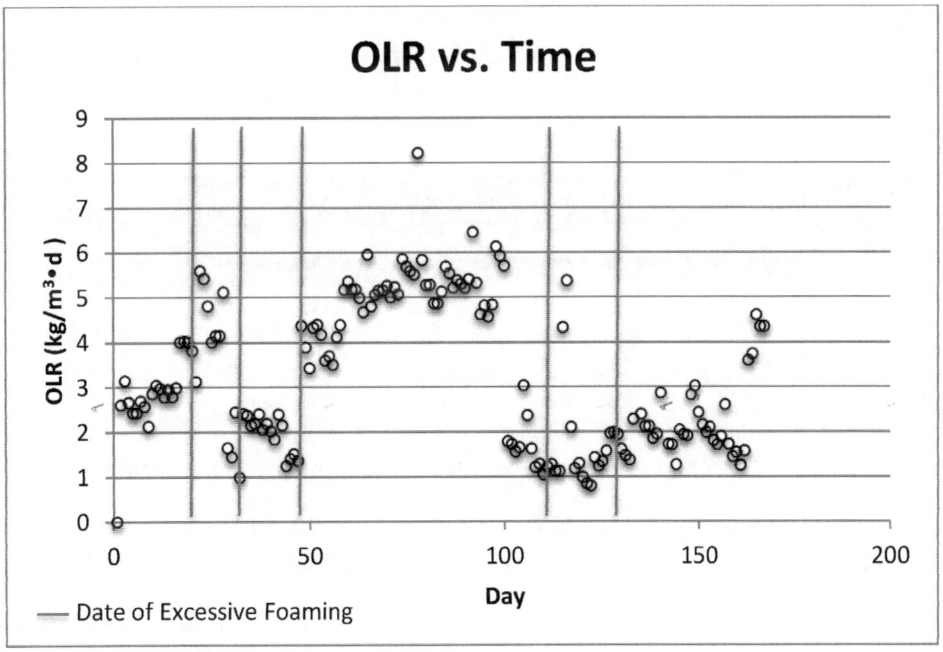


Figure 2: Organic loading rate (OLR) vs. time throughout the study period.

The excessive foaming on days 17, 28, 42, and 116 may be due entirely or partially to a shock in loading rate. On days 17, 28, 42, and 116, the OLR was raised by 135% (1.0 kg/m$^3$•d), 123% (1.0 kg/m$^3$•d), 131% (0.6 kg/m$^3$•d), and 124% (1.1 kg/m$^3$•d), respectively, which is quite a large increase in loading rate for a day-to-day basis. As the influent wastewater is composed of both a soluble sugar and particulate starch fraction, the shock load's soluble sugar portion would first be prioritized for bacterial anaerobic digestion. This would result in an abrupt accumulation of particulate starch COD, or in other words, surfactants, providing additional nucleation sites for gas bubbles in the sludge, and resulting in sudden foaming. The tendency of a reactor to foam has been noted to increase with an increasing suspended particulate concentration (Subramanian and Pagilla 2015). The only excessive foaming event where a spike in OLR was not exhibited prior to the foaming event was day 100, the day with the greatest amount of foaming. As it will be noted later, this day had many other factors contribute towards the quantity of foaming yielded.

### 3.1.2 Temperature

Reactor temperature has been found to have both a positive and negative effect on foaming propensity in an anaerobic digester. With a rapid increase in temperature, both the rate of gas bubble formation, and the rate of conversion of complex substrates to volatile fatty acids and biogas increase rapidly, increasing foaming tendency (Subramanian and Pagilla 2015). If a reactor were operated under thermophilic conditions, and compared to another reactor operating at mesophilic conditions, the thermophilic reactor would have a lower foaming tendency due to its lower viscosity providing better foam drainage and sludge breakup (Ganidi et al. 2009). A profile of the reactor temperature over the duration of the study may be viewed in Figure 3.

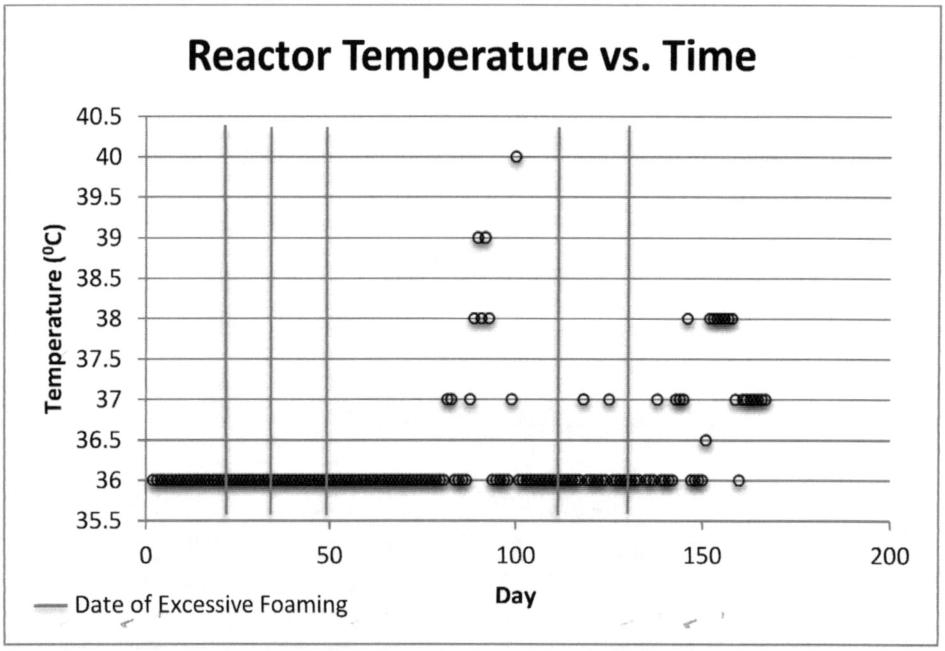


Figure 3: Reactor temperature vs. time throughout the study period.

As seen in Figure 3, there were a number of spikes in reactor temperature throughout the duration of the study, however excessive foaming was only exhibited when the reactor temperature jumped by 4°C over two days. It should be noted that when the reactor temperature increased to 39°C minor foaming events were present. However the temperature remained at 36°C for all other excessive foaming events. Due to other excessive foaming events taking place without a spike in temperature, and other minor temperature changes not causing excessive foaming events, it may be concluded that although an increased reactor temperature may contribute towards an increased foaming propensity, minor fluctuations in its value do not initiate reactor foaming. Large fluctuations may initiate a foaming event however, if other reactor characteristics are present to simultaneously contribute towards the foaming event.

### 3.1.3 Total Alkalinity

Reactor sludge alkalinity level has been proposed to be inversely proportional to reactor surface tension, therefore making sludge with a higher alkalinity concentration more surface active, and increasing its foaming propensity (van Niekerk et al. 1987). Reactor alkalinity was measured throughout the duration of the experiment, and may be viewed in Figure 4.

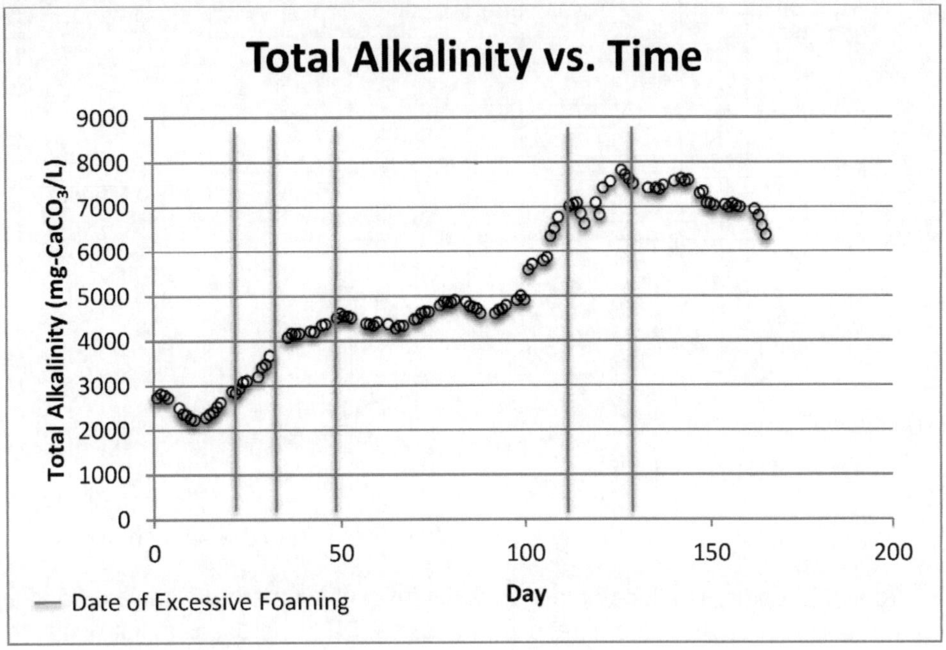


Figure 4: Reactor total alkalinity level vs. time throughout the study period.

Throughout the duration of the study, the total alkalinity level of the reactor was on a relatively steady increase, with there being a few periods where total alkalinity increased rapidly. Alkalinity accumulated in the beginning of the study because initial sludge wastage rate was very low (1 – 2 liters of sludge wasted per day). During days 30 – 100, sludge wastage rate was raised to an average of 21.1 L/d (± 10.1 L/d), which is why the alkalinity accumulation rate steadied out during this period. On day 100, the alkalinity level in the reactor was at its maximum of 4909 mg-$CaCO_3$/L when the largest excessive foaming event took place. Immediately following this event, sludge wastage rate was decreased by 62% (18 L/d), causing alkalinity to increase rapidly to 5583 mg-$CaCO_3$/L, and kept increasing for the next 23 days.

Although total alkalinity increases have been known to increase reactor foaming propensity, they do not appear to trigger excessive foaming events during the duration of this study. The largest foaming events did take place when alkalinity was at its highest levels in the reactors, but abrupt increases did not result in excessive foaming events. It would appear that larger foaming events take place when the reactor total alkalinity concentration is at its highest level, however further data should be collected to support this observation.

### 3.1.4 Ammonia Nitrogen

High ammonia nitrogen concentrations may act as surfactants increasing the foaming propensity of a reactor (Kale and Singh 2016). The ammonia nitrogen level in both the influent and effluent may be viewed in Figure 5.

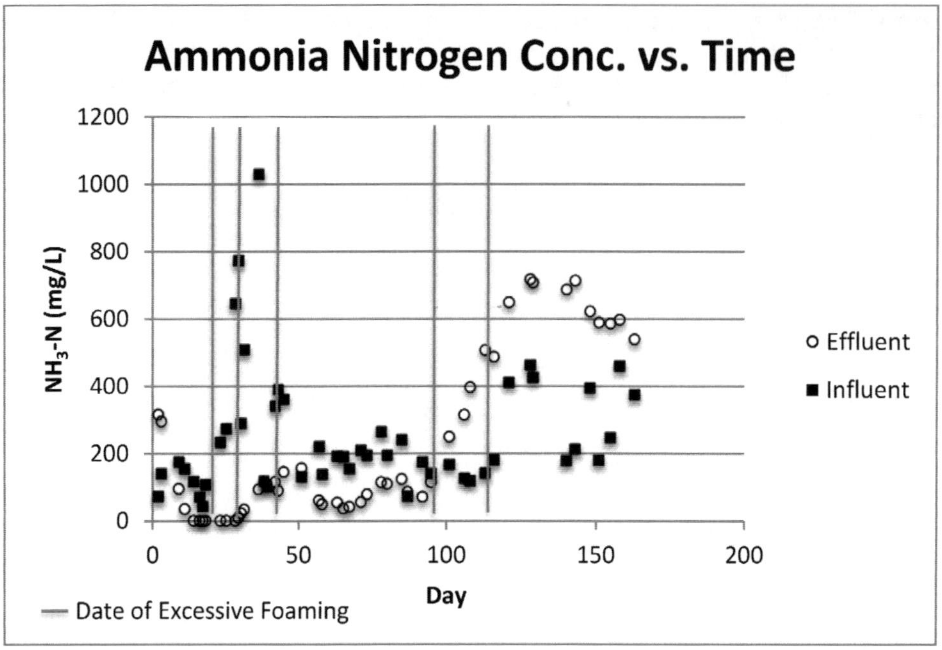


Figure 5: Reactor total alkalinity level vs. time throughout the study period.

The ammonia nitrogen concentration fluctuated in both the influent synthetic wastewater, and in the effluent leaving the reactor throughout the duration of the study. Early increases in influent ammonia nitrogen were due to an initially large retention time on the feed tank. As the wastewater sat in the tank before feeding, more of its nitrogen content was converted to ammonia nitrogen. On day 28 there was both excessive foaming and an elevated influent ammonia nitrogen concentration. This day also saw a spike in OLR of 1.0 kg/m$^3$•d, which would have resulted in a larger influx of ammonia nitrogen into the reactor as well. Although this shock load of ammonia nitrogen would have contributed towards the foaming event, the concentration within the reactor would have been 37.2 mg-$NH_3$-N/L, which is quite a low concentration and not likely the initiator of the foaming event.

On days 100 and 116, effluent ammonia nitrogen was quite high at levels of 249 and 500 mg-$NH_3$-N/L, respectively. This is during the period where sludge wastage was greatly reduced, meaning nitrogen began to accumulate in the system and convert to ammonia nitrogen. These concentrations were not in the range of general inhibitory levels for the overall anaerobic digestion process, however they may have been sufficient enough to contribute towards the excessive foaming events (Chen et al. 2007). The effluent ammonia nitrogen concentration did increase rapidly leading up to the excessive foaming events of days 100 and 116, however it also sat at an elevated level after these events, and no further excessive foaming events were noticed. For this reason, it may be concluded that an elevated reactor ammonia nitrogen concentration certainly contributes towards excessive foaming, but is not the factor that triggers it.

### 3.2 Foaming mitigation methodology

Using the experimental data collected thus far in regards to factors affecting reactor-foaming propensity, a foaming mitigation methodology was developed to apply to future operations with a similar reactor configuration. To develop this methodology, the trends in factors known to contribute towards excessive foaming events are summarized in Table 2.

Table 2: The percent-change values for each parameter leading into an excessive foaming event.

| Day of Excessive Foaming Event | %-Change in Organic Loading Rate (%) | %-Change in Temperature (%) | %-Change in Total Alkalinity (%) | %-Change in Influent $NH_3$-N (%) | %-Change in Effluent $NH_3$-N (%) |
|---|---|---|---|---|---|
| 17 | + 34.6% | - | + 4.9% | - 39.4% | - |
| 28 | + 23.4% | - | + 3.1% | + 136.3% | - |
| 42 | + 30.6% | - | + 1.2% | + 234.3% | + 5.5% |
| 100 | - 3.9% | + 7.5% | + 2.6% | + 19.3% | + 116.5% |
| 116 | + 24.3% | - | - 3.4% | + 27.5% | - 3.9% |

It would appear that a shock load in OLR has the greatest influence on reactor foaming propensity. Every day of excessive foaming, aside from day 100, had a %-increase in OLR greater than 23.4% exhibited. On day 100, there was a quick increase in reactor temperature, an increase in total alkalinity concentration, and an increase in both influent and effluent ammonia nitrogen concentration. All of these factors are known to increase foaming propensity, and happened simultaneously, which is likely the reason an excessive foaming event occurred on this day. The shock increase in temperature, and the additional volatile fatty acids and biogas formation associated with it, most likely initiated the foaming event, along with the increased alkalinity and ammonia nitrogen concentrations accelerated the event.

To prevent future events from occurring, the following steps may be followed to minimize reactor foaming propensity:

1. Ensure OLR increases on a day-to-day basis are below 23.4%. A lower increase in OLR would be favourable, if possible.

2. Have cooling mechanisms present for the anaerobic digester to prevent shock increases in temperature.

3. Be cautious of shock increases in ammonia nitrogen present in feed wastewaters. This may be accomplished through maintaining a low hydraulic retention time on feed tanks containing a high pH and a nitrogen concentration that may be readily converted to ammonia nitrogen.

4. When the sludge wastage rate is reduced rapidly, ensure nitrogen and total alkalinity concentrations supplied to the reactor are lowered as well to prevent their accumulation in-situ. For full-scale application, influent wastewater could be diluted or stripped of a portion of its nitrogen and alkalinity concentrations to ensure they do not accumulate in the reactor. After sludge wastage rate increases once again, the previously removed concentrations may be slowly re-added to new feed wastewater for treatment.

## 4 CONCLUSIONS

Foaming events were most likely catalyzed by a shock increase greater than 20% in OLR. The excessive foaming event on day 100 is the only excessive foaming event where a shock increase in OLR was not present, however on day 100 all other factors known to increase foaming propensity increased respectively, and there was also a rapid increase in temperature of 7.5%.

A mitigation priority list was developed to prevent future foaming events during operation. Reducing OLR increases to <20% is the greatest priority, as a shock increase in OLR was found to be the likely cause for most of the excessive foaming events observed. Cooling wraps to maintain a reactor temperature of 36°C during hot summer months, ammonia nitrogen control in the effluent wastewater, and alkalinity and ammonia nitrogen control in the influent wastewater when sludge wastage rate decreases rapidly are also important to reduce reactor foaming propensity.

## Acknowledgements

The authors would like to acknowledge and thank the funding agencies the Natural Science and Engineering Research Council (NSERC), the New Brunswick Innovation Foundation (NBIF), and ADI Systems Inc.

## References

Chen, Y., Cheng, J. J., Creamer, K. S. (2007). Inhibition of anaerobic digestion process: A review. *Bioresource Tehnology*, **99**: 4044-4064.

Eaton, A. D., Clesceri, L. S., Greenberg, A. E., Franson, M. A. H., American Public Health Association., American Water Works Association., & Water Environment Federation. (2005). *Standard methods for the examination of water and wastewater.* Washington, DC: American Public Health Association.

Ganidi, N., Tyrrel, S., Cartmell, E. (2009). Anaerobic digestion foaming causes – A review. *Bioresource Tehnology*, **100**: 5546-5554.

Hach Company. (2002). *Water Analysis Handbook*. Loveland, CO: Hach Company.

Kale, M. M., Singh, K. S. (2016). Foaming in sludge-bed anaerobic membrane bioreactor. *Journal of Environmental Engineering*, **142**(4): 1-3.

Kinnunen, V., Ylä-Outinen, A., & Rintala, J. (2015). Mesophilic anaerobic digestion of pulp and paper industry biosludge-long-term reactor performance and effects of thermal pretreatment. *Water Research,* **87**: 105-111.

Meng, F., Chae, S. R., Drews, A., Kraume, M., Shin, H. S., Yang, F. (2009). Recent advances in membrane bioreactors (MBRs): Membrane fouling and membrane material. *Water Research,* **43**: 1489-1512.

Ozgun, H., Kaan Dereli, R., Evren Ersahin, M., Kinaci, C., Spanjers, H., van Lier, J.B. (2013). A review of anaerobic membrane bioreactors for municipal wastewater treatment: Integration options, limitations, and expectations. *Separation and Purification Technology,* **118**: 89-104.

Pilli, S., More, T. T., Yan, S., Tyagi, R. D., & Surampalli, R. Y. (2016). Fenton pre-treatment of secondary sludge to enhance anaerobic digestion: Energy balance and greenhouse gas emissions. *Chemical Engineering Journal*, **283**: 285-292.

Skouteris, G., Hermosilla, D., Lopez, P., Negro, C., Blanco, A. (2012). Anaerobic membrane bioreactors for wastewater treatment: A review. *Chemical Engineering Journal*, **198-199**: 138-148.

Subramanian, B., Pagilla, K. R. (2015). Mechanisms of foam formation in anaerobic digesters. *Colloids and Surfaces B: Biointerfaces*, **126**: 621-630.

Van Niekerk, A., Kawahigashi, J., Reichlin, D., Malea, A., Jenkins, D. (1987). Foaming in anaerobic digesters: A survey and laboratory investigation. *Journal (Water Pollution Control Federation)*, **59**(5): 249-253.

Leadership in Sustainable Infrastructure
*Leadership en Infrastructures Durables*

Vancouver, Canada
May 31 – June 3, 2017/ *Mai 31 – Juin 3, 2017*

# GEOTEXTILE FILTRATION FOR IMPROVING SURFACE WATER QUALITY OF EUTROPHIC LAKES

Palakkeel Veetil, Dileep[1], Mulligan,. Catherine N[2,6], Eliaslara, Mariana[3], Castillo Arriagada, Esteban[4], and Bhat, Sam[5]
[1, 2, 3, 4, 5] Concordia University, Canada
[6] mulligan@civil.concordia.ca

**Abstract:** Eutrophication and impaired surface water quality are challenging problems due to the restriction in using many lakes and ponds for drinking and public recreational purposes. Lake Caron, located in the municipality of Sainte-Anne-des-Lacs in Quebec, is one such lake hit by algal blooms every year. This *on-site* study investigates the use of a floatable geotextile filtration unit for removing suspended particles, nutrients and algae to improve the lake water quality. Different combinations of three non-woven geotextiles (GTX-300, GTN-300, and GTN-350A), differing in apparent opening sizes, were tested with lake water and monitored real time for surface water quality parameters with a YSI EXO2 probe. Water quality was assessed by considering parameters such as suspended solids (SS), turbidity, total phosphorus (TP) and nitrogen (TN), and chlorophyll and blue green algae-phycocyanin (BGA-PC) concentrations in the lake water. Effective removal of SS by filtration resulted lowering TP concentration in the lake water, close to the Quebec provincial norms set for the protection of aquatic life. Overall, the filtration tests resulted in 85-98% turbidity, 89-99 % SS, 60-82% TP, 25-50% (TN), 71-78% chlorophyll and 93-97% BGA-PC removal.

***Key Words****:* Geotextiles, Filtration, Suspended solids, Phosphorus, and Eutrophication

## 1 INTRODUCTION

Eutrophication is an outcome of enrichment of nutrients, especially phosphorus and nitrogen, in aquatic systems and is characterized by dense algal blooms in freshwater, especially by cyanobacterial blooms (blue–green algae) (Correll 1998, Reeders et al. 1998). It poses serious threats to water quality, aquatic biodiversity, public health, and social life. The major consequences of eutrophication in surface water include high turbidity, reduced aquatic biodiversity, low dissolved oxygen and toxin release from certain types of cyanobacteria (blue green algae) (Shinder 2006, Chislock et al. 2013). Phosphorus (P) is considered as a growth limiting nutrient for primary production and its excess concentration can result in harmful algal blooms in lakes (Lewis and Wurtsbaugh 2008).

Lake Caron (Quebec, Canada) is a shallow artificial lake impacted by algal blooms every summer, and has been on the list of the MDDEP (Ministère du Développement Durable de l'Environnement et des Parcs (MDDEP), of eutrophic lakes in Quebec since 2008. The lake has been exposed multiple times to massive algal blooms in the past. In order to protect the inhabitants around the lake from cyanobacterial toxins, MDDEP restricted the use of lake for swimming. Many times, the total phosphorus concentration in the lake water was above the lower limit (30 μg/L) set for eutrophic lakes by the MDDEP. The considering factors for eutrophication in Lake Caron include surface runoff from the forest area around the lake, fertilizer use and septic tank discharges in the past, and sediment phosphorus release (Karim et al. 2013; Sarma et al. 2016).

Many remediation techniques have been practiced so far for the restoration of eutrophic lakes, which include source control, application of chemicals (use of alum, calcite, chemicals, adsorbents, algaecide)(Cooke et al. 1993; Xie et al. 2014), sediment dredging (Reddy et al. 2007) and capping with clean sediments. However, high transportation costs and lack of disposal sites associated with sediment dredging, and toxicity problems associated with chemical additions (Cooke et al. 1993) etc., restrict the use of these technologies for remediating small and shallow eutrophic lakes. Therefore, economical and environmentally acceptable restoration alternatives must be identified and tested for mitigating eutrophication in lakes.

Nutrients that enter surface water adsorb on to fine suspended particles including particulate organic matter, and sediment fines, and eventually settle on to the bottom sediments (Mulligan et al. 2009; Inoue et al. 2009). In eutrophic lakes, microbial degradation of algal biomass and other organic materials create anaerobic conditions in the bottom sediments which results in the release of sediment phosphorus in to the overlaying water (Chislock et al. 2013). Therefore, removing algal biomass together with fine sediment particles, and suspended particulate organic matter could reduce the total phosphorus in the water column (Mulligan et al. 2009, Inoue et al. 2009, Sarma et al. 2016).

Filtration is a solid –liquid separation technology, which separates solid particles from a liquid stream based on the differences in particle size. Geotextiles are strong, robust, and chemically and microbial resistant materials that are used for many environmental applications such as drainage, solid separation, filtration and soil reinforcement (Franke et al. 2012; Tota-Maharaj et al. 2012). Recent research studies addressed potential feasibility of using on-woven geotextiles as filter media for removing suspended solids, nutrients, heavy metals and organic matter from surface and storm water (Mulligan et al. 2009, Sarma et al. 2016, Frank et al. 2012, Inoue et al. 2009).

The present study is an extension of on-site filtration experiments that were conducted last year, and was to evaluate the reliability of geotextile filters for improving Lake Caron water quality under new lake circumstances with different combinations of selected non-woven geotextiles, prior to future in-situ filtration tests. The objective was to determine also the effect of sediment resuspension on water quality by conducting filtration tests with sediment to simulate actual lake conditions.

## 2 MATERIALS AND METHODS

### 2.1 Study Area

Lake Caron, an artificial, shallow, eutrophic lake, located in the municipality of Sainte-Anne –des-Lacs was selected for this study. The approximate lake surface area is 35,300 $m^2$ with average water volume of 46,400 $m^3$. The maximum and average depths are 2.6 and 1.4 m, respectively. Main sources of water in Lake Caron include precipitation, surface run off and snow. Sloped land with wild trees covers one side of the lake and surface runoff from this side significantly increases the load of nutrients and vegetative organic materials into the lake. In the past, trees on the shore were chopped down into the lake to expand the lake and thus this resulted in high organic matter in some parts of the lake. Lake Caron sediments in some places are rich in nutrients and organic matter. Sediment phosphorus concentration in some parts of the lake varied between 908-1310 mg/kg. Figure 1a shows algal blooms that occurred in Lake Caron in 2016.

### 2.2 Onsite floating filtration setup

Filtration experiments were performed in a plastic tank (35.6 cm height and 543 L volume) placed on the ground levelled with gravel, near the lakeshore. During the summer, algal blooms in the lake often accumulated at station 4 by the action of wind and currents. Therefore, this station was selected for the onsite and future in-lake experiments. In this study, two different kinds of filtration tests were performed: (i) filtration with lake water alone, and (ii) filtration with lake water and sediment. The filtration unit consists of a cylindrical filtration column, made of plexi glass, of 20 cm internal diameter and 25 cm in height, and a square shaped base, as a filter holder, with a circular hole at the centre with the exact diameter of the filtration column (20 cm). The filtration column was placed on the base to hold water and support a hydraulic head of 18 cm above the filter. Screws were used to fix and tighten the filtration column with the square base. A water level controller, which can turn on and off the pump automatically, was used at a height of 18 cm of the filtration column to avoid water overflow. For the filtration tests, the tank was first filled with about 300 L of lake water using a submersible pump. After this, the filtration unit, with selected geotextile filters, was placed on a piece of polystyrene foam to float the entire unit on the tank water. A circular hole with a 20 cm diameter was made in the centre of the foam to pass the filtered water into the tank.

For the filtration tests with sediment, the surface sediment was sampled at station 4 using a grab sampler and then stored in a closed container at 4°C for about 2 weeks. A calculated quantity of lake surface sediment was added into the tank filled with lake water to lay sediment over the bottom of the tank. Since, the addition of sediments resulted high initial turbidity, 2-3 days of settling time was given prior to starting the filtration tests in such cases. Four submersible pumps were used in the tank continuously circulate the water in the tank with the objective of filtering all water in the tank. However, three pumps with a very low mixing speed were used in the cases of the filtration tests with lake sediment to avoid sediment resuspension. A tarpaulin was used to cover the whole filtration set-up to protect it from rain and other external influences. A YSI EXO2 multiparameter (Hoskin) probe was placed in the tank to measure the physical and biological parameters of the lake water.

Figure. 1: (a) Algal bloom in Lake Caron (2016); (b) On-site floating geotextile filtration unit

### 2.3 Filter Media

Non-woven geotextile filters (TE-GTX 300, TE-GTN 300, and TE-GTN 350) custom developed and received from Titan Environmental Containment Ltd. MB, were used in this study. Different combinations of these filters, with varying numbers of filters (4-5), were tested in the field test. Table 1 shows some of the properties of selected geotextiles used in this study. The filter selection, combination, and number of layers were based on the previous field tests. The TE- GTX 300 filter is made of continuous filament polyester (PET) with an apparent opening size of (AOS: 110 µm) and has a characteristic white color that enables visual inspection of the filter for clogging and to take the decision of when to change the filters. The filters TE-GTN 300 (AOS: 90µm) and TE-GTN 350-A (AOS: 75 µm) are of black color and made of 100% virgin staple fiber polypropylene. The filters were cut in a circular shape with a 22 cm diameter and washed with deionized water overnight before placing them on the filter support. The filter with high AOS was placed on the top followed by other filters with decreasing AOS. During each filtration test, clogged filters were replaced by new filters and continued the test until the turbidity was reduced to a desired level. Lake water quality was monitored frequently by analyzing chemical, biological and physical parameters.

Table 1: Characteristics of the non-woven geotextiles used in this study

| Filters | Apparent Opening Size (AOS) µm | Flux (L/m²/min) | Mass/unit area (g/m²) | Material | Permittivity (sec⁻¹) |
|---|---|---|---|---|---|

| | | | | | |
|---|---|---|---|---|---|
| TE-GTX 300 | 110 | 3,900 | 300 | PET | 1.62 |
| TE-GTN 300 | 90 | 3,300 | 300 | PP | 0.75 |
| TE-GTN 350 A | 75 | 2,700 | 350 | PP | 0.56 |

PET: Polyester; PP: Polypropylene

### 2.4 Analysis

Water samples were collected from the tank before, after and during the experiments for measuring turbidity, total suspended solids (TSS), total phosphorus (TP), total nitrogen (TN), chemical oxygen demand (COD), nitrate ($NO_3^-$) and phytoplankton. The dissolved forms of phosphorus, nitrogen and COD were also measured. For this, collected water samples were filtered through a 0.45μm syringe filter and the filtrate was used for the analysis. Hatch test kits were used for the analysis of phosphorus, COD, $NO_3^-$ and TN in the water samples. The turbidity of lake water was measured using an Oakton turbidity meter. Physical and biological water quality parameters such as pH, temperature, turbidity, dissolved oxygen (DO), oxidation-reduction potential (ORP), chlorophyll *a* and blue green algae-phycocyanin (BGA-PC) were measured in-situ with a YSI EXO2 water quality sonde (Hoskin). The flow rate and head development were measured over time during the experiment.

## 3 RESULTS AND DISCUSSION

### 3.1 Water Quality

Table 2 shows the quality of Lake Caron water samples collected from station 4 over the period of July-Sept. 2016.

Table 2: Quality of Lake Caron water during July-Sep, 2016

| Parameters | Values |
|---|---|
| TP (μg/L) | 32-50 |
| COD (mg/L) | 30.8-46 |
| TN (mg/L) | 0.82-1.4 |
| Chlorophyll (μg/L) | 8-3.4 |
| BGA-PC (μg/L) | 0.12-0.7 |
| pH | 7.2-6.8 |
| Turbidity (NTU) | 10.4-32.4 |
| ORP (mV) | 199-351 |
| Temp (°C) | 22-29 |
| OD (mg/L) | 7-7.9 |

The general water quality parameters used to define the tropical state of a lake are TP, COD, chlorophyll a concentrations and transparency (secchi disk). TP concentration in the lake water (32-50 μg/L) was within the criteria set for eutrophic lakes (30-100 μg/L) by MDDEP and its concentration in the surface water must be less than 30 μg/L in order to control algal growth and protect aquatic organisms (MDDEP). The total dissolved phosphorus (TDP) concentration was between 7-9 μg/L. The pH (6.3-7.4) and DO (7.0-7.9 mg/L) were within the proposed limit to protect the aquatic organisms. The COD must be less than 20 mg/L for unpolluted surface water (Chapman and Kimstach 1996) and it was higher for the lake water. The chlorophyll a concentration in the lake water was close to the lower limit (8-25 μg/L) set for the eutrophic lakes by the MDDEP.

## 3.2 Filtration Tests

### 3.2.1 Filtration of lake water alone

A total of 5 filtration experiments with lake water alone were executed with different filter combinations. However, only the experiments with the best results obtained are discussed here. In this test, a combination of three TE-GTX 300 (110 µm) and a TE-GTN 350 A (75 µm) was used for filtering lake water with an initial turbidity of 14.9 NTU, TP of 34 µg/L, TCOD (total COD) of 36 mg/L and TN of 0.9 mg/L. The top and bottom filters, and some times the whole filter combination, were replaced by new filters due to clogging during the experiment. The experiment was run for 5 days continuously and showed significant removal for turbidity (97%), TSS (96%), TP (64%), COD (33%), chlorophyll a (84%), BGA-PC (91%). The TP concentration in the filtered water was about 12 µg/L, which is close to the recommended level for TP in surface water needed for aquatic organisms (MDDEP 2012). About 90% of the turbidity was removed within a day and that was reflected in the removal of TP as well (Figure 2a). TP removal was found to be directly related to the turbidity removal, as the TP in the lake water was mainly associated with suspended particles including fine sediments, particulate organic matter and algal biomass. Thus, removing turbidity or suspended particles by filtration can significantly reduce the TP level in the surface water (Sarma et al. 2016).

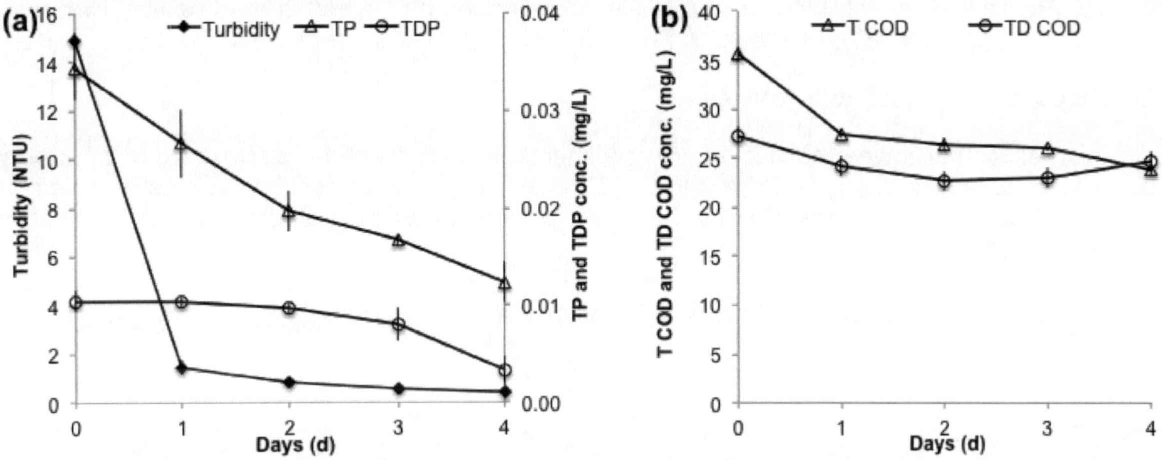


Figure 2: Removal of (a) Turbidity, TP, &TDP and (b) TCOD & TDCOD during filtration using 3xTE-GTX (110 µm) + 1x TE-GTN (75 µm) filter combination.

Figure 2b shows the TCOD concentration in the filtered water over the experimental period. As can be seen in Figure 2b, a reduction in TCOD was observed initially (within 24 h) because of the significant removal of algal biomass and suspended particles that contribute to the turbidity. Thereafter, TCOD concentration remained the same or changed slightly as the COD was mainly due to dissolved organic forms and thus resulted less COD removal (33%) by filtration. Similar to COD, the TN concentration decreased (by 25%) on the first day, and then remained almost the same throughout the experiment as TN was mainly in the dissolved form, which is difficult to remove by filtration (Figure 3a). No removal was observed for $NO_3^-$ as most of the $NO_3^-$ was in the dissolved form in the lake water. Besides, an increase in total and dissolved nitrate concentrations ($NO_3^-$ and $DNO_3^-$, respectively) was observed over time. This could be due to the oxidation of other reduced nitrogen species ($NO_2^-$ and $NH_4^+$) in the water. During filtration, the ORP increased from 299 to 415 mV due to the dynamic condition in the tank water, created by mixing and the filtration process. The increased ORP and high DO might have helped the oxidation of reduced chemical species in the water. Due to some analytical variability, dissolved nitrate ($DNO_3^-$) concentrations were found, to be sometimes, slightly higher than the total $NO_3^-$. Figure 3b shows the removal trend for chlorophyll a and BGA-PC over time. Both chlorophyll a and BGA-PC concentrations were found to decrease, gradually. Another experiment with the same combination of filters with lake water having a high initial turbidity of 32 NTU showed significant removal for TP (59%), turbidity (98%),

TCOD (41%), chlorophyll a (77%) and BGA-PC (93%). However, due to the high initial turbidity, more filters were used due to fast clogging.

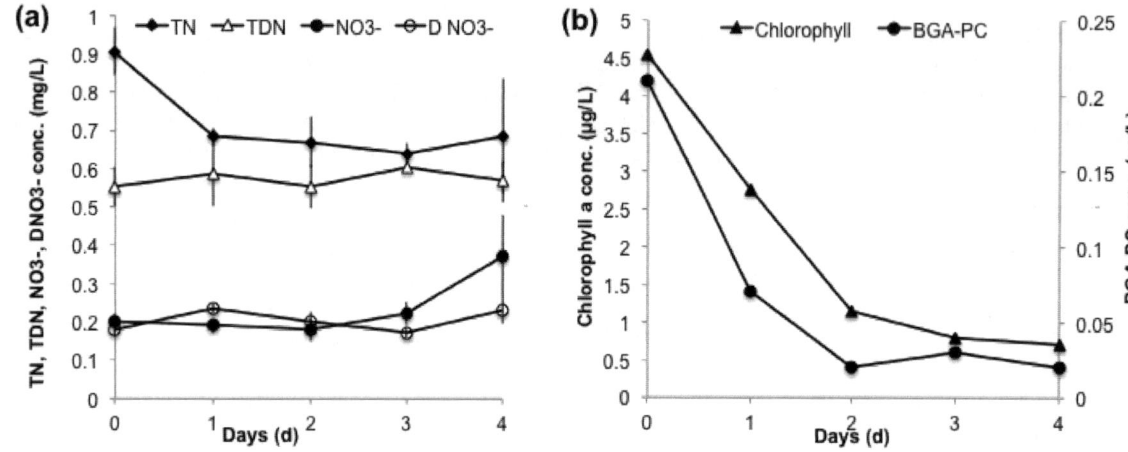


Figure 3: Removal of (a) TN, TDN, $NO_3^-$, & $DNO_3^-$ and (b) chlorophyll a and BGA-PC during filtration using 3xTE-GTX (110 µm) + 1x TE-GTN (75 µm) filter combination.

3.2.2 Filtration of lake water with sediments

Two filtration experiments were conducted with sediments. For the first test, approximately 3.8 kg dry sediment were added into the tank containing 300 L of lake water, which resulted in a solid content of about 1.25%. The filtration unit was idled for 3 days to settle re-suspended particles, after which filtration was started and continued for 19 days.

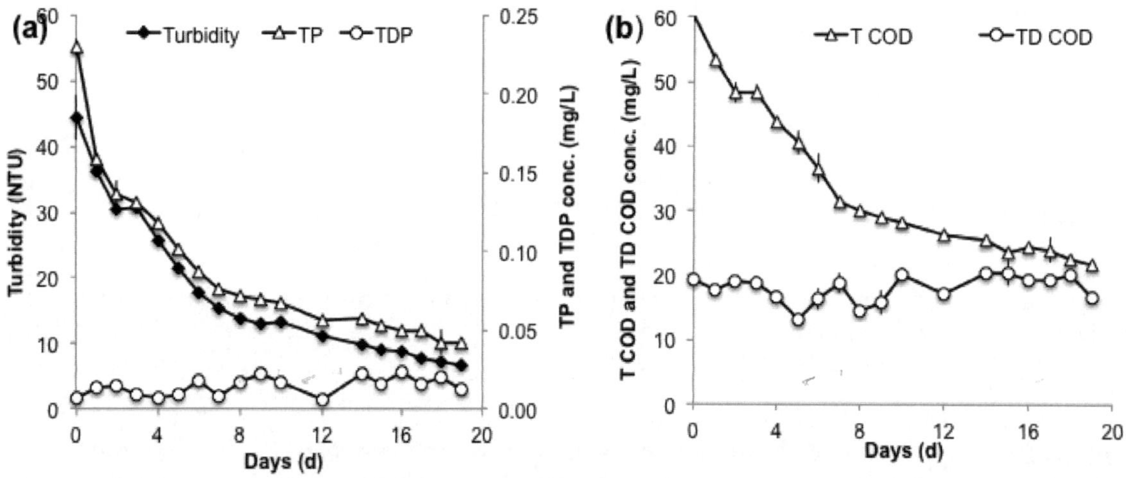


Figure 4: Removal of (a) Turbidity, TP, &TDP and (b) TCOD & TDCOD during filtration using 3xTE-GTX (110 µm) + 1x TE-GTN (75 µm) filter combination.

The initial turbidity, chlorophyll a and BGA-PC concentrations resulting from sediment addition were 202 NTU, 19.9 µg/L, 1.26 µg/L, respectively. Three days of settling allowed significant reduction in turbidity (49.8 NTU), chlorophyll a (3.75) and BGA-PC (0.21µg/L). The initial concentration of TP, TCOD, TN and turbidity in the lake water alone were 0.032, 38.5, 0.76 mg/L and 12.4 NTU, respectively. Addition of sediment resulted in low DO (4.13 mg/L) and pH 5.5. The filter combination used for the filtration test included three TE-GTX 300 (110 µm) and one TE-GTN 350 A (75 µm). Because of the high turbidity, the filters clogged rapidly and were replaced by a new set of filters with the same combination. Six sets of filters were used for the entire test. The test results observed for turbidity, TP and COD are given in

Figure 4a-b. The initial concentrations of TP (0.25 mg/L) and COD (60 mg/L) were relatively high due to the fine suspended sediment particles and particulate organic matter in the overlaying water, resulting from the sediment resuspension. As Figure 4a shows, the turbidity was reduced gradually and about 52% and 85% removals were achieved in 5 and 19 days, respectively. The TP removal trend was found to be similar to turbidity, which indicates that more than 70-75% of the P was in the particulate form associated with fine sediment particles, algae, and particulate organic matter. The removal mechanism for turbidity and TP includes both filtration and gravity settling. However dynamic conditions in the tank hindered settling of suspended particles, and therefore removal could be mainly due to filtration. The TDP concentration in the lake water varied between 7-23 μg/L, and its increase sometimes during the experiment could be due to either sediment phosphorus release or microbial degradation of settled organic matter (dead algae and sedimentary organic matter) in the sediment.

Figure 4b shows the removal of total and dissolved COD during the filtration test. Filtration was found to effectively remove particulate forms of organic matter, and there was a good correlation between TCOD and turbidity removal. The concentration of dissolved COD was almost the same throughout the experiment and indicates that filters alone cannot remove dissolved forms of COD, and phosphorus. Using filters incorporated with organic rich sediments could enhance the removal of dissolved forms of COD and P through adsorption on the sediment.

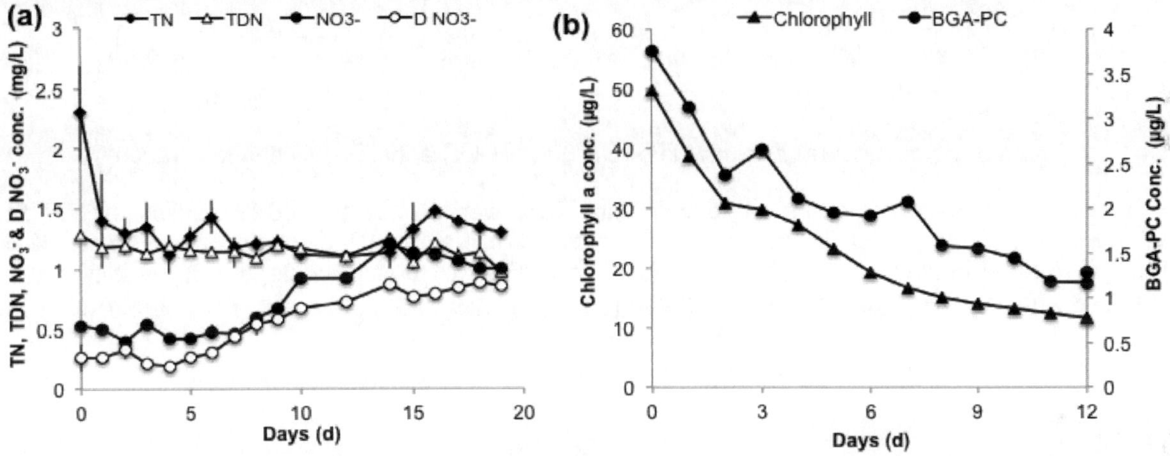


Figure 5: Removal of (a) TN, TDN, $NO_3^-$, & $DNO_3^-$, and (b) Chlorophyll a & BGA-PC during filtration using 3xTE-GTX (110 μm) + 1x TE-GTN (75 μm) filter combination.

Figure 5a shows the removal trends for TN, $TNO_3^-$, chlorophyll a and BGA-PC during the filtration test. As discussed before, the majority of TN was in the dissolved forms as $NO3^-$, $NO_2^-$ and other nitrogen forms. Like TN, $NO_3^-$ was mainly in the dissolved forms and therefore, no significant removal was observed for both TN and $NO_3^-$. A gradual increase in $NO_3^-$ concentration was observed (Figure 5a) and this could be due to the oxidation of reduced forms of nitrogen ($NO_2^-$ and $NH_4^+$) in the lake water. Chlorophyll a and BGA-PC removal trends during filtration are given in Figure 5b. Due to some technical problems with the YSI probe, the chlorophyll and BGA-PC concentrations were not monitored between days 12 to19. About 66% chlorophyll *a* and 78% BGA-PC removal was achieved by the first 12 days. Overall, turbidity, TP, and TCOD removal were 85%, 82% and 65%, respectively.

Another filtration test with sediment (solids content of 1.15%) was performed under the same conditions as mentioned before with different filter combinations as follows: two TE-GTX 300 (110 μm) and three TE-GTN 350 A (75 μm) (day 0), two TE-GTX 300 and three TE-GTN 300 (90 μm) (day 1), three TE-GTX 300 and 1 TE-GTN 300 (day 3) and one TE-GTX 300 and three TE-GTN 300 (day 9). A significant reduction in turbidity was observed when the number of filters with small opening sizes (75 and 90 μm) was increased. Using more filters with a smaller pore size could capture more solids and minimize carry over of undersized particles through the filters. However, these kinds of filter combinations are susceptible to fast clogging. Filters with a large opening size (110 µm) were less liable to clogging. However they were not very effective towards fine particle capture. Figure 6a-b shows the removal trends for turbidity, TP, and COD during the filtration. Actual concentrations of TP, COD, TN and turbidity in the lake water were 0.04, 32, 0.7 (mg/L) and 2.6 (NTU), respectively. Addition of sediment resulted in very high turbidity, and chlorophyll a concentration in the water. Two days of settling significantly reduced turbidity, chlorophyll a, and BGA-PC concentrations in the lake water to 46.5 NTU, 7.44 µg/L and 0.52 µg/L, respectively.

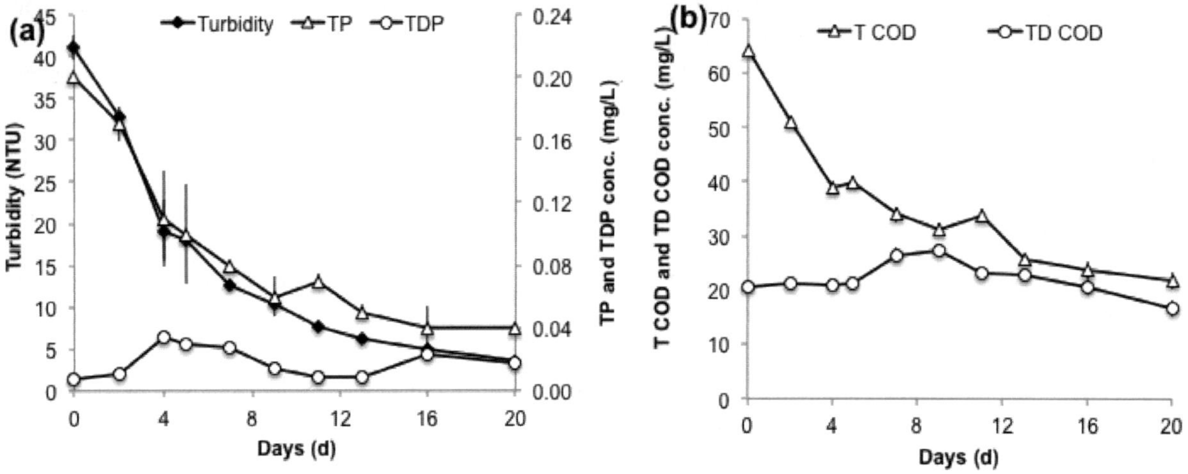


Figure 6: Removal of (a) turbidity, TP, TDP & (b) TCOD and TDCOD during filtration

The maximum removal obtained for turbidity, TP and COD were 91, 80 and 66%, respectively. TP and COD removal trends were similar to turbidity as they were mainly in the form of particulates, which contribute to turbidity in water. Significant reduction in TCOD was observed along with turbidity until all particulate organic matter was removed by filtration.

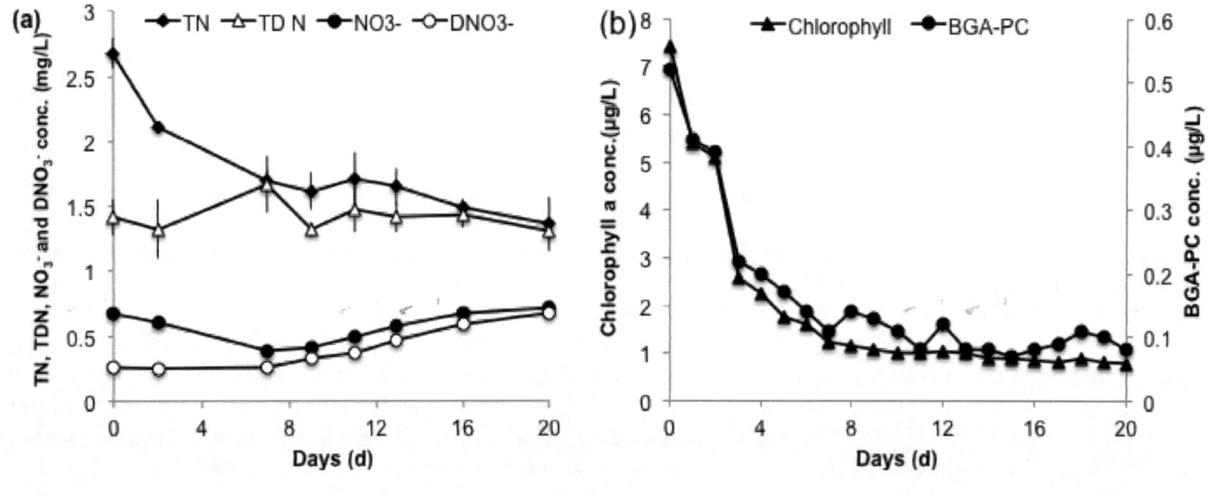


Figure 7: Removal of (a) TN, TDN, $NO_3^-$, & $DNO_3^-$, and (b) chlorophyll a & BGA-PC during filtration.

Figure 7a-b shows removal of TN, $NO_3^-$, chlorophyll a and BGA-PC during the filtration test. The ORP increased from 306 to 525 mV over time. Overall percent removals obtained for turbidity, TP, TCOD, chlorophyll a, BGA-PC were 93, 80, 66, 89, and 93 %, respectively.

In general, filtration using selected non-woven geotextile filters improved lake water quality in terms of turbidity, suspended solids, TP and chlorophyll a concentration in the filtered water. However, the filters are not effective for removing nitrogen (as TN, and $NO_3^-$) as it mainly exists in its dissolved forms in the lake water.

## 4 CONCLUSIONS

The test results from this study have shown that non-woven geotextile filters can be used for improving surface water quality. The geotextile filters allowed reduction of the TP and chlorophyll a concentration in Lake Caron water to the limit set for oligotrophic lakes by MDDEP. Since phosphorus in overlaying water is mainly in the particulate form, removing suspended particles by filtration could lower TP concentration and improve the aesthetics of the water. The geotextile filters resulted in the removal of 85-98% turbidity, 59-82% TP, 67-89% chlorophyll a, 78-93% BGA-PC and 33-66% COD depending on the initial water quality. Filter combinations with more (2-3) smaller size filters are more effective for suspended solids removal. This method is environmental attractive since it does not add any foreign materials in the lake water, and is simple and feasible to operate onsite. Since the filtration process generates clogged filters, reuse of the used filters must be addressed in the future work. In situ filtration tests will be performed in the next phase of this project.

## ACKNOWLEDGMENTS

The authors thank NSERC and Concordia University for financial support for this project. The authors are also grateful to their industrial partner, Titan Environmental Containment Ltd., for supplying geotextiles and providing financial support for this project.

## REFERENCES

Chislock, M.F., Doster, E., Zitomer, R.A., Wilson, A.E. 2013. Eutrophication: Causes, Consequences, and Controls in Aquatic Ecosystems. *Nature Education Knowledge*, **4**(4): 10.

Cooke, G. D., Welch, E. B., Martin, A. B., Fulmer, D. G., Hyde, J. B., Schrieve, G. D. 1993. Effectiveness of Al, Ca, and Fe Salts for Control of Internal Phosphorus Loading in Shallow and Deep Lakes. Paper Presented at the *Proceedings of the Third International Workshop on Phosphorus in Sediments,* 323-335.

Correll, D.L. 1998. The Role of Phosphorus in the Eutrophication of Receiving Waters: A review. *Journal of Environmental Quality*, **27**: 261-266.

Franks, C.A., Davis, A.P., Aydilek, A.H. 2012. Geosynthetic Filters for Water Quality Improvement of Urban Strom Water Runoff. *Journal of Environmental Engineering*, **138**(10): 1018-28.

Inoue, T., Fuke, M., Mulligan, C.N., Uehara, K. 2009. In-situ Removal of Contaminated Suspende Solids from a Pond by Filtration. *Ecological Engineering*, **35**:1249-1254.

Karim, R.A., Mulligan, C.N., Fukue, M. 2012. Preliminary Evaluation of a Sediment Resuspension Technique for Reduction of Phosphorus in Lake Water. Fifth ASTM Int. Symposium on Contaminated Sediments: Restoration of Aquatic Environment, Montreal, QC, Canada: 45-65.

Lewis, W.W., Wurtsbaugh, W.A. 2008. Control of Lacustrine Phytoplankton by Nutrients: Erosion of the Phosphorus Paradigm. *International Review of Hydrobiology*, **93**: 446-465.

MDDEP (Ministère du Développement Durable de l'Environnement et des Parcs. 2012. Qualité de l'eau de surface. Accessed in February, 2017.

Mulligan, C.N., Davarpanath, N., Fukue, M., Inoue, T. 2009. Filtration of Contaminated Suspended Solids for the Treatment of Surface Water. Chemosphere, **74**: 779-786.

Mulligan, C.N., Ramalingaih, T.N., Fukue, M. 2011. Removal of Suspended Solids and Phosphorus from a Eutrophic Lake. Environmental Pollution and Remediation, Ottawa, ON, Canada.

Quaranta, J. D., Tolikonda, R. 2011. Design of Non-woven Geotextiles for Coal Refuse Filtration. *Geotextiles and Geomembranes,* **29**(6): 557-566.

Reddy, K., Fischer, M., Wang, Y., White, J., James, R.T. 2007. Potential Effects of Sediment Dredging on Internal Phosphorus Loading in a Shallow, Subtropical Lake. *Lake and Reservoir Management*, **23**(1); 27-38.

Reeders, H.H., Boers, P.C.M., van der Molen, D.T., Helemerhorst, T.H., 1998. Cyanobacterial Dominance in the Lakes Veluwemeer and Wolderwijd, The Netherlands. *Water Science and Technology,* **37**(3): 85-92.

Sarma, S., Mulligan, C.N., Kim, K., Veetil, D.P. 2016. Use of Nonwoven Geotextiles for Removing Nutrients and Suspended Solids from a Eutrophic Lake. Environmental Speciality Conference , CSCE, London, ON, Canada: 1-10.

Schindler, D.W. 2006. Recent Advances in the Understanding and Management of Eutrophication. *Limnology and Oceanography*, **51**: 353-363.

Tota-Maharaj, K., Grabowiecki, P., Babatunde, A., Coupe, S.J. 2012. The Performance and Effectiveness of Geotextiles within Permeable Pavements for Treating Concentrated Storm Water. *International Journal of Water Technology*, **2**: 207-216.

Xie, F., Wu, F., Liu, G., Mu, Y., Feng, C, Wang, H. 2014. Removal of Phosphate from Eutrophic Lakes through Adsorption by In-situ Formation of Magnesium Hydroxide from Diatomite. *Environmental Science and Technology*, **48**: 582-590.

Leadership in Sustainable Infrastructure
*Leadership en Infrastructures Durables*

Vancouver, Canada
May 31 – June 3, 2017/ *Mai 31 – Juin 3, 2017*

# MEMORIAL UNIVERSITY OF NEWFOUNDLAND ON-CAMPUS CYCLING STUDY

Kashif Rasheed[1,2], and Cynthia A. Coles[1]
[1] Memorial University of Newfoundland, Canada
[2] kashifraheedarain@gmail.com

**Abstract:** Cycling meets the criteria of strong sustainability and is one of the most ideal modes of transportation because it is health promoting, low cost, low pollution emitting, and low greenhouse gas emitting. The purpose of this study was to investigate what was needed to increase the participation in bicycling at the university in St. John's and thereby help address issues such as health, sustainability, climate change, and biodiversity loss. This objective was achieved by, in sequence, studying places that have more advanced cycling cultures and how they achieved this, interviewing experienced cyclists at Memorial University of Newfoundland (MUN), surveying members of the university community, polling students and employees within the Faculty of Engineering and Applied Science since this is where the largest proportion of campus cyclists are located, and considering if there were any existing places in the engineering building where indoor bicycle parking could be an economical option. It was concluded that important improvements to support cyclists to the campus would include indoor secure bicycle parking, facilities to wash up, especially for those commuting long distances, more outside bike racks, and separate bike lanes with lighting for bicycling after dark. A few deficiencies were discovered in the survey that could be improved upon if future work is undertaken or other universities decide to follow this approach.

## 1 INTRODUCTION

Much has been written on the topic of sustainability and it is necessary to distinguish between weak sustainability, which is most prevalent in the literature, and strong sustainability. The former is especially related to incremental changes such technological improvements and increases in efficiency that will slow the rate of environmental degradation, and the latter is focused on transformative changes and low-consumption lifestyles that will bring the world to a place where resource constraints and the wellbeing of humans and other species that share the planet are respected (Lorek and Fuchs, 2013).

Using renewable energy, renewable resources and bicycles for mobility are all elements of strong sustainability. Bicycling is one of the most environmentally friendly forms of transportation and while it can make a great contribution to the health of its user it does not damage the health of others (Gatersleben and Appleton, 2007). This is important because environmental justice requires that those engaged in active transportation, such as walking and cycling, be given higher priority than those who employ more polluting means of transportation, and environmental justice is the most important aspect of strong sustainability (Pawlowski, 2008). A carbon tax beginning in Canada in 2018 should help address this current inequity.

Transportation along with energy and agriculture are the main areas where the habits of sustainability need to be introduced (Zhang et al. 2015). Bicycling is a key component of sustainable mobility and it is an area where progress has been made due to the persuasion of municipalities and universities though overall universities have played the greater role in introducing, maintaining and promoting bicycle use (Beuhler and Pucher, 2012). Universities need to be at the forefront of environmental change for their role in promoting knowledge and the large number of students they attract can be highly adaptable and forward thinking. Therefore this type of study might be considered relevant at other universities that have not yet done as much as they could have to promote active transportation.

The interest in cycling has been growing among governments and planners in recent years and the benefits of reduced traffic congestion and emissions have been a key reason. Electric bicycles or e-bikes are also seen as a better option than automobiles and are considered a potentially innovative means to reduce air pollution while being less physical demanding to use (Wolf and Seebauer, 2014).

Public health organizations and medical journals are recommending cycling to improve health and avoid the harmful impacts of car use including an excessively sedentary lifestyle in many developed countries (Giles-Corti et al., 2016). Many cities have been introducing infrastructure, programs and policies to promote cycling including providing separate bicycle paths for off-street and separate lanes for on-street cyclists. From 1990 to 2008 Washington DC tripled the use of bicycles after expanding its cycling infrastructure and Denmark and the Netherlands have increased bicycling among their populations (Beuhler and Pucher, 2012; Thomas and DeRobertis, 2013).

Pucher et al. (2011) studied nine cities in the United States and Canada to observe bicycling trends and found in Montreal, Vancouver, and Toronto the percentage of university students who bicycled were 10.7%, 10.1% and 5.2%. Large cities, city centers and university campuses had the highest bicycle densities and the cyclists were most often from the middle and upper middle classes (Pucher et al., 2011). The cities that had actively promoted bicycling ended up seeing the largest numbers of cyclists (Pucher et al., 2011).

Memorial University of Newfoundland (MUN) has taken some steps to help the bicycling community and encourage new bicycle users though there is still much more needed. In this paper, steps that have been taken and that can be taken to promote the bicycling culture at MUN are discussed.

## 2 BICYCLING BENEFITS

Bicycling is reliable, effective, faster over short distances, very convenient during heavy traffic, consumes almost zero energy, requires some of the lowest maintenance costs, improves the health of the user, causes no adverse effects on the health of others, and could significantly reduce GHG emissions (Gonzalez et al., 2015; Gatersleben and Appleton, 2007). The costs of operating all motorized vehicles in Canada are $14-36 billion per year and include environmental damage, deterioration of people's health, and GHG emissions (Börjesson and Eliasson, 2012). Cycling can also prolong health and cyclists may be able to continue into their 80's, 90's and beyond while inactive seniors will be more likely to spend the end of lives in wheelchairs. In the UK it was concluded that bicycling could be the most reliable and affordable form of transportation for most sectors of the population (Gatersleben and Appleton, 2007).

Compared to bicycles, motorized vehicles over their life cycles produce more air, noise, water and visual pollution. Carbon monoxide has a local impact while methane, ozone and carbon dioxide emissions have large global impacts. Engine acceleration, tire/road contact, braking, honking, and theft alarms are the sources of annoyance and discomfort while bicycling can create a more peaceful environment (Campbell and Wittgens, 2004). Bicycles can be a good option for tourists and do less damage to scenic, conservation and historically important areas. They can also help people adopt slower paced and higher quality lifestyles (Zhang et al., 2015).

Busses may also be more expensive than bicycles as a monthly transit pass in Montreal is $100. The Canadian Automobile Association estimates that maintaining a car costs about $6200 annually excluding fuel and parking costs (Willis et al., 2013). Bicycle transport also can reduces traffic congestion and delay. Every motor vehicle trip made can increase the time in traffic experienced by other vehicles which can be especially costly if goods are delayed. In Ontario congestion to cost approximately $6.4 billion annually in 2001. Without alternatives to the motor vehicle and increased investment, the time an average commuter takes in the Greater Toronto Area could increase 50 percent by 2021 (Campbell and Wittgens, 2004).

## 3 CYCLING IN SELECTED CITIES AND IN CHINA

### 3.1 Copenhagen, Denmark

Over the past 75 years' bicycles have been more dominant in Copenhagen than cars, since 2000 the city has been actively promoting cycling, in 2007 they set the goal of becoming the "best city for cyclists" and every two years they audit their progress towards this goal (Gössling and Choi, 2015). In Copenhagen there are plans to expand the network of three lane fast track "bicycle highways" which are able to narrow the travel speed gap between cyclists and automobile drivers (Gössling and Choi, 2015). Gössling and Choi (2015) considered 13 factors including environmental costs (air pollution, climate change, and noise pollution), health costs (fitness, fewer sick days, increased longevity and safety), vehicle and infrastructure costs (vehicle operation and road deterioration), and travel time costs and showed that bicycling was clearly the more economical alternative.

### 3.2 Seville, Spain

Seville, Spain, with a city proper population of 700,000 (or 1,500,000 in the metropolitan area) is impressive for its rapid recent growth in bicycling. Prior to 2006 less than 1% of the people cycled and there were only 12 km of bike paths. However, in 2006 a survey predicted that the cycling population could reach 89,000 (or greater than 10%) and this began infrastructure improvements for cyclists. By 2010 the bicycle network included 120 km and by 2013 it reached 164 km. In 2006 through to 2011 the millions of annual bicycle trips were 3.07, 6.48, 9.29, 14.0, 15.9 and 17.1 respectively and in 2013 Seville was considered the fourth most bicycle friendly city in the world (Marqués et al., 2015).

The proportion of cyclists among the Seville University population of 70,000 is now double that for the city as a whole and this is due partly to the large student numbers. Even before 2000 the university had provided enclosed and secure bike parking areas at all of its buildings for a total of 1,618 bicycles and free parking for 771 bicycles. From 2006 onward the university developed initiatives that were complimentary to those of the city of Seville including a bike lending program with 400 bicycles (Marqués et al., 2015).

### 3.3 St. John's, Canada

The City of St. John's has a Cycling Master plan in which bicycling has been accepted and recognized as a part of the transportation system, giving cyclists more status and legal support (MacDonald, 2009). Currently St. John's has very few on-road and off-road bicycle lanes and the province of Newfoundland and Labrador has the highest rates of heart disease, cancer, diabetes and stroke in Canada as explained in the Newfoundland and Labrador Department of Health and Community Services 2006-2008 Strategic Plan. The Cycling Plan is a healthy solution and would help fulfill residents' needs for physical and recreational activity (MacDonald, 2009). Since the province is currently experiencing financial hardship, promoting cycling could be particularly advantageous at this time.

### 3.4 China

China is one of the leaders in cycling. In the 1980's in the city of Fuzhou, with a population of 1,460,000, the bike modal share was 65.91%. Around the mid 1990's bicycle use peaked and then declined. However, more recently due to traffic congestion, accidents and environmental concerns, there is renewed interest in cycling and especially in electric bikes and bike sharing programs. From 1998 to 2010 the number of e-bikes grew from 30,000 to 30 million, and bike sharing programs have been linked to public transit so that each can use the same smartcard and bike share users can receive a discount. So far it appears that non-profit bike share may be the most successful approach (Zhang et al, 2014).

## 4 CURRENT BICYCLE FACILITIES AT MUN

MUN provides some basic facilities for cyclists. There are a total of 54 bike racks most of which are in sheltered areas, and 18 of which were installed in 2015. There are five bike lockers located between the Science and Chemistry buildings and they are available for rent by semester. Then there is the volunteer operated MUN Bike share program that aims to provide help to cyclists by lending tools and providing bikes for rent (personal communication, Mrs. Toby Rowe, Sustainability Co-ordinator, Department of Facilities Management, MUN, 2017).

## 5 MUN CYCLISTS WHO WERE INTERVIEWED

One science professor, one engineering professor and one administrative staff specialist in science were interviewed for their views on cycling conditions at MUN and what types of improvements could be made to make the university more cyclist friendly.

### 5.1 Dr. Len Zedel

Dr. Len Zedel, Associate Dean of Science (Research) has bicycled to MUN for 25 years all year round. The three most important improvements for him would be providing indoor parking facilities, more bike racks around the campus and wash-up facilities. Due to the unpredictable weather in St. John's outside bicycle racks, even the sheltered ones, are not very weather friendly and he suggested that some of the space in the underground parking in the Earth Science building could incorporate a fenced bicycle parking facility with a security camera for safety. Creating bike lanes and proper lightning around would help the existing cyclists and attract more people. There are virtually no shower facilities around the campus for cyclists and it gets especially messy after cycling in extreme weather. Snow and rain is a big challenge if there are no wash-up facilities. He also suggested that the showers in the Physical Education building for the gym users should be made available to cyclists. In a perfect world there should be cycling storage and lockers in the building by the showers so that cyclists could conveniently store their cycles and clean themselves up.

### 5.2 Mr. Kerry Hiscock

Mr. Kerry Hiscock, Facilities Coordinator in the Department of Earth Sciences, bicycles 8 km from Mount Pearl to the university every day all year round. It takes him 25 minutes one way, and he has been doing it for 27 years. He avoids rush hour traffic by arriving before and leaving after the majority of the other road users. The main problems he sees are not having enough bicycle storage and parking facilities for students and employees which creates bike security issues. He has had two of his bikes stolen from the campus. He is also concerned about bike lanes and poor lighting on campus and feels that cycling on and around campus is the most dangerous part of his commute.

### 5.3 Dr. Tariq Iqbal

Dr. Tariq Iqbal is an Electrical Engineering professor and the biggest problem he faced when cycling to the university was not having proper bicycle parking or storage available to him. He would have to store his bike in his office but the cleaning staff were telling him to leave it outside. He thinks this issue could be solved with indoor bicycle parking while taking care not to increase the burden too much on the cleaning staff. He recommends that the university provide incentives for people who cycle to work or school as a form of promotion, such as bookstore credit, a free bike or bike contests, and free bike repairing.

## 6 THE SURVEY AND POLL

A survey of 10 questions was administered to 40 individuals at MUN and it is in the Appendix. Its main purpose was to understand the problems of cyclists and determine what enhancements could make cycling more attractive. In an effort to target cyclists who might be the best qualified to answer the questions, the survey was administered near outdoor bicycle racks at the engineering building main and east side entrances, the science building main and west side entrances, and the library main and north side entrances.

Through the process of conducting the survey and analyzing and sorting the results some weaknesses in the questions were discovered and they are mentioned briefly below along with the findings. Occasionally when an answer was not provided or did not make sense then these responses were omitted from the analysis. Where questions were not answered exactly as intended, some changes were made in analysing the results to account for this variation, as explained below.

The bicycle poll was a single question directed at approximately 1000 undergraduate students, 650 graduate students, and 150 staff and professors or a total of 1,800 people in the Faculty of Engineering and Applied Science. The question was simply whether they would like to have indoor bicycle parking facilities in the engineering building.

### 6.1 Survey Results

Question 1 identified the participants and they consisted of 29 undergraduate students, 8 graduate students, and 3 faculty and staff. Of the total, 35 were male and 5 were female. From question 2 it was found that 92% of the participants owned a bicycle.

In answering question 3 about how they usually travelled to the university if only one mode of transport was selected, it was weighted 100%, if two modes were selected they were each weighted 50%, and if three modes were selected they were each weighted 33.3%. Only the first five categories were included as there was no explanation as to what an "other" category might have been. Among the five modes bicycling represented 40.6%, public transit represented 23.6%, walking was 18.5%, automobile was

17.3%, and taxi was 0%. These responses showed that a high proportion of those completing the survey were cyclists.

The results of question 4 indicated that 37% of those surveyed lived less than 1.5 km from MUN, 29% lived between 1.5 and 3 km from MUN, 10% lived 3 to 5 km away, and 24% had to commute more than 5 km. Some people were not sure of the exact distance and one improvement would have been to provide a map of St. John's with the question to assess accurate distances.

In answering question 5 where people had to estimate what percentage of the time they bicycled to MUN, 25% said never, 32.5% said up to 20% of the time, 35% said up to 50% of the time, and 7.5% said they bicycled up to 100% of the time. These statistics correspond well with bicycling being the mode of transportation 40.6% of the time (in question 3) for this group of individuals. It was noted (in question 2) that only 8% did not own a bicycle so this meant that 17% did not use their bicycles to travel to MUN.

Question 6 asked the participants to rank in order the three most important obstacles to cycling. Of the 40 respondents 18 ranked their choices, many selected three choices but did not rank them, and a small number considered only one of the choices important. A first choice was weighted 50%, a second choice was weighted 33.3%, and a third choice was weighted 16.7%. Unranked responses were equally weighted (as in question 3).

The most important reason for not bicycling was unpredictable weather at 26.6%, followed by poor facilities for washing up at 25.4%, living too far from MUN at 17.8%, fear of bicycle theft at 17.5%, cloths they would normally wear were not suitable for bicycling at 9.4% and not having a bicycle at 3.3%. Under "other", one person mentioned traffic. Leaving out "traffic" as one of the choices was unfortunately a weakness in the study. Some people who exercise indoors on stationary bicycles are afraid to use the main roads because of traffic and in this way motorists are depriving potential outdoor cyclists and their enjoyment of fresh air.

Question 7 asked what problems cyclists faced on campus. The answers were weighted as was done for question 3. Lack of bicycle lanes was most important at 30%, inadequate bicycle parking was 25.8%, bicycle theft was 24.6%, and having to park outdoors in winter was 19.6%.

In question 8 those surveyed were asked to rank, or in some cases respondents simply indicated, the improvements that MUN could make to promote cycling. Providing indoor parking facilities was the most important at 31.2%, and this was followed by washing up facilities at 18.4%, more bicycle parking racks at 18%, bicycle theft prevention at 14.0%, education on bicycle safety at 9.6%, and develop a network of bicycle pathways on campus at 8.8%.

If lack of campus "cycling lanes" was the most important deficiency in question 7 why was "developing a network of bicycle pathways on campus" the least important improvement in question 8? This might have been due to inconsistent wording among the two questions. Cycling lanes might be thought of as following the main access roads. A network of bicycle pathways might be thought of as including those along the main access roads as well as many interconnecting pathways on campus, which some may have thought was going beyond what was the most essential. There is also the possibility that in completing the survey some may have decided to focus on what they thought might have seemed to be the easiest and least expensive changes to make.

Question 9 asked individuals what they felt were the best benefits they were getting from cycling and 66.7% said health, 19.4% said saving money, and 13.9% said helping the environment. Ranking was not asked for in this question and people selected either one, two or all three of the benefits. Education on how bicycling improves health might be a good way to promote it.

Results from question 10 revealed that 56.8% of cyclists said that having high quality, secure and weather safe bicycle parking on campus would increase their frequency of bicycling to MUN, 35.1% said it might increase their frequency of bicycling, and 8.1% said it would not make them bicycle to MUN more often.

### 6.2 Poll Results

Of the 1,800 people polled, 36 or 2% said they would like to have indoor bicycle parking in the engineering building. Three people identified themselves as undergraduate students, 10 identified themselves as graduate students, 6 were employees, and it was unclear who the remaining 17 people were. It is expected that an email poll would get a lower percentage of responses, so this poll shows good support for indoor bike parking. It was mentioned earlier that the number of enclosed bicycle parking areas at Seville University was more than double the number of free bicycle parking places and they had been spread throughout the campus at each building.

## 7 RECOMMENDATIONS AND POTENTAIL BICYCLE PARKING FACILITES

### 7.1 Recommendations

MUN has a responsibility to ensure that its students and employees are safe and the university needs to provide the proper facilities and services just as the motorized parking facilities are properly managed and enforced. The Campus Enforcement and Patrol is an administrative body with its own regulations, penalties, full time staff and budget. The cycling society at MUN is less well attended and under-serviced. The survey was a useful tool in pointing out some of the problems and potential improvements.

These improvements include better bike parking facilities, proper storage with security cameras, wash-up facilities with storage nearby, and cycling lanes on campus with proper lighting. However, indoor parking facilities was the most emphasized improvement. Winter cycling can be very demanding with the snow, water, salt and the daily cycles of freeze-thaw that are rapidly wearing out bicycles and extra maintenance and care is required to keep the bikes functional.

Cyclists must protect themselves from the slush and wet weather and need good storage for bicycle clothing and wash-up facilities to clean up afterwards. Many bicycles that are stolen in St. John's, are stolen while parked at MUN. Secure indoor parking to keep bicycles safe and dry and provide a place for maintenance would be helpful and would attract more participants. Currently many cyclists are left to store their bikes in slush and their clothing in places where they are not meant to be stored. Showers and change facilities can also strongly attract cyclists and especially those who have a long commutes.

The showers at the gym in the Physical Education building might be brought into use for cyclists and a recommendation of bicycle racks or storage near the gym showers was made by quite a few respondents of the survey. There are showers available to the male and female professors and staff in the engineering building but it would be helpful if similar facilities could also be provided for the student population.

### 7.2 Potential bicycle parking locations 1 and 2

There are some unused spaces on the first floor of the engineering building, mostly under the staircases, that were assessed for bicycle parking but the City of St. John's Fire Chief, Mr. Mike Maher indicated that they would require blocking access from the staircase by putting a wall of the same material as other walls, and having a separate exterior access door. The pathways leading to the parking and storing facilities should be clear and not used for heavy foot traffic.

One of these locations was in the engineering building by exit door #25 on the north east corner of the building. There is some space here for the area to be converted into an indoor bicycle parking facility but the modifications required would be costly. On the other hand the general area would make an ideal location for bicycle parking. The floor area available is 16 feet by 8 feet excluding the space under the stairs, and the ceiling height is 7'-6" where an exterior entrance door would be required for this place to be used. This area is shown in Figure 1.

A second location on the northwest corner of the engineering building by exit door #2, is a mirror image of the first location, has the same dimensions, and so might be similarly modified.

Figure 1. The Engineering building at door number 25 showing outside and inside views.

### 7.3 Potential bicycle parking location 3

The third potential location in the engineering building is located on the southeast corner facing the business building. There is not heavy foot traffic here and the area is much wider than the total entrance door area and the corridor door area. The space is wide enough to put vertical wall mounted bicycle racks, on the right wall facing out or on the left wall facing in. Of the three locations this one presents a low cost solution for an area that is underutilized and would require the least amount of modifications. The location is shown in Figure 2.

This study has been brought to the attention of the Dean of Engineering, Dr. Greg Naterer and so there is a good possibility that some indoor bike parking will be considered in the engineering building. There are also future lab expansion renovations planned and this would be a good time to make room for indoor bicycle parking facilities with this new construction.

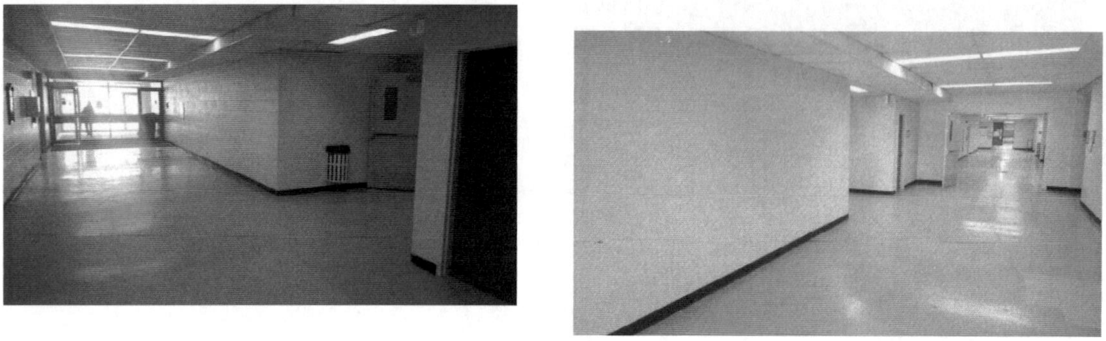

Figure 2. Inside the Engineering building at the south east entrance facing the business building.

## 8.0 Conclusions

Strong sustainability requires that resource consumption be adapted to long term resource availability. Most of the literature is focused on weak sustainability which will buy some time but will not resolve the most critical challenges. A transformation in attitudes and behaviors is required and increasing participation in bicycling is a part of strong sustainability. Some of the statistics from China and elsewhere are showing that a much higher proportion of cyclists would be possible in St. John's if they were only treated as fairly as automobile drivers and it seems clear that they should be considered an even higher priority.

The study showed that people are being held back by restrictive conditions and an outdated and expensive transportation system. Improvements which would be recommended for MUN are secure indoor bike parking which would also prevent theft and which was one of the concerns, facilities for washing up after arrival, more bicycle racks outside, and separate bikes lanes along the main access routes with lighting for travel after dark. A search though the engineering building found one less used location where indoor bicycle parking could be effectively and economically introduced.

Universities can play a key role in promoting sustainable mobility and students are likely to be highly receptive. The process outlined in this study was to survey the literature, consult with experienced cyclists, survey members of the university community and especially cyclists, and poll those who were potentially the most progressive. One benefit of the study was an increased awareness among the participants.

The automobiles on the road and their drivers are preventing people from participating in bicycling as long as the traffic policy does not fully accommodate cyclists on the road and does not make conditions safe and convenient for them. The numbers of cyclists could be greatly increased and conditions need to be made much more favorable. When a carbon tax is introduced in Canada it will help but both the introductory amount of the tax and the rate at which it will be increased are so small that it could take a long time to be really effective and other measures are needed at the same time.

**Acknowledgements**

An immense appreciation and hearty gratitude is extended to the Associate Dean of Science (Research) Dr. Len Zedel, Earth Sciences Facilities Co-ordinator Mr. Kerry Hiscock, Engineering Professor Dr. Tariq Iqbal, City of St. John's Fire Chief, Mr. Mike Maher, MUN Sustainability Coordinator Mrs. Toby Rowe, Mr. Patrick St. Croix from the Department of Health and Safety, and the Dean of Engineering, Dr. Greg Naterer for their valuable time, advice, support, encouragement and help in this project.

## References


Börjesson, M. and Eliasson, J. 2012. The value of time and external benefits in bicycle appraisal. *Transportation Research Part A,* **46**: 673–683.

Buehler, R. and Pucher,J. 2012. Cycling to work in 90 large American cities: new evidence on the role of bike paths and lanes. *Transportation,* **39**: 409–432.

Campbell R. and Wittgens, M. 2004. The business case for active transportation: the economic benefits of walking and cycling. Glouces-ter (ON): Go for Green, Google Scholar.

Gatersleben, B. and Appleton, K.M. 2007. Contemplating cycling to work: Attitudes and perceptions in different stages of change. *Transportation Research Part A,* **41**: 302–312.

Giles-Corti, B., Vernez-Moudon, A., Reis, R., Turrell, G., Dannenberg, A.L., Badland, H., Foster, S., Lowe, M., Sallis, J.F., Stevenson, M., and Owen, N. 2016. City planning and population health: a global challenge, *Lancet,* **388**(10062): 2912-2924.

Gonzalez, D.D., Nazelle, A.D, Hunter, T.C., Curto, A., Rodriguez, D.A., Mendez, M.A., Aymerich, J.G., Basagaña, X., Ambros, A., Jerrett, M., Nieuwenhuijsen, M.J. 2015. The Added Benefit of Bicycle Commuting on the Regular Amount of Physical Activity Performed. *American Journal of Preventive Medicine,* **49**(6):842–849.

Gössling, S. and Choi, A.S. 2017. Transport transitions in Copenhagen: Comparing the cost of cars and bicycles. *Ecological Economic,* **113**: 106-113.

Litman, T. 2016. Evaluating Active Transport Benefits and Costs. Guide to Valuing Walking and Cycling Improvements and Encouragement Programs. Victoria Transport Policy Institute.

Lorek, S. and Fuchs, D. 2013. Strong sustainable consumption governance – precondition for a degrowth path? *Journal of Cleaner Production,* **38:** 36-43.

MacDonald, H.M. 2009. City of St. John's. Cycling Master Plan.

Marqués, R., Hernández-Herrador, V., Calvo-Salazar, M., and García-Cebrián, J.A., 2015. How infrastructure can promote cycling in cities: Lessons learned from Seville, *Research in Transportation Economics,* **53**: 31-44.

Pawlowski, A. 2008. How many dimensions dose sustainable development have? *Sustainable Development,* **16:** 81-90.

Pucher, J., Buehler, R. and Seinen, M. 2011. Bicycling renaissance in North America? An update and re-appraisal of cycling trends and policies. *Transportation Research Part A,* **45**: 451-475.

Stutman, M. 1997. Planning for the Bicycle: Parking Issues and Facilities Planning at the University of Waterloo.

Thomas, B. and DeRobertis, M. 2013. The safety of urban cycle tracks: A review of the literature. *Accident Analysis and Prevention,* **52**: 219–227.

Willis, D.P., Manaugh, K., Geneidy, A.E. 2013. Uniquely satisfied: Exploring cyclist satisfaction. *Transportation Research Part F,* **18**: 136–147.

Wolf, A. and Seebauer, S. Technology adoption of electric bicycles: A survey among early adopters. *Transportation Research Part A,* **69**: 196–211.

Zhang, H., Shaheen, S.A. and Chen, X. 2014. Bicycle evolution in China: From the 1900s to the present. *International Journal of Sustainable Transportation,* **8:** 317-335.

Zhang, L., Zhang, J., Duan, Z. and Bryde, D. 2015. Sustainable bike-sharing systems: characteristics and commonalities across cases in urban China. *Journal of Cleaner Production,* **97:** 124-133.

## Appendix

**ON CAMPUS BICYCLE PARKING FACILITIES SURVEY**

**Name:** ____________________________________

**Gender:** ___ Male___Female

1. **Please indicate whether you are?**

___Undergraduate Student ___Graduate or PHD Student ___Faculty Member ___Staff

2. **Do you own a bicycle?**

___Yes ___No

3. **How do you usually travel to and from MUN?**

___Private Automobile ___Walk ___Public Transit ___Taxi ___Bicycle ___Other

4. **Approximately how far do you live from MUN?**

___Less than 1.5 km ___1.5 - 3 km ___3 - 5 km ___> 5 km

5. **Approximately what percentage of your trips to MUN have been made on bicycle?**

___0% ___1%-20% ___21%-50% ___More than 50%

6. **Please rank (in order of importance, 1 being most important) the three most applicable reasons why you do not ride a bicycle to MUN:**

___Do not own a bicycle ___Distance is too far ___Fear of bicycle theft

___Clothes not suited for cycling ___Poor wash up facilities ___Weather is unpredictable

___Other

7. **What problems do students face on campus who bicycle?**

___No Cycling lanes ___Outdoor parking in winter ___Less or improper parking

___Bicycle theft ___All of the above

8. **Please rank (in order of importance, 1 being most important) the three most beneficial improvements MUN can make to encourage students to bicycle?**

___Provide improved wash up facilities ___Provide indoor parking facilities

___Develop a network of bicycle pathways on campus ___Prevent bicycle theft

___Provide more bicycle parking racks ___Education on bicycle safety

9. **What is the best benefit you are getting from bicycling?**

___Environmental benefits ___Health benefits ___Money saving ___Other benefits

10. **Would the availability of high quality, Secure and weather safe parking increase the frequency with which you ride a bicycle to MUN?**___Yes ___No ___Maybe

Leadership in Sustainable Infrastructure
*Leadership en Infrastructures Durables*

**Case Study**

Vancouver, Canada
May 31 – June 3, 2017/ *Mai 31 – Juin 3, 2017*

# MODELING STRUVITE CRYSTALLIZATION IN A BATCH REACTOR USING POPULATION BALANCE MODULE COUPLED WITH 3-D CFD MODEL

Mousavi, Seyyed Ebrahim [1, 3], Rahaman, Md. Saifur [2]
[1,2] Concordia University, Canada
[3] ebrahim.mousavi65@gmail.com

Discharge of significant amounts of phosphate from nutrient-rich wastewater is a problematic issue considering the potential contamination of groundwater and eutrophication of recipient water bodies. On the other hand, phosphorus is also a limited, non-renewable resource with reserves that may be depleted in the near future. As a result, phosphorus recovery is receiving due attention as an environmental concern. The recovery of phosphorus using a crystallization technique is one proven solution for this issue. In this work, crystallization of Magnesium Ammonium Phosphate (Struvite) as an alternative method for the recovery of phosphorous is modeled in a batch stirred tank reactor. Different aspects of agricultural benefits of struvite have been discussed by the researchers, e.g. having low water solubility makes struvite a candidate for slow nutrient releasing fertilizer. Since the crystal size distribution is a cornerstone to control the quality of produced struvite, this study focuses on developing a model to predict the crystal size.

The aim of this work is to link Computational Fluid Dynamics (CFD) with Population Balance Modeling (PBM) to develop a 3-D CFD model which could predict Crystal Size Distribution (CSD) of struvite. In the simulation, the CFD flow field was solved through a Eulerian multiphase approach and RNG k-ε turbulence model, using a commercial CFD package, ANSYS Fluent 17.1. The population balance equation was solved using the discrete method, implementing 25 different size classes. The size-independent growth rate as a function of the supersaturation index (SI) was employed in the model through User Defined Function (UDF). A series of batch experiments were conducted utilizing synthetic wastewater in a 1-L glass stirred tank reactor agitated with a Rushton impeller to confirm the simulation results. Some CSD results presented in fig. 1

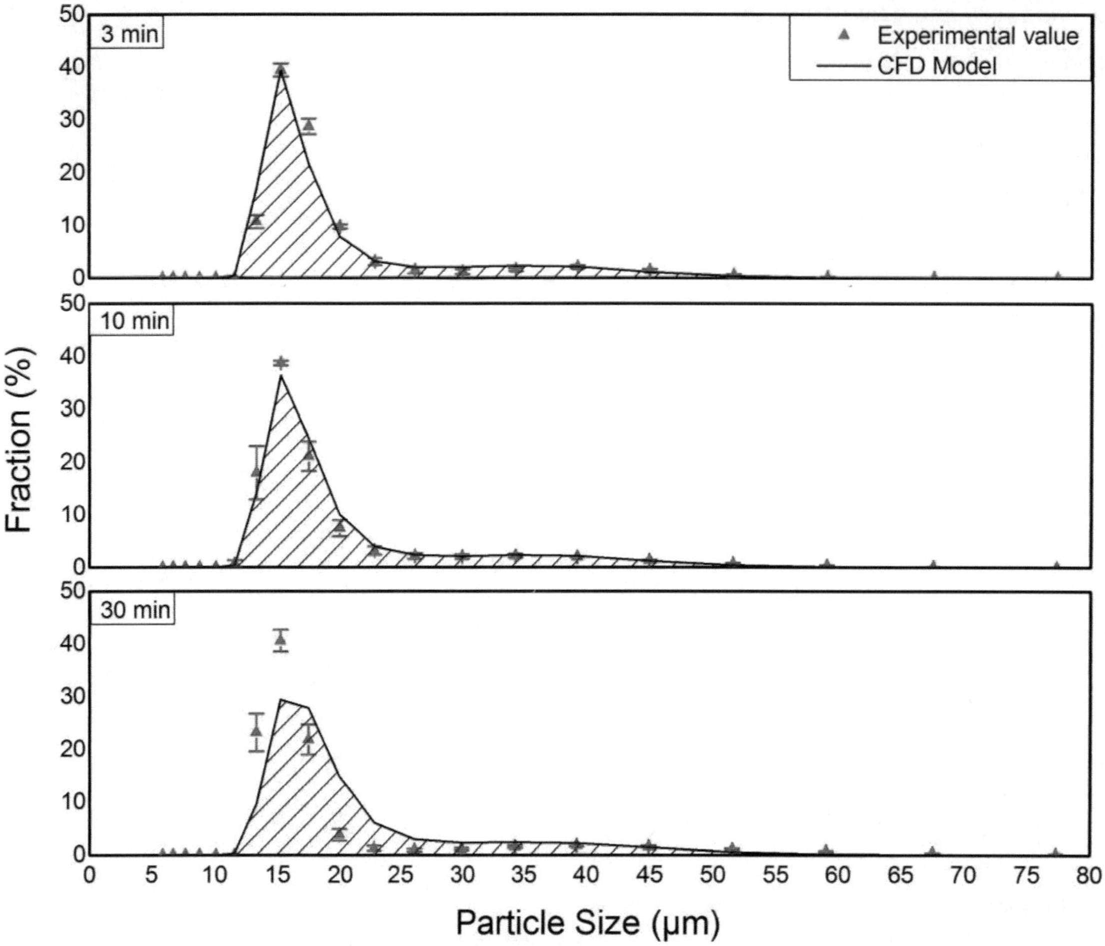


Figure 1 Experimental and CFD model predicted number based CSD at 3, 10 and 30 min

The model successfully predicted the variation of crystal size distribution as well as SI values with experiment time. Results showed that the general trend and the position of the peaks in CSD curves are in a good accordance with the experimental measurements.

KEYWORDS: Struvite, Crystallization, CFD Simulation, Population Balance Model, Stirred Reactor, Growth rate

Leadership in Sustainable Infrastructure

*Leadership en Infrastructures Durables*

**Case Study**

Vancouver, Canada

May 31 – June 3, 2017/ *Mai 31 – Juin 3, 2017*

# NOVEL ADVANCED OXIDATION PROCESS (AOPS) BASED ON PRE-MAGNETIZATION FE$^0$ FOR WASTEWATER TREATMENT

Yuwei Pan [1,2], Minghua Zhou [1,2,3], Xiang Li [1,2]

[1] MOE Key Laboratory of Pollution Processes and Environmental Criteria, College of Environmental Science and Engineering, Nankai University, Tianjin 300350, China

[2] Tianjin Key Laboratory of Urban Ecology Environmental Remediation and Pollution Control, College of Environmental Science and Engineering, Nankai University, Tianjin 300071, China

[3] zhoumh@nankai.edu.cn

**Abstract**

Novel advanced oxidation processes (AOPs) based on pre-magnetization $Fe^0$ was employed to enhance the degradation of organic pollutants and reduce reagents dosage. Compared with AOPs($Fe^0/H_2O_2$, $Fe^0/K_2S_2O_8$) based on conventional $Fe^0$, AOPs based on pre-magnetization $Fe^0$ could have 1.4–51.1 folds enhancement in the degradation rate of different refractory pollutants. Meanwhile, the feasibility of pre-magnetization $Fe^0/H_2O_2$ was also better in salty system and practical wastewater. AOPs based on pre-magnetization $Fe^0$ is more promising and highly efficient processes since it does not require any change of the present water and wastewater treatment processes, and does not need an extra energy source and complex equipment.

**Introduction**

Nowadays, advanced oxidation processes (AOPs) have been regarded as good alternatives for the treatment of various recalcitrant organic pollutants. AOPs could degrade recalcitrant organic pollutants into less toxic products through the generation of highly powerful hydroxyl radical (·OH) or sulfate radicals ($SO_4^{\cdot-}$) via Eq. (1) and Eq. (2).

$$Fe^{2+} + H_2O_2 \rightarrow Fe^{3+} + OH^- + \cdot OH \quad (1)$$

$$Fe^{2+} + S_2O_8^{2-} \rightarrow SO_4^{\cdot-} + Fe^{3+} + SO_4^{2-} \quad (2)$$

Therefore, over the last decades, a large number of AOPs based on $Fe^0$ have been reported to remove different organic and inorganic pollutants and showed that the release of $Fe^{2+}$ can accelerate the removal rate of pollutants. Therefore, some investigators adopted the integration of electricity, ultraviolet and ultrasonic to improve the generation of $Fe^{2+}$ inducing the degradation of pollutants. However, these technologies would increase the costs for construction and operation.

In our previous research, it was verified that these AOPs based on pre-magnetization $Fe^0$ based on "magnetic memory" which defined as a period in which particles can sustain their magnetization properties of certain intensity have miraculous improvement on organic pollutant degradation when comparing with conventional one, which would provide a more cost-effective alternative for environmental remediation[1]. The objects of this study is to (1) explore the characterizations of pre-magnetization $Fe^0$, (2) investigate the optimal value of f through response surface, (3) investigate the feasibility of AOPs based on pre-magnetization $Fe^0$ in salty system, (4) investigate the feasibility of AOPs based on pre-magnetization $Fe^0$ for re wastewater

**Results and Discussion**

As shown in Fig. 1, the application of pre-magnetization $Fe^0$ process had remarkable improvement. The degradation rate (*k*) of sulfadiazine(SD), tetracycline(TC), methylene blue (MB) and rhodamine B(RB) were improved greatly (Fig. 1) by pre-magnetization $Fe^0/H_2O_2$, and the value of *f* （rate constants between pre-magnetization process and conventional process） of these contaminants were 2.4 - 5.7. Therefore, pre-magnetization $Fe^0/H_2O_2$ could degrade a wide type of organic pollutants and accelerate the degradation of different pollutants.

Effect of common cations and anions in the pre-magnetization $Fe^0$/PS and $Fe^0$/PS processes were evaluated. Five cations($Mn^{2+}$, $Cu^{2+}$, $Mg^{2+}$, $Ca^{2+}$, $NH_4^+$) and five anions($HCO_3^-$, $CO_3^{2-}$, $NO_3^-$, $PO_4^{3-}$, $HPO_4^-$) were investigated. The corresponding results are shown in Fig. 2. As shown in Fig. 2, we could found that pre-magnetization $Fe^0$/PS process improved the degradation of RB in none of ions, each cations and each anions. Therefore, the feasibility of pre-magnetization $Fe^0$/PS process was better in different cations and anions.

**Acknowledgments**

Sino-Canada Cooperation Projects of Tianjin Binhai and National Undergraduate Training Program for Innovation and Entrepreneurship (201510055108)

**References**

[1] Pan, Y., Zhou, M., Li, X., Xu, L., Tang, Z., Sheng, X., Li, B., 2016. Highly efficient persulfate oxidation process activated with pre-magnetization $Fe^0$. Chem. Eng. J. in press.

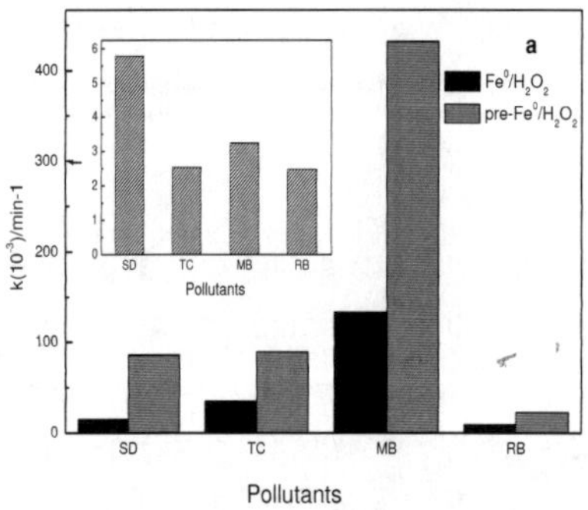


Figure 1: Pseudo-first-order rate constants of different pollutants by two processes

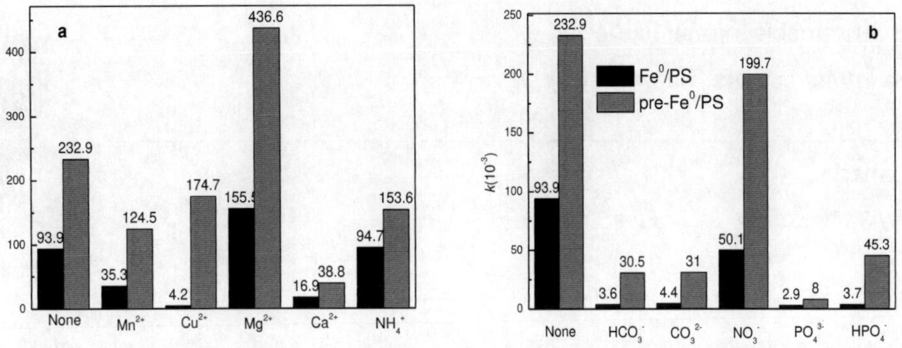


Figure 2: Pseudo-first-order rate constants at different cations (a) and anions (b)

Leadership in Sustainable Infrastructure
*Leadership en Infrastructures Durables*

Vancouver, Canada
May 31 – June 3, 2017/ *Mai 31 – Juin 3, 2017*

# OIL SANDS PROCESS-AFFECTED WATER (OSPW) MANAGEMENT USING BASAL DEPRESSURIZATION WATER (BDW) IN FORWARD OSMOSIS (FO)

Gamal El-Din, Mohamed[1,4], Zhu, Shu[2] and Li, Mingyu[3]
[1] University of Alberta, Canada
[2] University of Alberta, Canada
[3] University of Alberta, Canada
[4] mgamalel-din@ualberta.ca

**Abstract:** Forward osmosis (FO) is a promising membrane technology with various potential advantages including low energy input, high rejection efficiency, and high fouling reversibility. The feasibility of FO process has been examined previously in water and wastewater treatment, especially for oil and gas wastewaters. In the current study, to reduce the cost and energy input, FO was proposed to desalinate oil sands process-affected water (OSPW), using on-site waste basal depressurization water (BDW) as the draw solution. The results of long-term desalination experiments did not provide a clear difference between two operating conditions (FO and PRO mode) on membrane fouling because of the comparatively low water flux. Membrane cleaning research indicated that osmotic backwash using clean water efficiently recovered the initial water flux. High rejection rate (> 90%) of inorganic and organic species was achieved, especially, the rejection of naphthenic acids (NAs) reached above 94%. Meanwhile, the volume of OSPW was decreased >40%, corresponding to 1.4 times dilution of BDW in 24 hours. The obtained high rejection rates indicated the potential of safe discharge and/or reuse the diluted BDW after a sufficient processing period.

## 1 Introduction

Similar to other petroleum refinery and hydraulic fracturing wastewaters, oil sands process affected water (OSPW) contains both dissolved inorganic (i.e., heavy metals) and organic compounds (i.e., naphthenic acids; NAs) which are acute and chronic toxic to various aquatic organisms [1, 2]. To date, tailing ponds cover an area of 180 square kilometers and the surface is increasing yearly as new ponds are being developed. Recent research has shown great interests in an emerging membrane desalination technology — forward osmosis (FO) [3]. FO process requires less or no extra hydraulic pressure, which minimizes the membrane fouling potential and reduces the energy input. However, the selection of proper draw solution was is still under investigation from the perspective of cost reduction and recycle and/or safe discharge [4]. In the current study, we proposed a novel concept of using basal depressurization water (BDW), one of on-site brackish wastewaters, to drive OSPW desalination through semi-permeable membrane. BDW, the depressurized groundwater, was produced in open mining to control surface runoff and seepage water accumulation [5]. It is hypothesized that by employing FO process, the concentration of OSPW and dilution of BDW could be achieved at one process. Additionally, considering complex water characteristics of OSPW, it is expected that the membrane facing OSPW is more tended to foul.

## 2 Materials and methods

The schematic of FO system is shown in Figure 1. The commercial available cellulose triacetate (CTA) FO membrane was placed in a SEPA FO cell without using feed spacer or permeate carrier. Two speed-variable peristaltic pumps were applied to control the crossflow velocities on each side of the membrane. A digital balance (Scout Pro, Ohaus Corp., Parsippany, NJ, USA) connected to a computer was used to monitor the weight change of the draw solution. Membrane orientations including active layer facing the feed solution (FO mode) and active layer facing draw solution (PRO mode) were tested as an important factor affecting the membrane performance (i.e., internal concentration polarization).

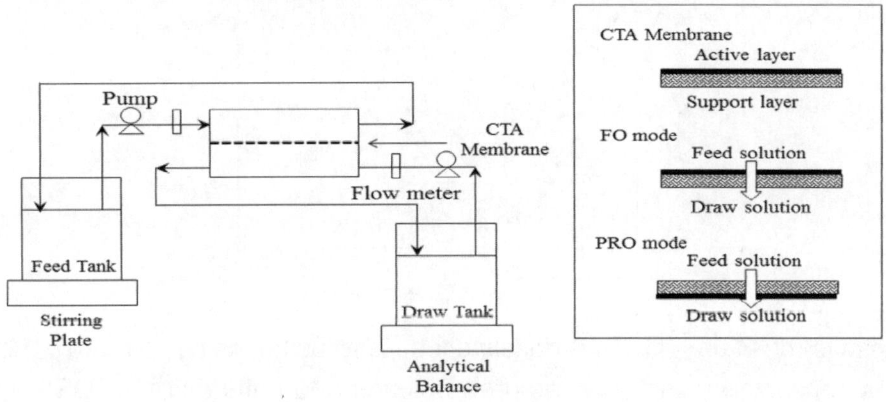


Figure 1: Schematic of FO system set-up.

## 3 Results

Figure 2 exhibits the water flux profiles under two operating modes. In FO mode, the initial water flux (1st day) started from 2.60 L/m² hr, which was lower than the initial water flux (2.95 L/m² hr) found in PRO mode. The difference of initial water flux on the two membrane orientation could be explained by the enhancement of internal concentration polarization (ICP) when BDW was directly contacting with the support layer (FO mode). The initial fluxes sharply declined to 1.92 L/m² hr (10.4%) and 2.57 (12.9%) observed on 2nd day in FO and PRO mode, respectively, and steady decreased afterwards. Immediate membrane pore blocking or scaling in the first 24 hours might be the reason caused the sharp drop of initial water flux. After 2nd day, foulants kept accumulating, thereby gradually decreasing the initial water fluxes in the following days. On 7th day, the initial water fluxes were reduced 36% in FO and 28% in PRO mode, respectively. Due to the comparatively weak driving force between OSPW and BDW, our results could not distinguish the fouling propensity of OSPW on the two membrane orientations because ICP phenomenon was more influential than membrane fouling under the low water flux condition throughout the experiment.

SEM images of the used membrane coupons further confirmed our previous assumptions. Figure 3 shows the SEM images of the active and support layers after 7-day desalination. In FO mode (Figure 3 a and b), scattered or clustered white foulants were observed on the side facing with OSPW while no surface foulants were found on the support layer against BDW. The same with FO mode, no fouling or precipitation were seen on the active layer surface against BDW but some accumulated precipitations were found surrounded by the pores on the support layer against OSPW in PRO mode. Due to the less smooth structure, particles or foulants were comparatively easier to deposit on the support layer and those smaller size foulants could flow into the inner pores, resulting in a continuous drop of permeate flux.

The overall salt rejection was above 90.0% and BDW was diluted approximately 1.5 times calculated by the conductivity before and after desalination in both modes (diluted BDW in FO mode: 12.9±0.1 mS/cm and in PRO mode: 12.2±0.1 mS/cm). As BDW diluted, chemical oxygen demand (COD) concentrations in BDW were decreased from 167.3 to 108 and 97.5 mg/L, respectively and total organic carbon (TOC) concentration was reduced from 12.9 to 6.13 and 5.8 mg/L in FO and PRO modes, respectively. The organic rejection quantified through NAs, TOC, and COD was above 94%.

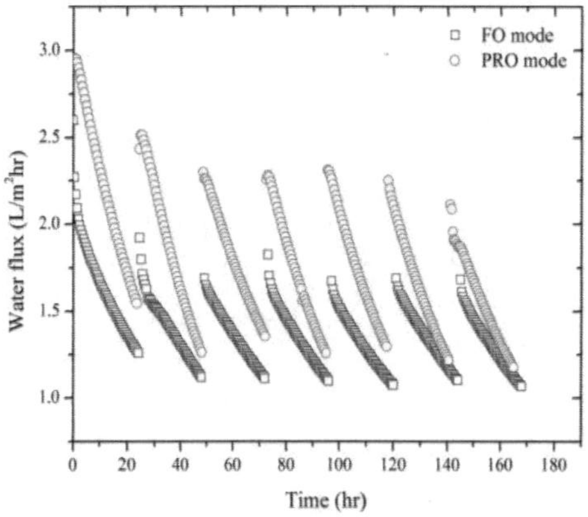


Figure 2： Results of 7-day OSPW desalination test using BDW as draw solution. The experiment was conducted using BDW as draw solution and natural settled OSPW as the feed solution. The crossflow velocity was maintained at 14 cm/s and two operating modes were tested.

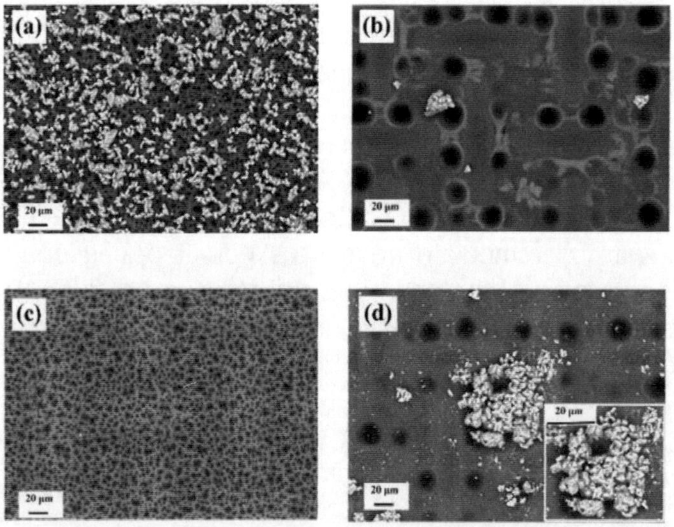


Figure 3： SEM images after 7 days desalination. FO mode (a) active layer (b) support layer. PRO mode: (c) active layer (d) support layer

## 4 Conclusion and recommendation

The feasibility of using BDW as the draw solution for OSPW desalination was confirmed by our investigation. The experimental results indicated that more than 40% volume of OSPW was reduced while the TDS concentration in BDW was diluted 1.4 times correspondingly. However, to apply it in practice, two main challenges need to be addressed: 1. Low driving force: BDW is less concentrated compared to other highly brackish water, to increase water flux, larger membrane area (i.e., using hollow fiber FO membrane) is recommended. 2. Membrane fouling: The impact of fouling can be diminished by carrying out pre-treatments (i.e., microfiltration, coagulation and flocculation) and optimizing operational condition (i.e., high crossflow velocity, using membrane spacer). By optimizing this process, it is expected that the environmental impact associated with large quantity of OSPW can be alleviated. The concentrated OSPW can be sent for further specific treatments, for example, advanced oxidation and reverse osmosis. The diluted BDW containing less total dissolved solids (TDS) and organic components can be used as process water (i.e., boiler feed) in oil sands industry after advanced treatment, for example, reverse osmosis. This process introduced both a cost and energy effective water management concept that not only aims to OSPW treatment but also the management of other process waters.

**Acknowledgements**

This research was supported by a research grant from a Natural Sciences and Engineering Research Council of Canada (NSERC) Senior Industrial Research Chair (IRC) Program in Oil Sands Tailings Water Treatment through the support by Syncrude Canada Ltd., Suncor Energy Inc., Shell Canada, Canadian Natural Resources Ltd., Total E&P Canada Ltd., EPCOR Water Services, IOWC Technologies Inc., Alberta Innovates - Energy and Environment Solution, and Alberta Environment and Parks. The financial support provided by the Helmholtz-Alberta Initiative (HAI) through the Alberta Environment and Parks' ecoTrust Program is also acknowledged. Also, the authors acknowledge the help from the technicians in the Department of Civil and Environmental Engineering from the University of Alberta.

**References**

Garcia-Garcia, E., et al., Commercial naphthenic acids and the organic fraction of oil sands process water induce different effects on pro-inflammatory gene expression and macrophage phagocytosis in mice. *Journal of Applied Toxicology*, 2012. **32**(12): p. 968-979.

Frank, R.A., et al., Toxicity assessment of collected fractions from an extracted naphthenic acid mixture. *Chemosphere*, 2008. **72**(9): p. 1309-1314.

Rastogi, N.K. and C.A. Nayak, Membranes for forward osmosis in industrial applications. *Advanced Membrane Science and Technology for Sustainable Energy and Environmental Applications*, 2011(25): p. 680-717.

Cath, T.Y., A.E. Childress, and M. Elimelech, Forward osmosis: Principles, applications, and recent developments. *Journal of Membrane Science*, 2006. **281**(1-2): p. 70-87.

Kim, E.S., et al., Desalination of oil sands process-affected water and basal depressurization water in Fort McMurray, Alberta, Canada: application of electrodialysis. *Water Science and Technology*, 2013. **68**(12): p. 2668-75.

Leadership in Sustainable Infrastructure
*Leadership en Infrastructures Durables*

**Case Study**

Vancouver, Canada
May 31 – June 3, 2017/ *Mai 31 – Juin 3, 2017*

# OPTIMIZATION OF AERATION IN A SINGLE-STAGE BIOLOGICAL AERATED FILTER FOR THE TREATMENT OF MUNICIPAL WASTEWATER

Ray, Rajan[1,5], Seth, Rajesh[2], Henshaw, Paul[3] and Biswas, Nihar[4]
[1,2,3,4] University of Windsor, Canada
[5] rray@uwindsor.ca

## 1 Project Overview

A biological Aerated Filter (BAF), is a secondary biological treatment process where microorganisms are used to the treatment of wastewater from primary sedimentation tank (Pujol et al., 1992). Lou Romano Water Reclamation Plant (LRWRP) uses an up flow single stage BAF system in secondary treatment where chemically enhanced phosphorus treatment (CEPT) is used in the primary stage. The plant treats municipal wastewater from both the city of Windsor and a portion from the city of La Salle. LRWRP the plant

of 16 cells and is the second largest BAF system in Canada. The plan area for each BAF cell is 140 $m^2$ with 5.94 m water depth. The average water velocity is 6 m/h for each cell, with a 21 ML/d per cell average treatment capacity. At present, the treatment plant has an ultimate capacity of 227 ML/d. The media in BAF has a nominal settled height of 3.9 m. The effluent target for 5-day Carbonaceous biochemical oxygen demand ($CBOD_5$) and unionized ammonia nitrogen ($NH_3$-N) is 10 mg/L and 0.08 mg/L respectively. Currently, there is no problem in achieving this target. BAF uses process airflow for achieving the required DO level for the treatment and the microorganisms to survive. The plant was initially designed to use a constant process airflow of 2.26 $m^3/h/\ m^3$ per cell. In BAF, process airflow can cost up to 70% of the total consumption of energy in the treatment plant (Hu and Wang, 2005). All the treatment plants are thriving for higher energy efficiency. A previous study revealed that the BAF effluent wastewater has a higher than normal dissolved oxygen (DO) concentration (≥5 mg/L) and the target $NH_3$-N level is achieved within the lower 50% of the cell height at 1.55 $m^3/h/m^3$ aeration rate (Ray et al., 2012). However, the study was limited to 1.55 $m^3/h/m^3$ and it suggested that a lower variable aeration rate be used in response to variable BOD loading and temperature.

## 2 Innovation

In the current study, two years' of data consisting of influent and effluent BOD loading, DO, pH, $NH_3$-N, and the temperature was analyzed for various air flow rates to understand the seasonal variable requirements of airflow while achieving the target limits of $NH_3$-N and $CBOD_5$.

## 3 Lessons Learned

The results show that the aeration can be varied and further reduced between 1.07 $m^3/h/m^3$ to 1.55 $m^3/h/m^3$ per cell in response to various influent $CBOD_5$ and temperature without affecting effluent limits. Figure 1 shows that the nitrification is still completed within 75% of the cell height after 120 min of steady state operation. Three sets of temperatures were selected (7.3 °C-11.0°C, 11.1°C-16.5°C and 16.6°C-23.4°C) in terms of year-round temperature variation. It was found that at 1.07 $m^3/h/\ m^3$ air flow rate $NH_3$ -N reduction rate is highest when the temperature is between 11.1°C-16.5°C. $NH_3$ -N reduction rate is almost similar up to 50% height of the cell for other temperatures. However, the rate reduces between 50-75% height of the cell for temperature between 16.6°C-23.4°C. It suggests that BAF shows better efficiency between 11.1°C-16.5°C at this specific air flow rate and aeration can be safely reduced when the water temperature is within the level. No significant variation was observed for 1.55 $m^3/h/\ m^3$ air flow rate. Effluent $NH_3$ -N varies between 3.2- 4.6 mg/L. BAF effluent DO concentration is ≥2 mg/L with the proposed variable aeration rate. Effluent $CBOD_5$ varies between 3-6 mg/L.

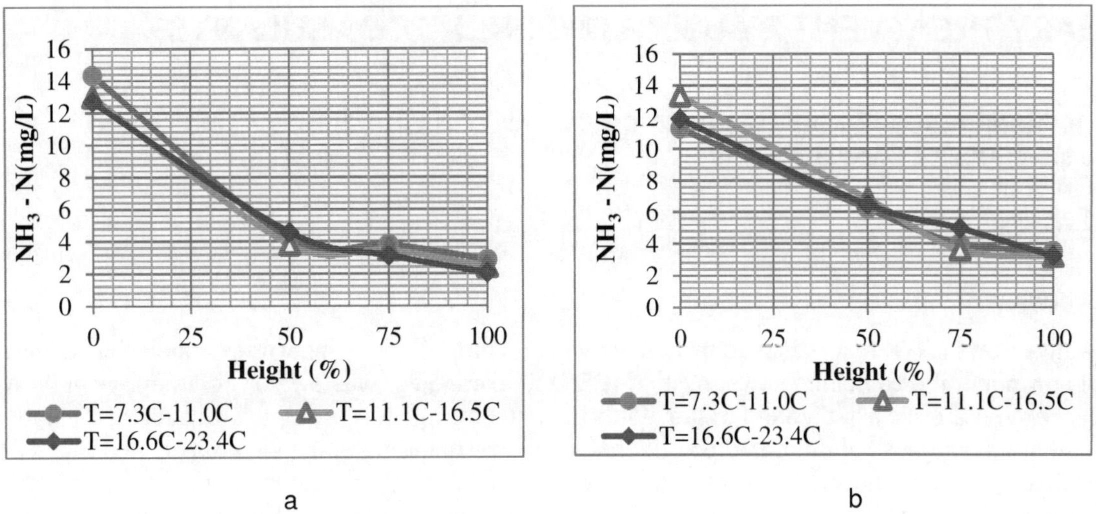


Figure 1: Change of $NH_3$-N for different water temperature (°C) after 120 min of operation a) at 1.55 $m^3/h/\ m^3$ per cell nominal air flow rates b) at 1.07 $m^3/h/\ m^3$ per cell nominal air flow rates (steady flow rate)

## Acknowledgements


This research was funded by NSERC, Canada. The City of Windsor provided data and access to the LRWRP.


## References


Hu, Y. and Wang, L. (2005) Effect of media heights on the performance of biological aerated filter. Journal of Environmental Science 17(2), 281-284.

Pujol, R., Canler, J. P. and Iwema, A. (1992) Biological aerated filters: an adopted biological process for wastewater from coastal areas. Water Science and Technology, 25 (12): 175-184.

Ray, R, Henshaw, P.F., Biswas, N., (2012), "Effects of Reduced Aeration in Biological Aerated Filter," Canadian Journal of Civil Engineering, 39, 432-438.

Leadership in Sustainable Infrastructure
*Leadership en Infrastructures Durables*

**Case Study**

Vancouver, Canada
May 31 – June 3, 2017/ *Mai 31 – Juin 3, 2017*

# OPTIMIZATION OF COMBINED SEWER OVERFLOW (CSO) ENHANCED PRIMARY TREATMENT BY USING DIFFERENT COAGULANTS

Shaikh, Mohammed Zakee[1], Mohammed, Abdul[2] and Gamal El Din, Mohamed[1,3]
[1] University of Alberta, Canada
[2] EPCOR Water Services, Canada
[3] mgamalel@ualberta.ca

## 1 Overview

Wet weather flows (WWFs) are caused by heavy rains and more frequent snowmelt conditions causing the influent flow out of the combined sewer system (CSS) to exceed the wastewater treatment plant (WWTP) capacity. WWFs are usually diverted to the North Saskatchewan River when this occurs, resulting in the release of large quantities of untreated pathogens and waste to the natural river ecosystem. WWFs out of CSSs consist of a wide range of physical, chemical and microbiological pollutants. In order to better treat WWFs before discharge, a system known as Enhanced Primary Treatment (EPT) has been initiated at a local WWTP whereof upwards of 1,200 MLD of CSO wastewater flow would undergo primary treatment before discharge. This EPT system has its drawbacks, since it does not treat the influent wastewater for pathogens, only enhancing the settling characteristics via the addition of polymer and alum coagulant. Coagulants such as Alum, polyaluminium chloride (PACl) and ferric chloride ($FeCl_3$) may help to achieve lower pathogen counts in the EPT effluent, thereby protecting the receiving river environment and meeting regulatory discharge limits. The coagulants and coagulant aid would assist in the removal of solids as well any biological pollution adsorbed on the solids via flocculation (Metcalfe & Eddy, fifth edition).

The objectives of this study are: (1) to determine the optimum mixing conditions that achieve maximum removal efficiency for important parameters such as chemical oxygen demand (COD), pH, turbidity, alkalinity, total suspended solids (TSS), total volatile suspended solids (TVSS), ortho-phosphate, coliforms and E. Coli colony counts, and (2) to qualitatively determine the best performing coagulant out of the three mentioned by observing the effect on the parameters mentioned above.

## 2 Methodology

A two-level factorial design was developed and implemented between the months of May 2016 and March 2017. Jar Test apparatus were used to implement the factorial design.

## 3 Results

The results obtained from the final sampling campaign are shown below in Figure 1.

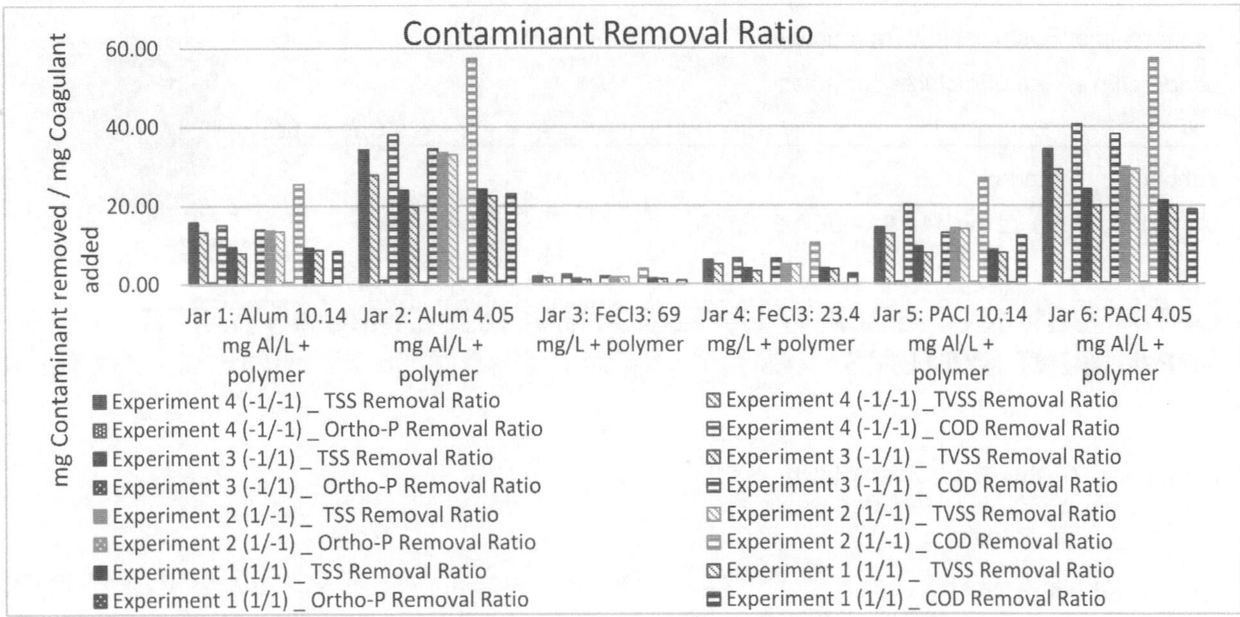


*Figure 1: Contaminant Removal Ratio for final sampling campaign*

Figure 1 is a representation of the four major parameters that were observed during the final sampling campaign, namely TSS, ortho-phosphate, TVSS and COD. The contaminant removal ratio was calculated based on how many mgs of contaminant was removed for every 1 mg of coagulant added. It was observed that the Alum and PACl low doses were most efficient at reducing each contaminant per mg of coagulant used. The $FeCl_3$, in comparison, was not effective. pH levels did not vary significantly throughout the campaigns and thus have been left out. Turbidity was mirrored effectively by TSS and has been left out for brevity. Nitrate, nitrite and the microbiological tests were inconclusive and have been left out for brevity.

## 4 Conclusions

It was concluded that the $FeCl_3$ was not an effective coagulant where as the Alum and PACl were most effective at lower doses. Out of the two, Alum is likely the most cost effective due to the relatively high price per ton of PACl.

## 5 Lessons Learned

During the course of the second set of experiments, it was observed that the high dose of PACl was too high when compared to the Alum dose range with respect to active Al species. The PACl doses were adjusted to match the Alum dose in regards to active Al species for the final sampling campaign.

## Acknowledgements

The authors acknowledge the financial support provided by the Natural Sciences and Engineering Research Council of Canada (NSERC) Collaborative Research and Development (CRD) grant and EPCOR Water Services. The author would like to thank Dr. Gamal El Din, Abdulrahim Al Omairi, Mirna Alameddine, Dr. Selamawit Messele and Dr. Pamela Chelme-Ayala for their support and assistance

## References

Metcalf & Eddy. 2014. Wastewater Engineering, Treatment And Resource Recovery. *Fifth edition*, AECOM.

Leadership in Sustainable Infrastructure
*Leadership en Infrastructures Durables*

**Case Study**

Vancouver, Canada
May 31 – June 3, 2017/ *Mai 31 – Juin 3, 2017*

# OPTIMIZATION OF SOLAR POWERED REVERSE OSMOSIS WATER TREATMENT SYSTEMS

Freire-Gormaly, M.[1, 3] and Bilton, A.[2]
[1,2] University of Toronto, Canada
3 mfreire@mie.utoronto.ca

## 1 Introduction

Over 663 million people lack access to improved drinking water and its associated health benefits (WHO 2015). These people rely on seasonally available water sources (cisterns, boreholes, and wells) that are often contaminated with chemicals and bacteria (WHO 2015). These remote communities also have limited access to electricity. The high environmental impact and fuel cost have limited diesel power's use for water treatment. Due to the relatively high levels of solar irradiance in many remote communities, photovoltaic reverse osmosis (PVRO) systems (Bilton 2013) are a promising option for these locations.

## 2 Project Overview

A community-scale PVRO system was installed through a joint-collaborative project between MIT, W.K. Kellogg Foundation, and PVPure in La Mancalona, Mexico in 2013. The community needed clean drinking water from a brackish (high saline and high mineral content well). The project scope required one cubic meter of water per day for the community and the water cost should be below the alternative supplies (purchased off a water truck). To improve the design tool for future PVRO installations, membrane fouling needed to be considered to ensure more robust and long-lasting systems.

## 3 Innovation

The innovation of the PVRO system was the use of minimal battery storage. Instead, the system operated only when the solar resources were sufficient and the end-product, clean drinking water, was stored in an inexpensive water tank. As well, the system was configured using an optimal design algorithm to select the system components (e.g. pumps, filters, and solar panel size) to maximize the water production. Industrial practice is to operate RO continuously since intermittent operation is thought to cause increased membrane fouling (Poovanaesvaran et al. 2011). Existing PVRO systems have not been designed taking into consideration membrane fouling. The goal of this work is to develop an innovative, automated design tool capable of considering the uncertainty in membrane fouling to recommend a cost-optimal PVRO system design. A preliminary study has shown the distinction between intermittently operated reverse osmosis systems compared to continuously operated systems at the lab-scale (Freire-Gormaly and Bilton 2017). The work here uses experimentally derived membrane fouling characteristics for different operating conditions (with anti-scalant and with or without rinsing of the membranes using 5L of clean water before shutting down the system) observed during intermittent operation. These correlations were used for the cost optimal design of operating conditions for PVRO systems that are robust to membrane fouling.

## 4 Lessons Learned

One of the key lessons learned is the trade-off between the cost of anti-scalant chemical usage and clean water rinsing can be quantified using optimization to determine the best operating conditions for a PVRO system. The optimization was performed using a genetic algorithm with membrane replacement time, type of anti-scalant used (anti-scalant A or anti-scalant B), and existence of a permeate rinse as design variables. Rinses were assumed to be performed with 5L of clean water produced by the system. Since the system is capable of producing approximately 1000L of water, it was selected as a sufficient amount to flush the groundwater out of the membrane housing, and piping. The membranes then sit in stagnant water with a very low mineral content.

The optimization uses a simulation model based on experimental measurements to estimate water production for different operating conditions. Two representative days of operation with anti-scalant usage and rinsing prior to shut-down (Figure 1a) shows the increased permeability at the start of each day due to flushing of the membrane. The 16 hours of shut-down are not shown. The optimization was performed using an hourly simulation over 5 years which determines the costs of the two anti-scalants ($/lb), and the cost of membrane replacement, and water cost. The operating costs were determined based on the total volume of water produced, the total mass of anti-scalants and their associated costs. The optimal system design used anti-scalant B with rinsing and membrane replacement every 2.6 years. The operating cost of this operating condition is $\$0.61/m^3$. The operating cost exponentially decreases to a minimum cost of $\$0.61/m^3$ at 2.6 years for the case of anti-scalant B with rinsing for different membrane replacement times (Figure 1b). Future PVRO systems can use this approach to select the appropriate anti-scalant, clean water rinsing and time to replace the membranes.

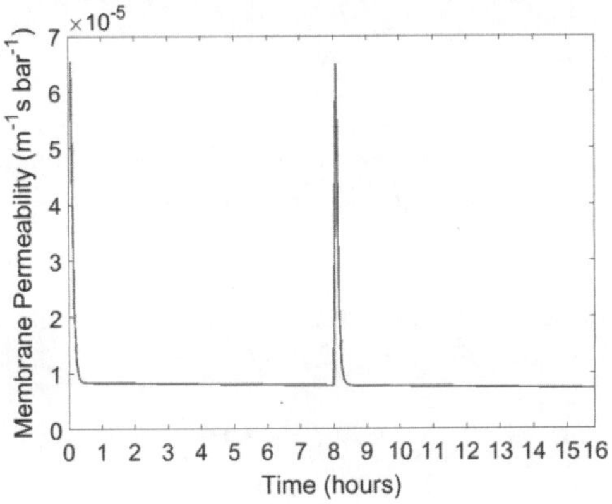


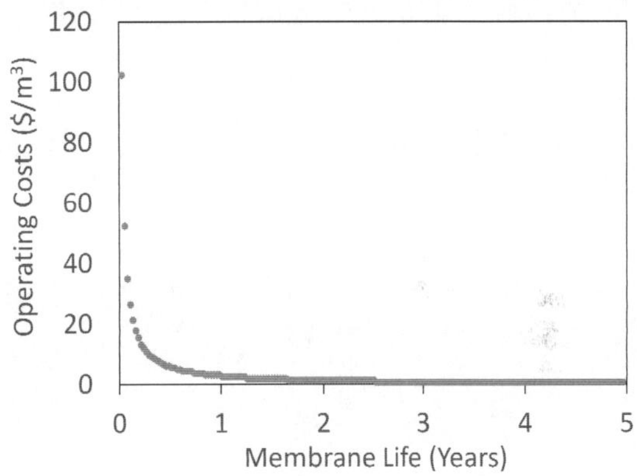


Figure 1: a) Two days of operation with increased permeability after 16 hours of shutdown. b) Operating cost as a function of membrane life for the case of anti-scalant B and rinsing.

**Acknowledgements**

The authors thank Prof. R.C. Andrews for his feedback on the experiments. The authors also thank research students Youngmok Ko and Francis Cruz. The financial support of MIE at UofT, OGS, Hatch and the NSERC Scholarships are also greatly appreciated.

**References**

Bilton, A. 2013. "A Modular Design Architecture for Application to Community-Scale Photovoltaic-Powered Reverse Osmosis Systems." PhD Thesis, Massachusetts Institute of Technology.

Freire-Gormaly, M, and Bilton, A. 2017. "An Experimental System for Characterization of Membrane Fouling of Solar Photovoltaic Reverse Osmosis Systems under Intermittent Operation." *Desalination and Water Treatment.*

Poovanaesvaran, P., M.A. Alghoul, K. Sopian, N. Amin, M.I. Fadhel, and M. Yahya. 2011. "Design

Aspects of Small-Scale Photovoltaic Brackish Water Reverse Osmosis (PV-BWRO) System." *Desalination and Water Treatment* 27 (1–3): 210–23. doi:10.5004/dwt.2011.1824.

World Health Organization and UNICEF Joint Monitoring Programme (JMP). 2015 Progress on Drinking Water and Sanitation, 2015 Update and MDG Assessment.

Leadership in Sustainable Infrastructure
*Leadership en Infrastructures Durables*

**Case Study**

Vancouver, Canada
May 31 – June 3, 2017/ *Mai 31 – Juin 3, 2017*

# OZONATION OF SYNTHETIC COMBINED SEWER OVERFLOW: IMPACT ON THE FATE OF SILVER NANOPARTICLES

Alameddine, Mirna[1], Messele, Selamawit Ashagre[1], Mohammed, Abdul[2] and Gamal El-Din, Mohamed[1,3]
[1] Department of Civil and Environmental Engineering, University of Alberta, Edmonton, AB, Canada
[2] EPCOR Water Services, Canada
[3] mgamalel-din@ualberta.ca

## 1 PROJECT OVERVIEW

A Wastewater Treatment Plant (WWTP) in Alberta is working in collaboration with a research team at the University of Alberta on improving the treatment of its Combined Sewer Overflow (CSO) to maximize the removal efficiency of common regulated parameters, microorganisms, micropollutants and nanoparticles before the final discharge. All efforts employed by the WWTP to optimize its current CSO Enhanced Primary Treatment (EPT) converge to reduce the emission of any potentially toxic material into the North Saskatchewan River and ultimately protect the environment and public health.

## 2 INNOVATION

Recent developments in nanotechnology allowed the introduction of many nanomaterials in regular consumer products. Among those nanomaterials, Silver Nanoparticles (AgNPs) are the most commonly used because of their size, high surface area and antimicrobial properties, yet not the most abundantly produced (Vance et al., 2015). They are generally found in textile and medical and household products and eventually find their way to the wastewater network. Researchers were able to ascertain the bioavailability of AgNPs in wastewater to fish like juvenile rainbow trout (Bruneau et al., 2016) and to depict a dose-dependent toxic effect of AgNPs in zebrafish embryos (Asharani et al., 2008). This potential risk of AgNPs necessitates the consideration of treatment techniques that guarantee their removal from effluents before discharge. Accordingly, optimizing the EPT of CSO at the Alberta WWTP by adding a treatment step to the current coagulation/flocculation/settling process, can either jeopardise or boost the removal of AgNPs. This study, will test in particular the addition of ozonation to the EPT. A preliminary bench scale experiment is designed to test the impact of ozone ($O_3$) application on AgNPs in raw and treated synthetic CSO.

## 3 METHODOLOGY

A sample of raw primary influent was collected from the Alberta WWTP, under dry weather conditions. The sample was diluted with deionized water to mimic a wet weather CSO and is denoted as synthetic CSO. A jar test was conducted using 2 liters of synthetic CSO per jar and a coagulation/ flocculation/settling process was initiated with a blank for reference, and 3 different coagulants: Alum at 10.4 mg Al/L, Ferric Chloride ($FeCl_3$) at 20.95 mg Fe/L and Polyaluminum Chloride (PACl) at 10.4 mg Al/L that were added with 1.25 mg/L cationic polymer. Fast mixing was performed at 150 rpm for 1 min, followed by slow mixing at 15 rpm for 10 min and settling for 1 hr. Samples were collected from the supernatant and stored in dark glass bottles at 4°C. Equal volumes of raw and treated synthetic CSO were added in 23 mL amber glass reactors and topped up with a fixed dose of applied $O_3$ (17.26 mg $O_3$/L). Upon the addition of $O_3$ stock solutions, all reactors were subjected to 20 min shaking at 203 rpm, after which $O_3$ residual was measured with the Indigo Method. Samples were then stored in amber glass vials at 4°C for AgNP analysis by Inductively Coupled Plasma – Mass Spectrometry (ICP-MS) and results are reported as concentration of $^{107}Ag$.

## 4 RESULTS AND DISCUSSION

Experimental results unveil the ability of the primary treatment of synthetic CSO to enhance the removal of AgNPs. In fact, the concentration of AgNPs dropped by 15% through settling alone, by 80% through the addition of $FeCl_3$, and by 89% through the addition of either alum or PACl. On the other hand, ozonation of raw and treated synthetic CSO considerably increased Ag concentrations in all tested samples (by 86% in synthetic CSO, 93% in blank jar, 99% in all 3 jars treated with alum, $FeCl_3$ and PACl) with a utilized $O_3$ dose equivalent to applied $O_3$. This increase does not coincide with the results of similar ozonation experiments performed on synthesized AgNP stock solutions in MilliQ water that showed merely 5% decrease in AgNP concentration. This small drop can be probably attributed to the inability of silver oxide, the likely product of AgNP ozonation, to settle out of solution (feature of its solubility product), and thus the persistence of Ag for detection by ICP-MS. Based on the observed impact of ozonation on samples of synthetic CSO, it can be speculated that both high organic matter present in wastewater matrices and flocs obtained during primary treatment processes form strong complexes with colloidal Ag and settle out as aggregates leading to less detectable Ag in solution. It is also possible that chlorine ions ($Cl^-$) that were added with $FeCl_3$ and PACl might have disturbed the stability of AgNP and simulated the formation of silver chloride precipitate decreasing the detectable Ag in solution. In a complex wastewater matrix, $O_3$ might have preferentially oxidized the organic matter releasing embedded Ag and thus provoking higher Ag concentrations as measured by ICP-MS.

## 5 LESSONS LEARNED AND FUTURE WORK

This preliminary experiment showed that the ozonation of synthetic CSO, in a raw condition or after primary treatment, increases the final concentration of silver. While the scope of this work did not provide evidence on the exact AgNP transformations, it still gave some insight on newly explored impacts of wastewater ozonation that involve nanoparticles. As such, the mechanisms for AgNP transformations evoked during ozonation along with the kinetics, competing agents in wastewater matrices and oxidation by-products will be the core of future studies.

### Acknowledgments

The authors acknowledge financial support for this project provided by the Natural Sciences and Engineering Research Council of Canada (NSERC) Collaborative Research and Development (CRD) grant and EPCOR Water Services.

### References

Asharani, P. V., Lian Wu, Y., Gong, Z., & Valiyaveettil, S. (2008). Toxicity of silver nanoparticles in zebrafish models. *Nanotechnology, 19*(25), 255102. doi: 10.1088/0957-4484/19/25/255102

Bruneau, A., Turcotte, P., Pilote, M., Gagne, F., & Gagnon, C. (2016). Fate of silver nanoparticles in wastewater and immunotoxic effects on rainbow trout. *Aquat Toxicol, 174*, 70-81. doi: 10.1016/j.aquatox.2016.02.013

Vance, M. E., Kuiken, T., Vejerano, E. P., McGinnis, S. P., Hochella, M. F., Jr., Rejeski, D., & Hull, M. S. (2015). Nanotechnology in the real world: Redeveloping the nanomaterial consumer products inventory. *Beilstein J Nanotechnol, 6*, 1769-1780. doi: 10.3762/bjnano.6.181

Leadership in Sustainable Infrastructure
*Leadership en Infrastructures Durables*

Vancouver, Canada
May 31 – June 3, 2017/ *Mai 31 – Juin 3, 2017*

# POTENTIAL AIR QUALITY IMPACTS DUE TO HYDRAULIC FRACTURING IN WESTERN NEWFOUNDLAND

Shareef Mohammed Mujtaba [1,3], and Tahir Husain [2]
[1,2] Memorial University of Newfoundland, St. John's, NL, Canada
[3] mms515@mun.ca

## ABSTRACT

Due to presence of potential Hydraulic Fracturing (HF) reserves in Western Newfoundland, there has been an increased interest in exploring these. However, due to the environmental concerns associated with HF a study was undertaken to identify environmental implications of HF in Western Newfoundland. This paper presents environmental and human health impacts associated with HF activities on air quality. Several studies have recognized that HF activities may lead to an increase in the type of air pollutants generally found at conventional operations, including those other pollutants specific to fracturing, such as silica sand, fracturing chemicals, and flowback wastewater. This study identified key air pollutants and their health impacts due to HF activities. An air quality modelling is conducted to study the dispersion of pollutants emissions as a result of HF activities. The emissions of pollutants were estimated based on EPA emission factors and other criteria, and using these emissions modelling of selected criterial pollutants were conducted in two domains in Western Newfoundland. Modelling exercise revealed that peak concentrations of $NO_2$ were relatively high while other pollutants were predicted low. Relative high concentrations of $NO_2$ is a cause of concern due to its health impact as well as its role in formation of tropospheric Ozone.

Keywords: - Hydraulic Fracturing; Newfoundland; Air quality; Air Dispersion Modeling

## 1.0 INTRODUCTION

Depending on the characteristics of the rock, hydrocarbon reservoirs are classified as either conventional or unconventional. In conventional reservoirs, layers of sandstone and fluid typically allow oil or natural gas to flow readily into wellbores. On the other hand, in unconventional reservoirs, due to the low permeability of the rock formation, the hydrocarbon cannot easily flow out unless a path is created artificially, for example, by hydraulic fracturing (HF), which is linked with the horizontal drilling of wells (GNLDNR, 2014 a).

Historically, oil and gas (O&G) have been largely extracted from the ground by drilling a vertical well into a hydrocarbon reservoir, called conventional reservoirs, and allowing the oil or gas to flow into the well by natural pressures. As conventional resources become limited, the exploitation of unconventional resources such as shales, tight sands, and coalbed methane is gaining recognition. Tapping shale reserves is not economically feasible for producing O&G without sophisticated technologies. To resolve this issue, research and innovation by O&G industries have produced advanced techniques for extracting the O&G trapped in shale formations. A key technology for accessing unconventional hydrocarbon resources is hydraulic fracturing, or fracking. HF involves pumping a mixture of water, sand, and chemicals into shale at a high pressure to create fissures or fractures in a tight rock formation. These fractures lead to an increase in the flow of natural gas and other hydrocarbons from the formation to the well. Fracking has gained popularity due to its capability to increase the exploration and production of shale O&G (PTAC, 2012). However, this expansion of fracking has also increased concerns from federal, provincial, and local agencies, including the public, about its associated potential environmental impacts on land, water, and air. Scientific research and investigations continue to look for an environmentally friendly use of HF. The HF operation consumes significant energy, which also brings new environmental management challenges for the O&G industry.

Due to the presence of potential HF reserves in Western Newfoundland, there has been an increased interest in exploring these reserves. However, due to the environmental concerns associated with HF a study was undertaken to identify the environmental implications of HF in Western Newfoundland. The main goal of this paper is to present the environmental impacts on air quality associated with the HF activities. The objectives of this paper include: identifying the main air pollutants and their health risks, evaluating the meteorology of the Western Newfoundland, estimating the air pollutants emissions, performing air dispersion modelling, and assessing the impacts.

## 2.0 AIR POLLUTANTS AND SOURCES

Conventional O&G operations have been known to create harmful air emissions. These pollutants include diesel particle emissions, hydrocarbons, VOCs (e.g., benzene), particulate matter (PM), and greenhouse gases (GHGs). The emissions of air pollutants from fracturing operations is similar to that from conventional O&G operations, but it is of a higher magnitude around fracturing sites because of the greater effort required (Green, 2014). The United States Environmental Protection Agency (USEPA) is also concerned about air pollutants that result from HF; according to USEPA (2014), there have been well-documented air quality impacts in areas with active natural gas developments, with increases in the emissions of methane ($CH_4$), VOCs, and hazardous air pollutants (HAPs). The main regional air emission issue is the generation of Ozone ($O_3$), which in some circumstances could adversely affect air quality (CCA, 2014).

Studies recognize that fracturing activity may lead to an increase in the type of air pollutants generally found at conventional O&G operations, including those other pollutants specific to fracturing, such as silica sand, fracturing chemicals, and flowback wastewater (Srebotnjak and Rotkin-Ellman, 2014). In general, the air emissions involved in fracturing operations can be categorized as:

- On-site criteria pollutants and their precursors such as carbon monoxide (CO), lead, nitrogen oxides ($NO_x$), $O_3$, PM, sulfur dioxide ($SO_2$) and VOCs.
- Air toxics and other HAPs, including fugitive emissions from mixing chemicals, spills, and flowback fluids (which can also include VOCs); and
- GHG emissions such as carbon dioxide ($CO_2$) and $CH_4$.

The sources of these emissions include combustion engines, powering on-site and transportation equipment, drilling process wastewater and condensate tanks, underground/downhole sources such as flowback fluids, and fugitive emissions from sand, dust, mixing chemicals, spills, or other uncontrolled gas releases (Tyner et al., 2014). Table 1 summarizes the potential activities/equipment and possible air pollutant emissions from a typical fracturing operation.

### 2.1 Criteria Pollutants

HF activities releases thousands of tons of air pollutants in the atmosphere, for example in 2012 fracturing activities in US emitted 13,000, 170,000, 250,000, 23,000 and 600 tons of PM, $NO_x$, CO, VOCs and $SO_2$ respectively (Ridlington and Rumpler, 2013). This potentially will pose significant impact on the local air quality in Western Newfoundland. The following sections elaborates on various criteria pollutants and their impact on health.

#### 2.1.1 Particulate Matter (PM)

Particulate emissions originate from the combustion engines of heavy trucks and machinery used during well-site preparation, drilling, and production. PM is a complex mixture of very small particles and liquid droplets found in the air, including dust, dirt, soot, and smoke. The size of the particulate is directly linked to its potential for causing health problems. PM that is 10 micrometres (μm) in diameter or smaller ($PM_{10}$) poses a health concern because it can pass through the throat and nose and accumulate in the respiratory system. PM less than 2.5 μm in diameter ($PM_{2.5}$) is believed to pose the greatest health risk because it can get deep into a person's lungs and even into the bloodstream (USEPA, 2010). Total particulate matter (TPM) is the term applied to any particle suspended in the atmosphere, but typically it is limited to PM less than 44 μm in diameter. PM larger than 10 micrometres in diameter is typically associated with a nuisance rather than a health issue (DECNL, 2013). According to USEPA (2010), each well requires on average of between 2 and 5 million gallons of water per HF event. Water is generally transported by diesel trucks, each of which has an approximate capacity of 3,000 gallons (USEPA, 2011). It has been estimated that approximately 2,300 trips by heavy-duty trucks are required for each horizontal well during the early stages of shale gas development (USEPA, 2011). With thousands of such wells concentrated in high-development regions, the levels of truck traffic and diesel-associated air pollution will increase in these areas. In addition to diesel PM, other pollutants prevalent in diesel emissions, such as $NO_x$ and VOCs, react in the presence of sunlight and produced ground-level $O_3$.

Table 1: Typical HF activities, sources, and air emissions from HF

| Activity and Associated Equipment | Emissions & Sources | Potential air pollutants |
|---|---|---|
| Well site preparation and road construction | Trucks and heavy machinery | Diesel PM, $NO_x$, $CO_2$, CO, BTEX, PAH, and dust |
| Well drilling, hydraulic fracturing, and well completion | Drilling | Diesel PM, $NO_x$, $CO_2$, CO, BTEX, PAH, $CH_4$, volatile drilling mud fluids, and volatile hydrocarbons from drill cuttings |
| | Hydraulic fracturing | Silica dust, volatile fracturing chemicals, BTEX, other volatile hydrocarbons, PM, $NO_x$, $CO_2$, and CO |
| | Flowback and produced water | Volatile fracturing fluids, BTEX, other volatile hydrocarbons, and hydrogen sulfide ($H_2S$) |
| Production | Produced water | BTEX, other volatile hydrocarbons, and $H_2S$ |
| | Gas flaring/venting | $CH_4$, $NO_x$, $CO_2$, CO, PM, $H_2S$, BTEX, and other volatile hydrocarbons |
| | Work-over and maintenance | Diesel PM, $NO_x$, $CO_2$, CO, $CH_4$, BTEX, PAH, and other volatile hydrocarbons |
| Processing and storage | Gas venting | $CH_4$, $H_2S$, BTEX, and other volatile hydrocarbons |
| | Separators and condensate tanks | $CH_4$, BTEX, and other volatile hydrocarbons |
| | Compressors | diesel PM, $NO_x$, $CO_2$, CO, BTEX, PAH, and other volatile hydrocarbons |
| Transmission | Pipelines | $CH_4$, BTEX, and other volatile hydrocarbons |
| | Compressor stations | Diesel PM, $NO_x$, $CO_2$, CO, BTEX, PAH, and other volatile hydrocarbons |
| | Gas venting | $CH_4$, $H_2S$, BTEX, and other volatile hydrocarbons |
| Well abandonment and site rehabilitation | Trucks and heavy machinery | Diesel PM, $NO_x$, $CO_2$, CO, BTEX, and PAH |
| | Abandoned orphaned wells | $CH_4$ |

Compiled from Srebotnjak and Rotkin-Ellman, 2014; Tyner et al., 2014

### 2.1.2 Nitrogen Oxides ($NO_x$)

Nitric oxide (NO) and nitrogen dioxide ($NO_2$) are collectively referred to as $NO_x$. These are generally produced due to combustion activities. In Canada, the main sources of $NO_x$ are from transportation (50%), petroleum industry (22%), electric power generation (10%), natural source (8%) and other industrial and non-industrial sources (10%).

### 2.1.3 Carbon monoxide (CO)

CO is a colourless and odourless gas which reduces the delivery of oxygen ($O_2$) to the body's organs. Generally, incomplete oxidation of fuel results in the formation of CO. However, if sufficient $O_2$ is not present to complete the combustion of the hydrocarbon fuel, the oxidation to $CO_2$ and water ($H_2O$) is not completed, and hence CO is emitted.

According to Natural Resources Canada (NRC) (2012), shale gas has a low $CO_2$ content, similar to typical conventional gas production. Therefore, as more shale gas development occurs, the GHG emission per unit of shale gas produced and consumed should be similar to that from conventional natural gas production and use.

### 2.1.4 Volatile organic compounds (VOCs)

Fracturing fluids can contain VOCs such as benzene, which can be released into the atmosphere when the fluid evaporates (Colborn et al., 2011). The wellheads themselves vent VOCs such as benzene and toluene that can combine with combustion by-products to create smog (Conrad et al., 2010). However, according to Bunch et al. (2014), shale gas production activities did not result in community-wide exposure to concentrations of VOCs at levels that would pose a health concern.

### 2.1.5 Ozone ($O_3$)

$O_3$ is a secondary pollutant that is formed in polluted areas by atmospheric reactions involving two main types of precursor pollutants: VOCs and $NO_x$. CO from the incomplete combustion of fuels is also an important precursor for O3 formation. The formation of $O_3$ and other oxidation products (e.g., peroxyacyl nitrates and hydrogen peroxide), including oxidation products of the precursor chemicals, is an extremely complex reaction that depends on the intensity and wavelength of sunlight, atmospheric mixing and interactions with cloud and other aerosol particulates, the concentration of VOCs and $NO_x$ in the air, and the rates of all chemical reactions. The majority of ground-level $O_3$ is formed when the $O_3$ precursors $NO_x$, CO, and VOCs react in the atmosphere in the presence of sunlight (Conrad et al., 2010). VOC sources can come from ponds, condensers, and other gas-processing equipment and compressor-transmission operation (Conrad et al., 2010).

## 2.2 Greenhouse Gas (GHG) Emissions

HF operations require a significant amount of energy and, as HF locations are generally remote, that energy has to be generated on-site (Green, 2014). In most cases, conventional power generators fueled by diesel fuel, natural gas, or other fossil fuels are used. The combustion of such fuel leads to the emission of GHGs (Green, 2014). In addition, improperly drilled wells may leak $CH_4$ and other GHGs to the atmosphere during and after the production period of the well. The literature claims that HF would increase natural gas emissions to the atmosphere due to leakage during the HF process and at the beginning of gas recovery (Green, 2014). $CH_4$ is considered to be one of the most potent GHGs. It is estimated that 100 million tons of $CO_2$ equivalent is released since 2005 due to HF activities in US, thus contributing to global warming (Ridlington and Rumpler, 2013).

More recently, reduced emissions completions (RECs), or "green completions," which capture and separate natural gas during well completion and workover activities, have become a key technology to limit the amounts of $CH_4$, VOCs, and HAPs that can be vented during the flowback period without the disadvantage of flaring. RECs use portable equipment that allows operators to capture natural gas from the flowback water. After the mixture passes through a sand trap, a three-phase separator removes natural gas liquids and water from the gas, which is then sent to sales pipelines for distribution. REC operations have been found to be very cost-effective even with low natural gas prices (Green, 2014).

According to Natural Resources Canada (NRC) (2012), "most prospective shale gas developments have low CO content, similar to typical conventional gas production. Therefore, as more shale gas development occurs, the greenhouse gas emissions per unit of shale gas produced and consumed should be similar to that from conventional natural gas production and use."

Numerous other cost-effective technologies have been developed to reduce natural gas leakage, such as plunger lift systems, dry seal systems, and no-bleed pneumatic controllers. Through the use of these technologies and practices, nearly 90 percent of the natural gas leakages could be addressed (PTAC, 2012). To further reduce the emissions impacts at well sites in densely populated areas, electric motors could be used instead of internal combustion engines (Clark et al., 2013).

# 3.0 POTENTIAL HEALTH RISKS

The specific health effects due to air contaminants and their extent are dependent on a variety of factors such as the type and length of exposure to a contaminant as well as the health status and lifestyle of the exposed individual. Along with the concern of increased GHG emissions, the impact of air emissions from HF has become a debated issue in the environmental movement. Due to a low emission rate, some emissions have been acknowledged to be relatively harmless to human health. Naturally occurring radioactive materials (NORM), one identified source of emissions, are brought to the surface during shale gas production but remain in such places as rock pieces with the produced water (USDOE, 2009). As the radiation from these NORMs is weak, it cannot penetrate dense materials or cause extreme risks from exposure. However, radiation hazards must be evaluated so that it does not exceed regulatory standards concentrations.

Other chemicals have been detected in drilling locations that are highly detrimental, particularly to air quality. For example, with benzene, a carcinogen that typically causes leukemia, health concerns arise when its level reaches 1.4 parts per billion (ppb). In 2009, air samples from a Targa Resources compressor station outside Decatur, Texas, revealed that the level of benzene reached 1,100 ppb; a sample from a nearby Devon Energy well revealed 15,000 ppb (Hawes, 2009). Of the 300 air samples taken from 30 facilities in north central Texas, 50 exceeded the Texas Commission of Environmental Quality's standard for long-term health risk (WISE, 2010).

Air emissions can also be attributed to the equipment used to extract natural gas by HF. Millions of gallons of water are commonly transported by tanker trucks; for instance, over 1,000 truck trips were required for one fracture (WISE, 2010). Each truck trip could stir up dust and release PM, $NO_x$, and $CO_2$ into the air. Diesel engines needed to run the drilling equipment use large amounts of fuel that also produce a significant amount of emissions.

A health impact assessment conducted by Witter et al. (2008) for O&G development concluded that air quality is most likely affected during well-pad construction and well completion and through truck traffic. Fugitive emissions from production equipment are another possible long-term source of air contamination that needs to be controlled. Table 2 lists the potential health and environmental effects from criteria air pollutants and BTEX respectively, released from HF operations.

A human health risk assessment of air emissions carried out in a region of Colorado with a shale gas development near a rural population detected several different air emissions in close proximity to the development. The study found that the highest air pollution concentrations occurred during well development and completion (McKenzie et al., 2012). Overall, two-thirds more hydrocarbons were detected during well completion than during the production phase. The range of concentrations detected for several VOCs and BTEX during completion was large. For instance, the minimum detected concentration of m-xylene/p-xylene was 2.0 $\mu g/m^3$ of air, whereas the maximum was 880 $\mu g/m^3$ of air (McKenzie et al., 2012). Health Canada's tolerable concentrations over a lifetime for xylene isomers (m-xylene/p-xylene) are 180 $\mu g/m^3$ and 348 $\mu g/m^3$ respectively (Ruth, 1986; Health Canada, 1996).

## 4.0 METEOROLOGY OF WESTERN NEWFOUNDLAND

### 4.1 General Climate

The western Newfoundland area is located between the latitudes 46°36' and 51°38'. Due to the influence of ocean, the area experiences slightly warmer winter and cooler summer than the inland areas. The average temperature in summer remains in the 20-25°C range, while winter temperatures are below zero. Geographically, the whole area is located in westerlies winds (30-60° of latitude), the dominating wind direction prevail from west. However, with the influence of ocean currents the wind direction of the area varies.

### 4.2 Winds

The winds were analyzed based on five-year meteorological data (2010-2014) obtained from National Centers for Environmental Information (NCEI) for various weather stations in western Newfoundland. Three stations i.e., Stephenville, Deer Lake and St. Anthony were selected to analyze the wind patter. Due to the influence of ocean currents, wind direction varies in different seasons of western Newfoundland. The dominant winds tend to prevail from W sectors for six stations with the exception of Wreck House, which the dominate winds tend to prevail from N and S.

Five-year wind rose plots indicated that the wind speed of areas on the coast is generally will be higher than it of inland areas. The dominate wind speed of Deer Lake is 3.00-5.70 m/s in summer and 5.70-8.80 m/s in winter, while it is 5.70-8.80 m/s in summer at St. Anthony and Stephenville, and over 8.80 m/s in winter at these two stations. Wind speed in summer could be 1-2 classes lower than it in winter. The dominate wind direction of all three stations are prevail from W in winter and SW in summer. In recent years, the trend of wind direction of St. Anthony is tending to SW in winter and SE in summer and varies in spring and fall. At Deer Lake station, the dominate wind direction prevail to WSW and NE all year round. In spring, NE is the dominate wind direction, while WSW in summer and winter. The trend of the wind direction of Deer Lake is moving to SW and SSW in summer. At Stephenville station, the dominate wind direction prevail from W and E. WSW is the dominate wind direction in summer, while E in spring. W and WNW are the dominate wind direction in winter. Recent years trend indicates that in summer the dominate wind direction prevail to WSW, while W in winter and ENE in spring.

Table 2: Air contaminants associated with HF and their effects

| Substance | Potential Health Effects | Environmental and Climate Effects |
|---|---|---|
| Particulate Matter (PM) | Non-fatal heart attacks<br>Irregular heartbeat<br>Aggravated asthma<br>Reduced lung function<br>Increased respiratory symptoms (e.g., coughing, difficulty breathing)<br>Premature death in people with heart or lung disease | Impairs visibility, adversely affects ecosystem processes, and damages and/or soils structures and property. Variable climate impacts depending on particle type. Most particles are reflective and lead to net cooling, while some (especially black carbon) absorb energy and lead to warming. Other impacts include changing the timing and location of traditional rainfall patterns |
| Oxides of Sulfur ($SO_x$) from process unit (EESI, 2011) | Aggravate asthma, leading to wheezing, chest tightness and shortness of breath, increased medication use, hospital admissions, and ER visits; very high levels can cause respiratory symptoms in people without lung disease. | Contributes to the acidification of soil and surface water and mercury methylation in wetland areas. Causes injury to vegetation and local species losses in aquatic and terrestrial systems. Contributes to particle formation with associated environmental effects. Sulfate particles contribute to the cooling of the atmosphere. |
| Nitrogen Oxides ($NO_x$) | Irritated respiratory system aggravated asthma, bronchitis, or existing heart disease<br>Combines with VOCs to form $O_3$ | Contributes to the acidification and nutrient enrichment (eutrophication, nitrogen saturation) of soil and surface water.<br>Leads to biodiversity losses. Impacts levels of $O_3$, particles, and $CH_4$ with associated environmental and climate effects. |
| Carbon Monoxide (CO) | Exacerbation of cardiovascular disease behavioural impairment<br>Reduced birth weight increased daily mortality rate | Contributes to the formation of $CO_2$ and $O_3$, GHGs gases that warm the atmosphere. |
| Volatile Organic Compounds (VOCs) | Carcinogen (some VOCs)<br>Leukemia and other blood disorders (benzene)<br>Birth defects (some VOCs)<br>Eye, nose, and throat irritation (some VOCs)<br>Adverse nervous systems effects | Contributes to $O_3$ formation with associated environmental and climate effects.<br>Contributes to the formation of $CO_2$ and $O_3$, GHGs that warm the atmosphere. |
| Ground Level Ozone ($O_3$) (Smog) | Reduced lung function<br>Aggravated asthma or bronchitis<br>Permanent lung damage | Damages vegetation by visibly injuring leaves, reducing photosynthesis, impairing reproduction and growth, and decreasing crop yields. $O_3$ damage to plants may alter ecosystem structure, reduce biodiversity, and decrease plant uptake of $CO_2$. $O_3$ is also a GHG that contributes to the warming of the atmosphere. |
| Methane ($CH_4$) | Asphyxiation in confined spaces.<br>May cause rapid breathing, rapid | |

| Substance | Potential Health Effects | Environmental and Climate Effects |
|---|---|---|
| | heart rate, clumsiness, emotional upset, and fatigue. At greater exposure, may cause vomiting, collapse, convulsions, coma, and death. | |

Source: Fierro et al., 2001; USEPA, 2013a, 2013b, 2013d; McKenzie et al., 2012

# 5.0 DISPERSION MODELLING

## 5.1 Dispersion Models

Several air dispersion models have been developed so far, however three main models are widely used, namely AERMOD (AERMOD, 2004), CALPUFF (Scire et al., 2000 a) and CMAQ (CMAS, 2015). AERMOD is a steady-state plume dispersion model designed to predict near field concentration of pollutants, while CALPUFF is a lagrangian Gaussian puff dispersion model for both near and far field applications. CMAQ is a 3D grid-based photochemical air quality model and it specialized in simulating $O_3$ and photochemical oxidants. These models have been developed by US EPA and used as regulatory model for several purposes. CALPUFF is selected for modelling a hypothetical scenario in Western Newfoundland, as it is recommended model for all regulatory applications in NL (NL Guideline for Plume Dispersion Modelling, 2012).

## 5.2 Modelling Scenario

In order to study the extent of impact of air emissions, two hypothetical release scenarios are developed and modelled. Table 3 shows the drill rig emissions per well for criteria pollutants, HAPs, and GHG.

The estimates are based on the following assumptions:

- It requires approximately 600 hours of operation (approximately 25 days at 24 hour/day).
- Drill rig horse power is 1, 500 hp
- Diesel fuel sulfur content 0.0005 % (EPA standard)
- AP-42 emission factors are used.

The emissions from a single well activity is very low, however HF drilling are performed in large numbers for sustainable production. In US, over 13,000 wells are drilled per year.

Two modelling scenarios were undertaken in this report, one assuming 500 wells will be drilled per year in the Western Newfoundland (hereafter referred to as South-West NL run), and the second scenario assumed 80 wells per year in a small domain near Port au Port bay (hereafter referred to as Port au Port run). Table 4 illustrates the calculation of number of wells that needs to be drilled simultaneously. It is assumed that the drilling operations are executed only during summer months due to adverse weather conditions in other months.

## 5.3 Study Area and Modelling Domain

HF activities potentially could occur in entire Western Newfoundland; hence study must cover the geography of Western Newfoundland. Further, small areas of interest may be selected and fine resolution modelling could be done to study the local impacts of the air pollutants. This study selected two domains: Domain 1, and Domain 2. The pollutants modelled were $PM_{10}$, $PM_{2.5}$, $NO_x$ CO and $SO_2$. However, similar study must be done for other domains and pollutants. The meteorological Domain 1 was 130 km x 130 km with a resolution of 4 km as shown in Figure 1a. This domain was scaled down to 17.5 km x 17.5 km near Port au port with a resolution of 500 m to study the local impacts in the area as shown in Figure 1b.

## 5.4 Meteorological Modelling

The CALPUFF simulation model requires meteorological data to represent the transport and dispersion of pollutants in the domain. The meteorological characteristics over the domain vary both spatially and temporally. The CALMET diagnostic model was used to provide CALPUFF model the spatial and temporal meteorological parameters. The WRF prognostic meteorological model was used to generate input for CALMET as described briefly in the following sections.

Table 3: Drill Rig Emissions Per Well of Criteria Pollutants and VOC (Adopted from Drill Rig Emissions, 2012)

| Species | Drilling Rig Emission Factor | Per Well (lb/hr) | Per well (g/s) |
|---|---|---|---|
| Criteria Pollutants & VOC | | | |
| $NO_x$ [a] | 0.0152 | 9.12 | 1.1490 |
| CO [a] | 5.73E-03 | 3.44 | 0.4330 |
| $PM_{10}$ [a] | 4.00E-04 | 0.24 | 0.0302 |
| $PM_{2.5}$ [b] | 4.00E-04 | 0.24 | 0.0302 |
| $SO_2$ [b] | 4.05E-06 | 2.43E-03 | 0.0003 |
| VOC [a] | 2.20E-03 | 1.32 | 0.396 |
| HAPs | | | |
| Benzene [d] | 5.82E-06 | 3.49E-03 | 1.05E-03 |
| Toluene [d] | 2.11E-06 | 1.26E-03 | 3.79E-04 |
| Xylenes [d] | 1.45E-06 | 8.69E-04 | 2.61E-04 |
| Formaldehyde [d] | 5.92E-07 | 3.55E-04 | 1.07E-04 |
| Acetaldehyde [d] | 1.89E-07 | 1.13E-04 | 3.40E-05 |
| Acrolein [d] | 5.91E-08 | 3.55E-05 | 1.06E-05 |
| Naphthalene [e] | 9.75E-07 | 5.85E-04 | 1.76E-04 |
| Total PAH [e, f] | 1.59E-06 | 9.54E-04 | 2.86E-04 |
| Greenhouse Gases | | | |
| $CO_2$ [b] | 1.16 | 696 | 207 |
| $CH_4$ [b, c] | 7.05E-04 | 0.423 | 0.127 |

[a] Emission factors for Tier II non-road diesel engine emission standards from dieselnet.com (NOX, CO, VOC and PM) - Tier II emission standards are not set for VOC (listed as Hydrocarbons), so the Tier I Standard is used - Tier II or Tier I emission standards are not set for PM2.5, so thePM10 emission factor is used

[b] AP-42 Volume I, Large Stationary Diesel Engines Tables 3.4-1 and 3.4-2 Diesel Fuel, 10/96. VOC emission factor represents total Hydrocarbon Emissions

[c] CH4 Emission Factor listed in notes of AP-42 Table 3.4-1 as 9% of Total Organic Compounds

[d] AP-42 Volume I, Large Stationary Diesel Engines Table 3.4-3

[e] AP-42 Volume I, Large Stationary Diesel Engines Table 3.4-4

[f] PAH (Polycyclic Aromatic Hydrocarbons) includes naphthalene and are a HAP because they are polycyclic organic matter (POM)

Table 4: Calculation of number of wells to be drilled simultaneously

| Item | 500 Wells/Year | 80 Wells/Year |
|---|---|---|
| Number of wells to be drilled per year | 500 | 80 wells |
| Number of days required per well | 25 | 25 days |
| Total drilling days | 500 x 25 = 12,500 | 80 x 25 = 2, 000 |
| Available days for drilling (May, June, July, August, September) | 150 days | 150 days |
| Simultaneous drilling | 12,500/150 ~ 83 wells | 2,00/150 ~ 13 wells |

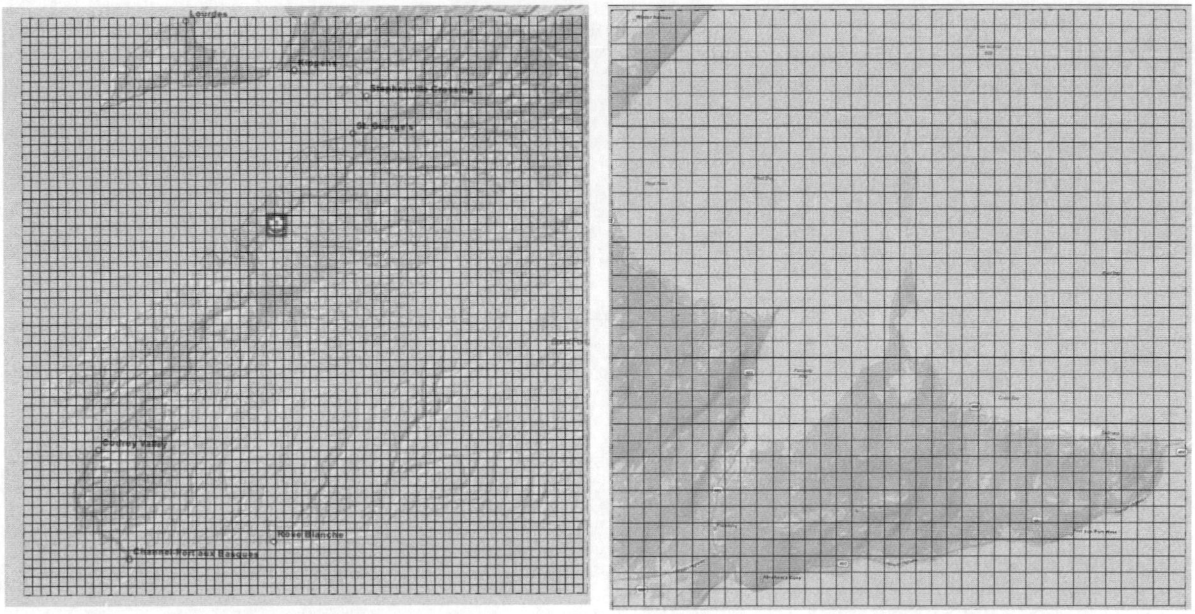

(a) Southwest (b) Port au port

Figure 1: Meteorological grids for dispersion modelling

### 5.4.1 CALMET Ready WRF Data

WRF version 3.4.1 (next-generation mesoscale model) is a prognostic meteorology model developed in partnership by the National Center for Atmospheric Research, the National Oceanic and Atmospheric Administration, the Air Force Weather Agency, the Naval Research Laboratory, the University of Oklahoma, and the Federal Aviation Administration. The model is a hydrostatic, terrain-following, eta-coordinate model designed to simulate or predict mesoscale and regional-scale atmospheric circulation. WRF was primarily developed using FORTRAN coding, is used as a community model in over 150 countries and is updated on an annual basis. For the current simulation, CALMET ready files were obtained from Lakes Environmental based on the WRF model for 2014 with horizontal resolution of 4 km.

### 5.4.2 CALMET

CALMET is a diagnostic meteorological model that produces 3-D wind fields based on parameterized treatments of terrain effects such as slope flows, terrain blocking effects, and kinematic effects. Using available sources of meteorological and geophysical information it produces a spatially varying wind field that is consistent with the local terrain and land use features, as well as atmospheric stability conditions and mixing height values necessary for the dispersion modeling.

The domain size configured in CALMET for South-West run was 128 km x 128 km, slightly less than WRF domain. The horizontal domain was downscaled to 2 km to study the dispersion of pollutants in detail. It consisted of 64 x 64 grid cells with a Universal Transverse Mercator (UTM) map projection of 21N and datum of WGS-84. Similarly, for Port au port run, the domain was 17.5 km x 17.5 km with a resolution of 500m and 36 x 36 grid cells.

In order to properly simulate pollutant transport and dispersion, it is important to define the vertical profile of meteorological parameters such as wind speed, temperature, turbulence intensity and wind direction with the atmospheric boundary layer. Eleven vertical levels are centred at 10, 30, 60, 120, 240, 480, 820, 1250, 1850, 2600, and 3500 meters above the ground. The cell face heights corresponding to these grid points are 0, 20, 40, 80, 160, 320, 640, 1000, 1500, 2200, 3000 and 4000 meters.

The resolution of terrain and land use data used in the modelling were 23 meters and 25 meters respectively. The source of terrain data was from Canada Digital Elevation Data (CDED, 1999) and the source of land use was Earth Observation for Sustainable development (EOSD, 2003. CALMET derived wind rose diagram during summer 2014 at the center of domain at an elevation of 10 m is shown in Figure 2 for South-West and Port au Port domain. As observed, winds in South-West were either blowing from south-east or north-west, however were predominant from south-east. The wind speed was most of the time in the range of 5-8 m/s with about 1% calm conditions. This shows that the pollutants are expected to disperse in all direction with a higher tendency to move towards north-west.

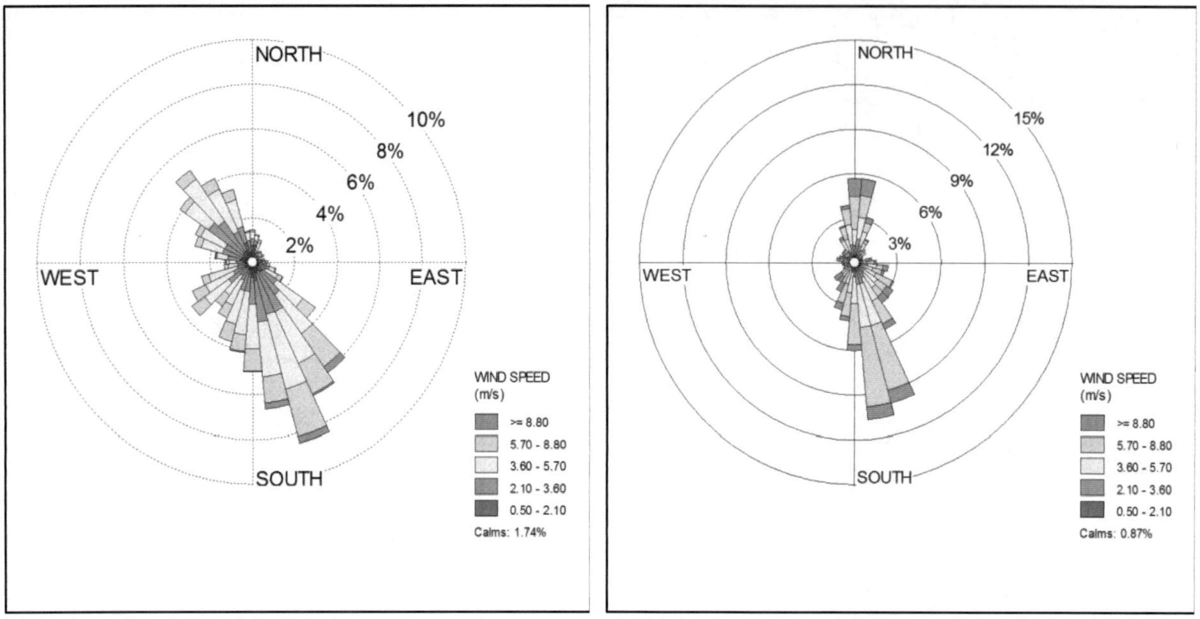


(a) Southwest (b) Port au port

Figure 2: CALMET derived wind rose at 10 m elevation at the center of the domains in summer 2014 (winds directions are blowing from)

### 5.4.3 CALPUFF Modelling

In this modelling study, the CALPUFF simulations were conducted using these model options: Gaussian near-field distribution, Transitional plume rise, Stack tip downwash, Turbulence based dispersion coefficients, Transition of $\sigma y$ to time-dependent (Heffter) growth rate, Partial plume path adjustment for terrain, Modeling of dry deposition, Consideration of chemical transformations and, No consideration of wet deposition.

CALPUFF View™, the Lakes Environmental graphical user interface was used to process the input and output. CALPUFF Version 7 was applied on five months (May-Sep) CALMET data. Continuous release of five criteria pollutants from 81 wells for five months was assumed. For the South-West run, the wells were assumed to be in the middle of the domain as shown in Figure 3a, while wells are scattered over the coastline for Port au Port run as shown in Figure 3b. The CALPUFF computational grid was same as CALMET, species modelled for both runs were $NO_x$, CO, $PM_{10}$, $PM_{2.5}$, and $SO_2$. The input emissions are provided to CALPUFF by variable point emission sources files, these are prepared by specialized processors in the format PTMAERB. The format is defined in the CALPUFF user guide (Scire et al., 2000 a).

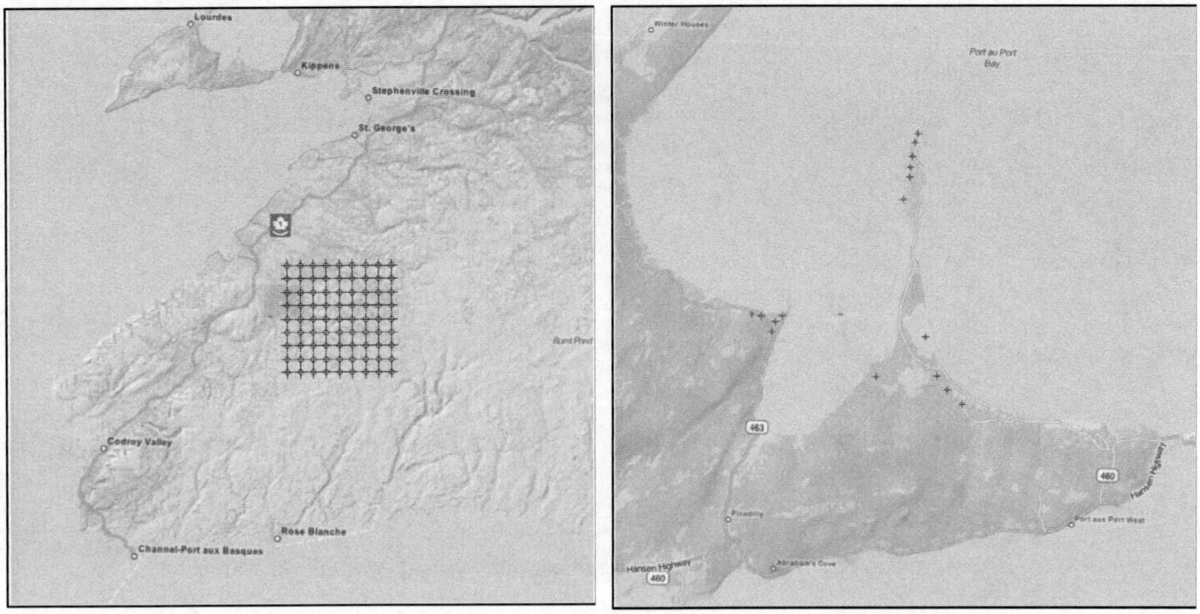


(a) Southwest (b) Port au port

Figure 3: HF drilling locations

### 5.5 Dispersion Results

The results of South-West and Port au Port runs are summarized in Table 5. The peak concentrations in the domain are compared with NL ambient air quality standards. The peak concentrations of all the pollutants are within the standards, however $NO_2$ shows significant high values. The peak concentrations of $NO_2$ were 168.71 µg/m$^3$ and 115.15 µg/m$^3$ for the South-West and Port au Port runs respectively. Due to highly varying winds the pollutants dispersed in all directions. The location of peak for the Port au Port run was towards south near the peninsula, while the peak for South-West run were in the vicinity of the assumed drilling location in the center of the domain. The spatial variation of the pollutants is presented as contour plots. Figures 4 and Figure 5 show contours plots for modelled pollutants for $NO_2$ in the two domains.

Table 5: Modelled peak concentrations of criteria pollutants

| | NL Air Quality Standard (µg/m$^3$) | Peak Concentration (µg/m$^3$) | |
|---|---|---|---|
| | | South-West Run[1] | Port au Port Run[2] |
| $NO_2$ | | | |
| 1-hour | 400 | 168.71 | 115.15 |
| 24-hour | 200 | 13.76 | 12.81 |
| Annual | 100 | 2.59* | 1.92* |
| $PM_{10}$ | | | |
| 24-Hour | 50 | 0.37 | 0.36 |
| Annual | | 0.07* | 0.05* |
| $PM_{2.5}$ | | | |
| 24-hour | 25 | 0.37 | 0.36 |
| Annual | - | 0.07* | 0.05* |
| CO | | | |
| 1-hour | 35,000 | 64.51 | 42.04 |
| 8-hour | 15,000 | 14.56 | 8.63 |
| $SO_2$ | | | |
| 1-hour | 900 | 0.04 | 0.02 |
| 24-hour | 300 | 0.004 | 0.003 |
| Annual | 60 | 0.0006* | 0.0005* |

[1] Based on drilling 500 wells/year (81 simultaneous) spread over western NL domain (300 km x 300 km)
[2] Based on drilling 80 wells/year (15 simultaneous) spread in Port au Port bay domain (17.5 km 17.5 km)
* Values are run length average (5 months)

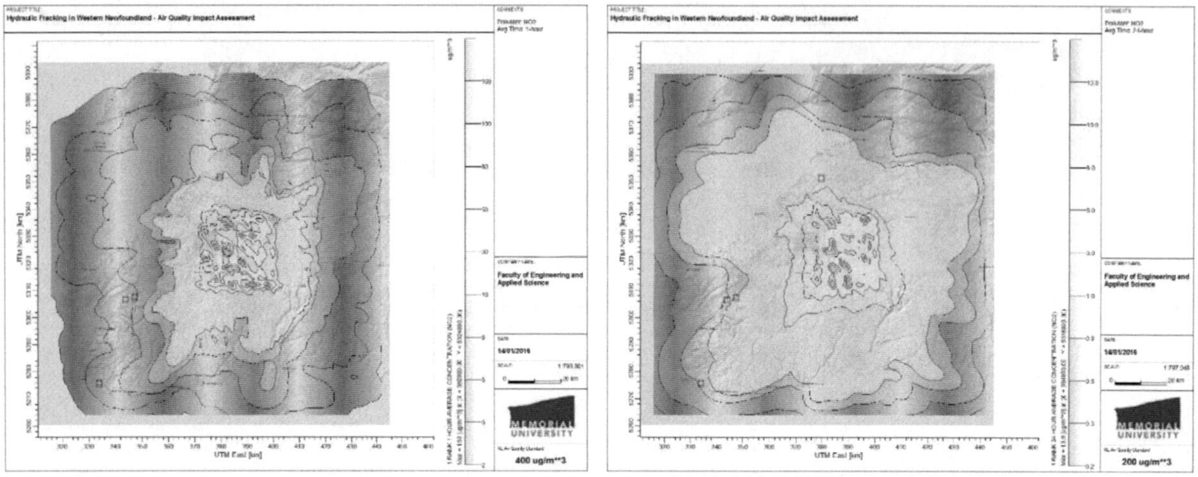

(a) 1-hour average (b) 24-hour average

Figure 4: Southwest run – Peak 1-hour and 24-hour average $NO_2$ concentrations

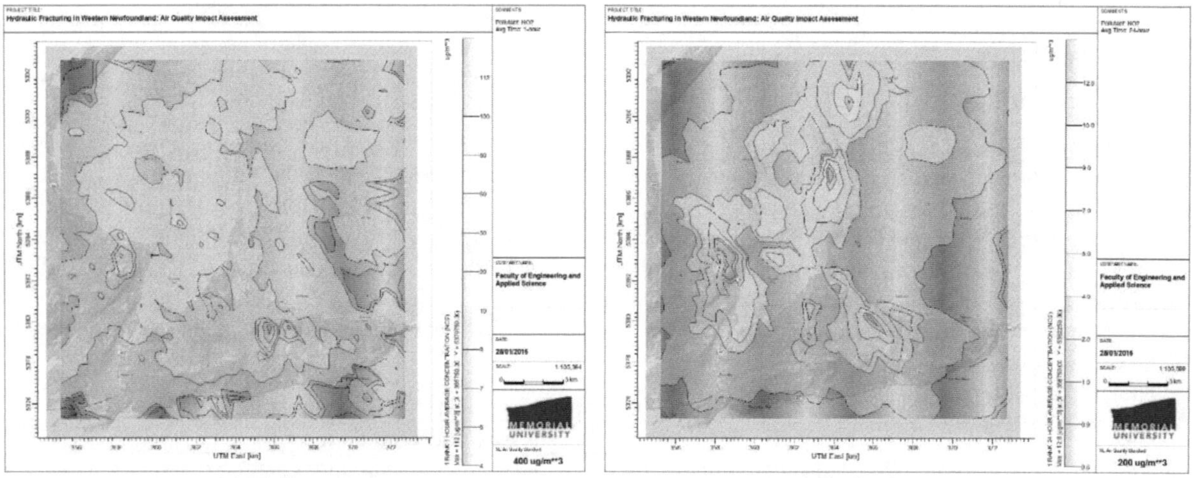

(a) 1-hour average (b) 24-hour average

Figure 5: Port au port run – Peak 1-hour and 24-hour average $NO_2$ concentrations

## 6.0 CONCLUSIONS

Although state-of-the-art technologies are employed in current HF operations, these technologies are not without any environmental footprint, including that on air quality. The main sources of air-quality pollutants during HF operations are emissions from trucks and heavy machinery, flowback and produced water, gas flaring venting activities, compressor stations, and separators and condensate tanks. The primary pollutants are particulate matter (PM), nitrogen compounds, volatile organic compounds (VOCs), and benzene, toluene, ethylbenenze, and xylene (BTEX), which are harmful to humans as well as to flora and fauna. Thousands of tons of these pollutants are known to be released due to HF activities. Methane ($CH_4$) which is one of the potent greenhouse gases (GHGs) may be leaked due to improperly drilled wells. Dispersion of pollutants depends largely on the meteorology of the area. Average summer temperatures in the area were in the range of 20-25 $^0$C while winters are below zero. The wind blows predominantly from west as geographically this area is located in westerlies, however due to the influence of ocean currents the wind direction varies highly in certain areas. The emissions of criteria pollutant per well were estimated based on EPA emission factors. Using these emissions, a preliminary air dispersion study of selected criterial pollutants was conducted in two domains. The first domain was a large area covering south-west NL, and the second one a small and high resolution area covering Port au Port peninsula. The modelling exercise revealed that the peak concentrations of $NO_2$ were relatively high while other pollutants were predicted very low. All modelled pollutants were below NL ambient air quality standards however, relative high concentrations of $NO_2$ is a cause of concern due to its health impact as well as its role in the formation of $O_3$. Therefore, a detailed photochemical modelling is highly recommended to study the regional $O_3$ formation in addition to detailed dispersion study for other criteria pollutants including BTEX which is an occupational health hazard. Appropriate air quality mitigation measures include avoiding venting, taking measures to reduce greenhouse gases, and conducting periodic site-specific air-quality monitoring.

## 7.0 REFERENCES


AERMOD: Description of Model Formulation (2004). United States Environmental Protection Agency. Available at: http://www3.epa.gov/scram001/7thconf/aermod/aermod_mfd.pdf

Bunch, A. G., Perry, C. S., Abraham, L., Wikoff, D. S., Tachovsky, J. A., Hixon, J. G., Urban, J. D., Harris, M. A., and Haws, L. C. (2014). Evaluation of impact of shale gas operations in the Barnett Shale region on volatile organic compounds in air and potential human health risks. *Science of the Total Environment*, 468–469, 832–842.

CCA (Council of Canadian Academies). (2014). *Environmental Impacts of Shale Gas Extraction in Canada*. Ottawa, ON: The Expert Panel on Harnessing Science and Technology to Understand the Environmental Impacts of Shale Gas Extraction, Council of Canadian Academies. ISBN 978-1-926558-78-3.

CMAS, (2015). Community Modeling & Analysis System (https://www.cmascenter.org /cmaq/).

Colborn, T., Kwiatkowski, C., Schultz, K., and Bachran, M. (2011). Natural Gas Operations from a Public Health Perspective. *International Journal of Human and Ecological Risk Assessment*, 17(5), 1039-1056.

Conrad D.V, Michanowicz, D., Christen, C., Malone, S., and Ferrer, K. (2010). "Potential Shale Gas Extraction Air Pollution Impacts," FracTracker—Marcellus Shale Data Tracking, Foundation for Pennsylvania Watersheds, 24 Aug. http://www.fractracker.org/2010/08/potential-shale-gas-extraction-air-pollution-impacts/.

DECNL (Department of Environment and Conservation Newfoundland and Labrador). (2013). *Ambient Air Monitoring* Report 2013 – February 2014. Available at: http://www.env.gov.nl.ca/env/publications. Link accessed on June 18, 2015.

GNLDNR 2014 a (Government of Newfoundland and Labrador, Department of Natural Resources). *The Green Point Shale of Western Newfoundland: A Review of Its Geological Setting, Its Potential as an Unconventional Hydrocarbon Reservoir, and Its Ability to Be Safely Stimulated Using the Technique of Hydraulic Fracturing*. Available at: http://www.nr.gov.nl.ca/nr/energy/pdf/green_point_shale_west_nl.pdf.

Green, K. P. (2014). Managing the Risks of Hydraulic Fracturing. Fraser Institute. Available at: http://www.fraserinstitute.org/.

Hawes, C. (2009). "Barnett Shale air study reveals alarming results." WFAA Dallas-Fort Worth. Available at: http://www.wfaa.com/home/related/More-Known-about-Barnett-Shale-Air-Quality-Study-73645207.html.

McKenzie, L. M., Witter, R. Z., Newman, L. S., and Adgate, J. L. (2012). Human health risk assessment of air emissions from development of unconventional natural gas resources. *Science of the Total Environment*, 424(1), 79-87.

NL Guideline for Plume Dispersion Modelling. (2012). Guideline for Plume Dispersion Modeling, Government of Newfoundland and Labrador, Department of Environment & Conservation. Available at: http://www.env.gov.nl.ca/env/env_protection/science/gd_ppd_019_2.pdf.

NRC. (2012). Shale Gas. Government of Canada. Available at: http:/ /www.nrcan.gc.ca/ energy/natura 1-gas/5687.

PTAC (Petroleum Technology Alliance Canada). (2012). The Modern Practices of Hydraulic Fracturing: A Focus on Canadian Resources.

Ridlington, E., Rumpler, J., (2013). Fracking by the numbers. Environment Americal Research & Policy Center. Available at: http://www.environmentamerica.org/sites/environment/files/reports/EA_FrackingNumbers_scrn.pdf

Ruth, J. H. (1986). Odor thresholds and irritation levels of several chemical substances: A review. *American Industrial Hygiene Association Journal*, 47(3), A142-A151.

Scire, J.S., D.G. Strimaitis, and R.J. Yamartino,. (2000a), *A User's Guide for the CALPUFF Dispersion Model (Version 5).* Earth Tech, Inc.Concord, MA.

Srebotnjak, T. and Rotkin-Ellman, M. (2014) Fracking Fumes: Air Pollution from Hydraulic Fracturing Threatens Public Health and Communities. Natural Resources Defense Council (NDRC).

Tyner, R., Johnson, M., Jamin, Y., and Picard, D. (2014). Evaluation of Air Emission Associated with Hydraulic Fracturing. Project Report to Petroleum Technology Alliance of Canada and Natural Resources Canada, March 20.

USDOE. (2009). Modern Shale Gas Development in the United States: A Primer Work Performed Under DE-FG26-04NT15455. Available at: http://energy.gov/sites/prod/files/2013/03/f0/ShaleGasPrimer_Online_4-2009.pdf.

USEPA (U.S. Environmental Protection Agency). (2010). Hydraulic Fracturing Research Study. Available at: http://www.epa.gov/safewater/uic/pdfs/hfresearchstudyfs.pdf.

USEPA. (2011). *Plan to Study the Potential Impacts of Hydraulic Fracturing on Drinking Water Resources*. Available at:
http://water.epa.gov/type/groundwater/uic/class2/hydraulicfracturing/upload/hf_study_plan_110211_final_508.pdf.

USEPA. (2014). *Natural Gas Extraction - Hydraulic Fracturing*. Government of the Unites States. Available at: http://www2.epa.gov/hydraulicfracturing#air.

WISE. (2010). Regulatory Options & Challenges in Hydraulic Fracturing, by Phi Nguyen, Texas Christian University.

Leadership in Sustainable Infrastructure
*Leadership en Infrastructures Durables*

Vancouver, Canada
May 31 – June 3, 2017/ *Mai 31 – Juin 3, 2017*

# POWER GENERATION POTENTIAL FROM CYANIDIZED WATER USING OSMOTIC PROCESSES.

Vives, Luis[1, 2], Quiñones, Edgar[1] and Bustos, Carlos[1]
[1] University of Cartagena, Colombia
[2] lvivesp@unicartagena.edu.co

**Abstract:** Substances such as sodium cyanide are highly toxic and are indiscriminately dumped into rivers and watersheds by mining industries, exposing communities to a high level of environmental risk. Looking for alternatives to mitigate the impact, the objective of this research was to estimate the electrical power potential in a lab scale osmotic power generator through the use of sea water and cyanided water gradient concentration (25 and 700 mg/L NaCN) using flat polymeric membrane. 60 tests were performed in two osmotic generators, using a dead-end filter operation scheme, one with vertical and another with horizontal membrane orientation. By using a solution of 700 mg/L, water concentrations similar to those found at the tailing of South Bolivar were achieved. This allowed the GE Osmonics SE membrane to develop a maximum achieved power density of 2.45E-04 W/m$^2$ at a value of 1.9 cm of water for the reactor with the vertical configuration. Seawater tests were performed with concentrations hundred times that of cyanide, at 71 g/L, a power density of 2.27E-04 W/m$^2$ and a head pressure of 3.3 cm were found for the reactor with the vertical configuration. The results show higher yields under the same operating conditions when replacing NaCl found in seawater by cyanide salts, thereby decreasing the concentration by 70.3 g. Through this research, an efficient alternative for the reuse of cyanide salt contaminated water is shown.

## 1 Introduction

The Renewable Energy (RE) sources is in continuous search for cost-effective relation and has taken many dimensions and directions. Trying to mitigate the current urgency of dwindling world supplies of conventional fossil fuels, increasing oil prices and the most important, climate change (Sharif, Merdaw, Aryafar & Nicoll. 2014). Nowadays, a carbon plant emits more than one billion metric tons of greenhouse gases every year (Elimelech, Yin Yip, Hoover, Phillip, & Tiraferri, 2011). The salinity gradient is one of the resources that use osmosis and electro dialysis. The power production by pressure retarded osmosis (PRO) and waste water treatment or desalination by forward osmosis (FO) may be economically feasible option to overcome this inadequacy problem of energy and water supply as compared to other available techniques. In an ideal PRO process, pure water permeates through a semipermeable membrane from dilute feed solution into a pressurized and concentrated draw solution; the resultant highly pressurized solution is further split into two streams. Fresh water enters the membrane modules; part of fresh water is transferred by osmosis across membrane into sea water. Then, the pressurized sea water comes out as brackish water. Generating power through osmosis between river and sea water has been known since 1970s (Attarde, Jain, Chaudhary, Gupta. 2015).

Water is a resource that has been approved for the generation of electric energy as the example of hydroelectric plants, which have become a sustainable source of energy throughout the world. Although in some cases it is questioned by the environmental impact it entails. In Colombia, the large number of rivers make the utilization of this type of energy an enticing prospect. The rivers make up 65% of the energy supply of Colombia, more than half of the national supply. However, there are zones that do not have electricity service. Zones affected by illegal armed groups, insular or ones that are hard to access are in need of a source of energy and presently are considered non-interconnected zones (ZNI). The unsatisfied energy demands in the ZNI in 2011 increase to 65.0 GW/h with a growth rate of 35.6% compared to 2010. This meant that by 2014, the unmet demand reached 80 GW/h (Ministry of Mines and Energy, 2011-2012).

Colombia is an example of the necessity for the creation of this type of plant since, there are areas where water is sometimes not properly supplied or arrives contaminated due to the human activities at the source of supply. The most common case being mining, which corresponds to 25% of the general mining activities throughout the country (De La Hoz, 2009). South of Bolivar is a prime example of the effects of mining on the quality of water. The continuous mining activities in the region have led to an increase in the concentrations of cyanide salts in the region of 700 to 800 mg/L. These high values of cyanide salts allow the production of osmotic energy due to its high degree of salinity. The potential of power generation however, comes with a drawback in the form of the toxicity of the water and poses a threat to the general health of the population (Ramos Avilez, 2012). Cyanide in some forms, is highly potent and a fast-acting toxicant. Its toxicity is related to its physicochemical specification (Hilson & Monhemiusb, 2006) and depends on its particular form. Thus, while free cyanide is a very dangerous metabolic poison metal complexes of cyanide vary in toxicity and the function of the ease with which cyanide releases them.

A large number of theoretical research has opened up a positive outlook for potencies between 2.7 and 5.1 $W/m^2$ for waters with salt concentrations between 35 and 60 g/l using flat-sheet cellulose membranes (Achilles, Cath, & Childress, 2009). The general consensus is that in order to produce energy at a competitive price, a minimum power of 5 $W/m^2$ should be reached. The world's first 4 kW osmotic plant was inaugurated by Statkraft on 24 November 2009 in Tofte, Norway. It is estimated that each year 1,600 TW/h could be generated worldwide, and with only 12 TW/h in Norway, would be enough to meet 10% of total electricity demand (Statkraft Company, 2009). Likewise, the possibility of rejection of sodium cyanide by polymer membranes is shown, by increasing the pressure up to a range of 15-20 atm; the maximum separation obtained was about 95% (removal of cyanide). These results show a pattern in the possibility of obtaining energy through sodium cyanide (Murthy & Gupta, 1999).

The implementation of the pilot plant using cyanide water allowed for the evaluation of the potential that this type of water can provide. The experiment took place in the civil engineering sanitation laboratory of the University of Cartagena and modeling laboratory of chemical engineering of the University of Ryerson. The reactor was developed using two stainless steel reactors, the first one with a horizontal configuration and capacity of 16 L and the second with a vertical configuration with a capacity of 10 L. A set of GE Osmonics SE polymer membranes generated the required pressure and a metering tube was used to indicate the potential generated by the water in each test. The results showed the existence of the potential that can be obtained when reusing the cyanide waters and also consolidated the important source for the advancement in this type of energy.

## 2 Osmotic power generation via the PRO process

The principle of this type of energy is based on the use of the pressure generated in the natural phenomenon of osmosis. To have an idea this phenomenon can be considered as an inverse process of the separation energy that the desalination plants use. Instead of using energy to separate a mixture in its constituent elements, the energy in this case is released when the elements are combined. When two solutions with different degree of salinity (salinity gradient) are mixed, a free energy of mixture is released (Elimelech, Yin Yip, Hoover, Phillip, & Tiraferri, 2011).

The performance of the PRO membranes is usually evaluated in terms of the energy density W, which is defined as the output force per unit area of membrane ($W/m^2$). A PRO membrane with a high W is essential because this determines the requirements between the membrane area and the size of the PRO plant to guarantee an energy production capacity. Mathematically, W is determined by:

[1] $W = Jw \Delta P$

Where ΔP is the applied effective hydraulic pressure and $Jw$ the water permeate flow through the membrane. Without considering the effect of polarization, the ideal $Jw$ through a perfect semipermeable membrane can be calculated by the ideal condition equation:

[2] $Jw$=A*($\Delta\pi$−$\Delta P$)

Equation [1] shows that *Jw* is determined as the product of the system permeability to water, *A*, and the net trans-membrane driving pressure, which is the net difference between the osmotic pressure, $\Delta\pi$, and the net hydraulic pressure, $\Delta P$. The density of the power obtained from the PRO process, *W*, can be estimated as the product from multiplying the water flux by the hydraulic pressure:

[3] $W$=Jw*$\Delta P$

[4] $\Delta P = \frac{P_{DS_{in}}+P_{DS_{out}}}{2} - \frac{P_{FW_{in}}+P_{FW_{out}}}{2}$

And similarly for $\Delta\pi$:

[5] $\Delta\pi = \frac{\pi_{DS_{in}}+\pi_{DS_{out}}}{2} - \frac{\pi_{FW_{in}}+\pi_{FW_{out}}}{2}$

Where, the subscripts DS and FW refer to the Draw Solution (the high concentration side) and the Feed Water (the low concentration side), respectively, To determine $\Delta\pi$, the osmotic pressures of the diluted and concentrated solutions must be determined, giving approximate values using the Van't Hoff equation:

[6] $\pi$=$Vi\ Ci\ RT$

Where $Vi$ is the number of ions formed with the solution of solute ($i$ is 2 for NaCL and NaCN), $Ci$ the molar concentration of solute, R is the ideal gas constant equal to 0.082 liters atm/mol, T the absolute temperature in (K), $Ci$ is the total concentration of ions in kmol/m3. On the other hand, using differential equations of equation [3] respect to $\Delta P$, the maximum power density ($Wmax$) occurs when $\Delta P$ is equal to half of the osmotic pressure difference ($\Delta\pi$/2) across the membrane.

[7] $Wmax$=$A$×$\Delta\pi$24

## 3 Experimental Setup

The GE Osmotic membrane (YMAGSP3001) was selected to achieve the stated objectives for the present research. The membrane is a 305-305 mm flat sheet - thin film commonly used for acid concentration, starch/sugar concentration and laundry and dye house wastewater reclamation. These membrane characteristics allows the operation of the osmotic generator at pressures higher than 60 MPa, cyanide concentrations up to 1000 mg/L at different pH ranges. The thin film membranes were further characterized using a JSM-6380LV Electron Microscope Scanner at magnifications of 100x, 1000x, 10000x, and 100000x to analyze the cyanide effects on membrane surface structure. Five membranes samples of 30 x 10 mm size were first immersed for 24 hours in cyanide solutions of 0.0, 25, 50, 100 and 200 NaCN mg/L, respectively. Then, all membrane samples were dried in a vacuum desiccator for 24 hours. Before SEM analysis, Samples were subjected to bombardment of gold ions in a Denton Vacuum Desck IV for two minutes. In addition to the surface study the chemical composition of the membranes was determined using the microscopic scanner and X-ray spectroscopy.

Two osmotic generators were designed and constructed at lab scale, one with a horizontal membrane and another with a vertical membrane orientation as shown in Figure 1. Similar operation schemes have been used in previous studies (Hernández Sanchez, Ríos Ponce Vicencio & Medina, 2010). The two reactor were made of stainless steel with a total volume capacity of 12 L for the vertical and 25 L for the horizontal. The surface area of the membranes were 0.05 $m^2$ for the horizontal and 0.03 $m^2$ for the vertical generator.

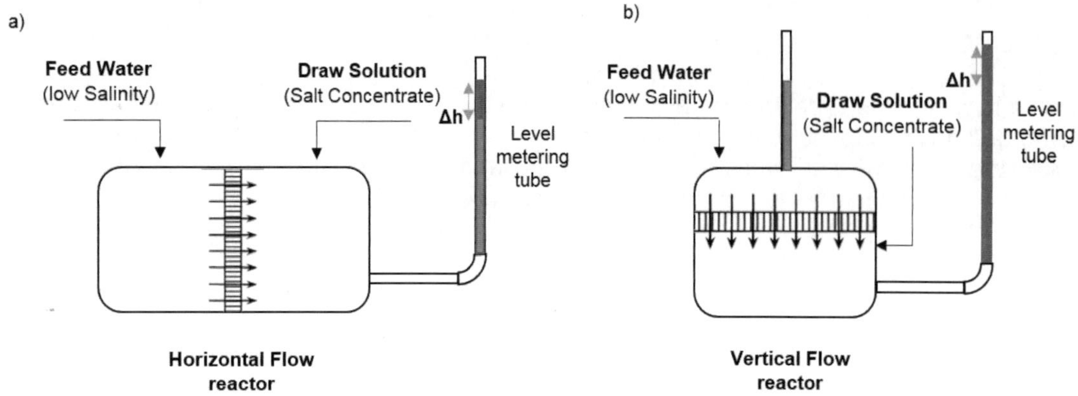


Figure 1: Schematic diagram of the Reactor a) Horizontal Flow configuration b) Vertical Flow Orientation.

For all tests in the vertical reactor five liters were used for feed water and seven liters for draw solution, the feed water tank was one meter high with reference to the base of the reactor. In the case of the horizontal configuration, a constant volume of 25 L was used on the draw solution side and for the feed water side a 500 L supply tank was used, using valves to control the flow to the reactor. In addition, it is important to note that the height difference between the outlet of the fresh water tank and the reactor was approximately 40 cm.

The first process that was performed after verification of the correct operation of the osmotic power generator was the calibration of the conductivity meters YSI and Consort C5010, which allowed reliability in the values thrown during the tests. The calibration consisted of using 15 concentrations of salt and 7 of cyanide, from which a graph was constructed that relates the conductivity and the concentration, allowing to establish equivalences of salinity and relation between NaCL and cyanide.

In the two reactors were initially carried out with clean water tests to determine the behavior of the membranes against it and thus have a point of comparison and calibration. On the other hand, it was also decided to perform tests with synthetic salt water (NaCL), which allowed to simulate the type of water normally used in PRO (seawater) systems with concentrations of 35, 17.5 and 70 g/L. Finally, cyanide was used for the tests, as planned, taking 5 different concentrations of 700, 200, 100, 50 and 25 mg/L.

# 4 RESULTS AND DISCUSSION

## 4.1 Water and membrane characterization

Measurements of the water were made to determine the approximate values of conductivity and therefore of salinity of the samples studied. Fifteen salt concentrations and six for cyanide by the risk with their manipulation. When comparing the conductivities, the NaCN samples are slightly higher than those of NaCl. Taking a sample of 500 mg/L, for cyanuric waters the conductivity was 1283µS/cm and for salt water 1023µS/cm; about 20% higher for the vertical configuration. Taking the same concentration for the horizontal configuration, samples had a conductivity of 1291 µS/cm for cyanided waters and 1060 µS/cm for the salted, approximately 18% higher. These two situations provide a positive outlook for obtaining good results in testing PRO waters with cyanide as the process of osmosis depends on the concentration differences.

With the help of the Microscopic Electron Scanner and specifically with the X-ray spectroscopy test, a series of images can be obtained that allowed the physical chemical analysis of the membranes according to the different concentrations at which the samples were exposed.

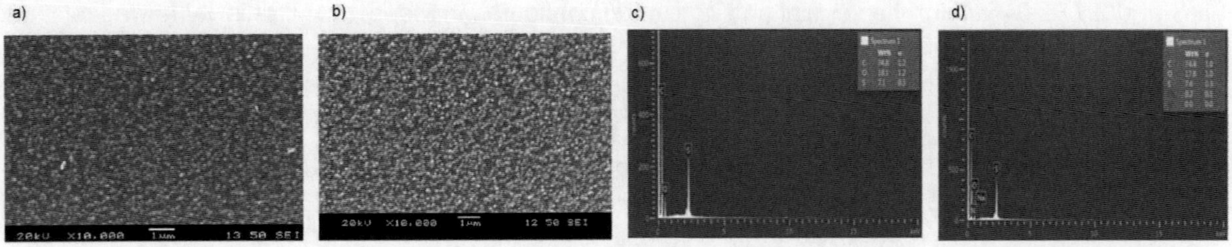

Figure 2: Microscopic Scanner a) white b) 200mg/L NaCN and Dispersive Energy X-ray Spectroscopy c) white d) 200mg/L NaCN for reference samples, active membrane layer.

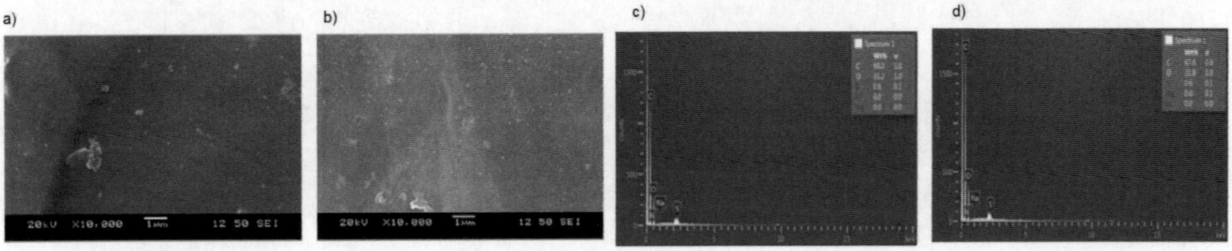

Figure 3: Microscopic Scanner a) blank b) 200mg/L NaCN and Dispersive Energy X-ray Spectroscopy c) blank d) 200mg/L NaCN for reference samples, Support membrane layer.

Figures 2 and 3 show that there were no significant changes in the membrane composition between the blank and the membrane exposed to the different concentrations of salts that were plated in the tests, showing it is not hydrophilic as assumed. Moreover, it should be noted that these samples were only submerged in the cyanide salts and were not subjected to pressure; an additional phenomenon that was not taken into account at this stage of the investigation but to which the membranes have to be exposed to. Low percentages of sulfur were also found. This is because, in order to improve the properties of the polymers, amounts of sulfur between 5 and 10% were added, allowing the polyamides to develop thermostability properties. Besides helping with the elastic and permeability properties, it has been established that with this addition, the aromatic structure of the polymer can withstand 70°C, a 42-68% increase from temperatures they are normally exposed to. This makes it more resistant to high temperatures (Shockravi, Abouzari, and Javadi, 2009).

## 4.2 Osmotic scale reactor operation

### 4.2.1 Fresh Water

Table 1: Fresh water test in an osmotic scale reactor

| Solution | Flow Configuration | Concentration (g/L) | Average Initial h (cm) | Average Final h (cm) | Δh prom (cm) | Time (s) |
|---|---|---|---|---|---|---|
| Distilled Water | Vertical | 0 | 39.1 | 32.33 | -6.8 | 189.5 |
| Tap Water | Horizontal | 0 | 6.4 | 22.55 | 16.2 | 583 |

As show in Table1, in both configurations an immediate increase in the water reference level was observed. In the case of the vertical configuration, around three minutes the system reached a maximum average equilibrium value of 32.33 cm, while for the horizontal configuration, around nine minutes the average height in the measuring tube was 22.55 cm. Therefore in the following tests it was considered that the water alone experiences a pressure head. In addition it was obtained that the flow through the membrane was 3.58E-

04 $m/s$ and 2.77E-04 $m/s$ for the vertical and horizontal configuration respectively. This value was relevant for tests with concentrations of salt and cyanide.

### 4.2.2 NaCL Solution

Table 2: NaCL Solution test in an osmotic scale reactor

| Solution | Flow Configuration | Concentration (g/l) | Average Initial h (cm) | Average Final h (cm) | Δh Average (cm) | Time (s) |
|---|---|---|---|---|---|---|
| NaCL Solution | Vertical | 17.5 | 39.5 | 38.1 | -1.4 | 187 |
| | | 35.5 | 40.0 | 40.74 | 0.7 | 189 |
| | | 71 | 39.8 | 43.1 | 3.3 | 189.6 |
| NaCL Solution | Horizontal | 17.5 | 6.5 | 31.5 | 25.1 | 574.0 |
| | | 35.5 | 6.4 | 34.3 | 27.9 | 576.0 |
| | | 71 | 6.5 | 37.2 | | 584.0 |

The variation of the concentrations by 17.5, 35.5 and 71 g/L show that by increasing salt concentrations the measurement of the meter tubes also increases. By raising the minimum concentration by four times in the case of the reactor with vertical configuration, the head is increased by about 5 cm, while in the horizontal the increase was just under 6 cm. In the same way in the observed water differences, that is to say the variation between the final and initial level with respect to the concentrations handled, a notable increase was observed for both configurations.

Comparing the results obtained from water head, it is identified that for the horizontal configuration these are drastically superior for each one of the concentrations compared with the values obtained in the vertical. In spite of the great difference existing on the basis of this variable it is not possible to define which of the two has a greater potential, since this strictly only depends on the power of density.

Table 3: Membrane permeability and Power density calculation for NaCl solution test

| Flow | Con (g/l) | Ci (MCon) | Δπ (bar) | Δp (bar) | Jw (m/s) | Δπ-Δp (bar) | A (L/m2·h·bar) | W (W/m²) | W max. |
|---|---|---|---|---|---|---|---|---|---|
| Vertical | 17.5 | 0.043 | 12.6 | 0.1 | 2.04E-03 | 12.54 | 1.63E-04 | 2.04E-04 | 4.90E-02 |
| | 35.5 | 0.087 | 26.3 | 0.1 | 2.16E-03 | 26.17 | 8.25E-05 | 2.16E-04 | 5.18E-02 |
| | 71 | 0.174 | 56.1 | 0.1 | 2.27E-03 | 56.01 | 4.05E-05 | **2.27E-04** | **5.45E-02** |
| Horizontal | 17.5 | 0.043 | 12.6 | 0.04 | 5.49E-04 | 12.6 | 4.36E-05 | 2.20E-05 | 1.32E-02 |
| | 35.5 | 0.087 | 26.3 | 0.04 | 5.96E-04 | 26.23 | 2.27E-05 | 2.38E-05 | 1.43E-02 |
| | 71 | 0.174 | 56.1 | 0.04 | 6.37E-04 | 56.07 | 1.14E-05 | 2.55E-05 | 1.53E-02 |

Table 3 shows variable values for the permeability coefficient, varying according to the concentrations. However, a representative value must be found for this membrane and specifically for the solute studied. For this reason, the slopes of the curve are calculated by means of linear regressions, which are described by the function that forms the points.

In the same sense when analyzing the power density produced by the different configurations, it is observed that the vertical is the most effective and that as shown in Graph 1.a. There is a linear proportionality between the osmotic pressure and the generated power, although this power in turn depends directly on the permeability coefficient, a fundamental parameter for the reached results. The maximum values obtained are 2.27E-04 W/$m^2$ for a concentration of 71 g/L in vertical flow, while for the horizontal configuration the maximum value under the same concentration was 2.55E-05 W/$m^2$.

**NaCN Solution**

Table 4: NaCN Solution test in an osmotic scale reactor

| Solution | Flow Configuration | Concentration (mg/l) | Average Initial h (cm) | Average final h (cm) | Δh prom (cm) | Time (s) |
|---|---|---|---|---|---|---|
| NaCN Solution | Vertical | 25 | 43.0 | 40.3 | -2.7 | 186 |
| | | 50 | 43.9 | 41.57 | -2.3 | 190 |
| | | 100 | 41.2 | 42.13 | 0.9 | 184 |
| | | 200 | 43.5 | 44.47 | 1.0 | 188 |
| | | 700 | 43.7 | 45.53 | 1.9 | 186 |
| NaCN Solution | Horizontal | 25 | 6.37 | 30.87 | 24.5 | 570 |
| | | 50 | 6.43 | 32.53 | 26.1 | 568 |
| | | 100 | 6.43 | 33.73 | 27.3 | 578 |
| | | 200 | 6.43 | 34.83 | 28.4 | 580 |
| | | 700 | 6.57 | 35.77 | 29.2 | 578 |

Table 5: Membrane permeability and Power density calculation for NaCN solution test

| Flow | Con (mg/l) | Ci (MCon) | Δπ (bar) | Δp (bar) | Jw (m/s) | Δπ-Δp (bar) | A (L/m2·h·bar) | W (W/m²) | W máx. |
|---|---|---|---|---|---|---|---|---|---|
| Vertical | 25 | 7.29E-05 | 0.0195 | 0.1 | 2.17E-03 | -0.081 | -2.70E-02 | 2.17E-04 | 5.21E-02 |
| | 50 | 1.46E-04 | 0.0388 | 0.1 | 2.19E-03 | -0.061 | -3.58E-02 | 2.19E-04 | 5.26E-02 |
| | 100 | 2.92E-04 | 0.0772 | 0.1 | 2.29E-03 | -0.023 | -1.01E-01 | 2.29E-04 | 5.50E-02 |
| | 200 | 5.83E-04 | 0.1534 | 0.1 | 2.37E-03 | 0.053 | 4.44E-02 | 2.37E-04 | 5.69E-02 |
| | 700 | 2.04E-03 | 0.5263 | 0.1 | 2.45E-03 | 0.426 | 5.75E-03 | **2.45E-04** | **5.88E-02** |
| Horizontal | 25 | 7.29E-05 | 0.0195 | 0.04 | 5.42E-04 | -0.021 | -2.64E-02 | 2.17E-05 | 1.30E-02 |
| | 50 | 1.46E-04 | 0.0388 | 0.04 | 5.73E-04 | -0.001 | -4.86E-01 | 2.29E-05 | 1.38E-02 |
| | 100 | 2.92E-04 | 0.0772 | 0.04 | 5.84E-04 | 0.037 | 1.57E-02 | 2.34E-05 | 1.40E-02 |
| | 200 | 5.83E-04 | 0.1534 | 0.04 | 6.01E-04 | 0.113 | 5.30E-03 | 2.40E-05 | 1.44E-02 |
| | 700 | 2.04E-03 | 0.5263 | 0.04 | 6.19E-04 | 0.486 | 1.27E-03 | 2.48E-05 | 1.49E-02 |

Comparing the values of average hydraulic height reached by the salt and cyanuric waters in the vertical configuration has shown that salt water with one hundred times greater concentration produced a little less than half compared to the cyanide solution. With a concentration of 71g/L for salt water and 700 mg/L for cyanide solution, a height of 3.3 cm and 1.9 cm was produced respectively, showing the potential that exists in the cyanide water. Likewise for the horizontal configuration, values obtained were even more encouraging with the maximum concentration of cyanide of only 700 mg/L and a head of 29.2. In contrast to 30.7 cm for the experiments with common salt at 71 g/L, the most outstanding results in these tests are the proximity of the heads of water and the remoteness of the concentrations used. However, it should be noted that the membrane was selected for use in more aggressive conditions, yet it is a reliable and very telling point of comparison.

A)

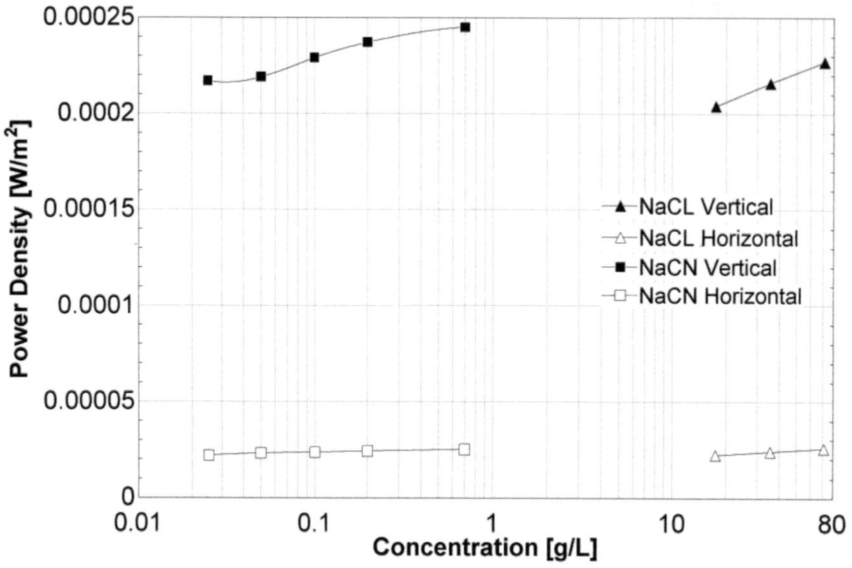


B)

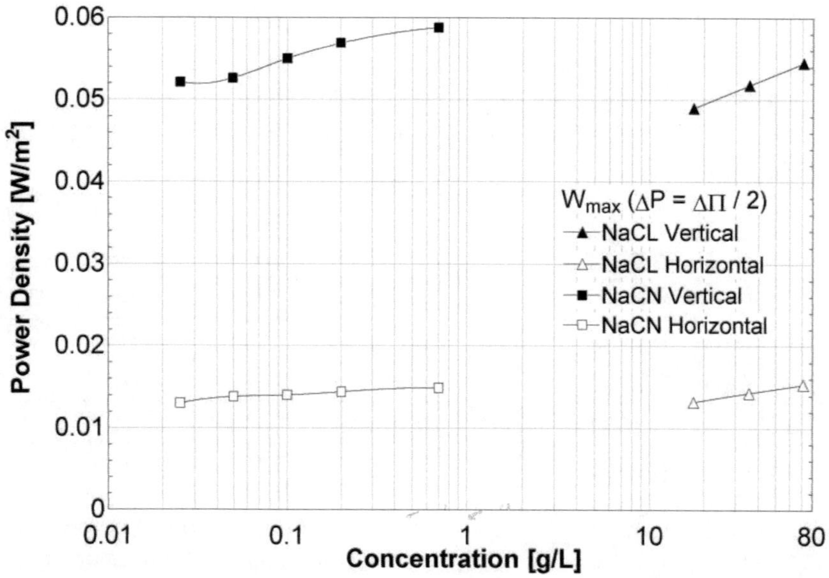


Graphic 1: Variation of the concentration influence on the Power density produced. A) Experimental condition

The average values of energy production according to the maximum concentration used, which corresponds to the concentrations found in the aquifer tailings in the south of Bolivar were 2.45E-04 W/m2 for the vertical configuration. The maximum power density calculated for the system was 5.88 E-02 W/m$^2$. It was possible to determine the partial concordance in the values as the coefficient of permeability in the studied references. These were around 1.87E-09 $ms\text{-}1Kpa\text{-}1$ and 2.02E-09 $ms\text{-}1Kpa\text{-}1$ using operating pressures between 7 and 15 bar in different combinations of sea water, fresh water and brine (Achilli, Cath, & Childress, 2009) (Gang, Sui, Xue, & Tai-Shung, 2015). Also, these studies used modified thin film membranes to improve performance in PRO and FO tubular hollow configuration membranes (Helfer, Lemckert & Anissimov. 2014), reaching densities of up to 3.8 W/m2. Figure 1 shows the results calculated for the tests with both solutions in the two different studied configurations; vertical and horizontal flow. Clearly it is seen in both solutions that the vertical configuration demonstrates a better performance than the horizontal system.

Comparing results, the most significant power density was 5.88 E-02 $W/m^2$ achieved in the vertical configuration, with a thin film membrane designed to operate in RO conditions, under very low pressure and flow times of less than 10 minutes. The calculated density is still far from meeting the needs of the system, as its viability is estimated at 5 W / m2 [Helfer, Lemckert and Anissimov, 2014]. The results expressed in the research shows how small amounts of cyanide salts have a higher performance than that generated using NaCl solutions. The NaCl solution consists of approximate concentrations of brine, which is the source with the highest concentration currently evaluated for this technology. This provides a new possibility to take advantage of contaminated water with cyanide.

**Conclusion**

Estimating the variation of the electric potential as a function of the osmotic pressure, it was found that there was a linear proportionality between them, for both salty and cyanuric waters. In addition, a maximum power density of 5.88E-02 $W/m^2$ was tested, using the GE Osmonic SE polymer membrane with solutions of cyanide concentrations of 25, 50, 100, 200, 500 and 700 mg/l for osmotic membrane reactors of different configurations (vertical and horizontal). This project can act as a preamble for the generation of eco-friendly energy and sustainable development, reusing contaminated water from mining activities in the south of Bolivar in Colombia.

Determining the physicochemical characteristics of the GE Osmonics SE membrane implemented in the research was found to be non-hydrophilic. This means that it does not allow the adhesion of cyanide particles and thus, being suitable to withstand the abrasion to the different concentrations of cyanuric waters to which it was exposed. In addition to the expected carbon and oxygen chains, sulfur content in small proportions of 5 -10% was also found. This guarantees the elasticity of the membrane, avoiding ruptures and allows the obtaining of a high energy potential while maintaining good efficiency in the system.

The reactors built in the University of Cartagena and the University of Ryerson allowed to generate a potential of energy represented in head of pressure by means of the use of a polymeric membrane and based on different configurations, through the tests carried out it was possible to identify that the reactor of vertical configuration is more effective for the use of the osmotic pressures. This is due to the force offered by gravity during the filling process in addition to the immediate contact with the membranes of the water sheets with different concentrations, allowing a more effective osmosis process.

The pilot plants demonstrated good performance under different conditions. A range of power densities was obtained at varying concentrations with the highest value of 2.45E-04 $W/m^2$ being generated at a concentration of 700 mg/L of NaCN. This represents the energetic potential of the developed system. In addition, through good monitoring and management the permeability coefficient of the membrane used was 3.26x10^-5ms^- 1kpa^-1. This is close in order of magnitude to that obtained in previous investigations between 1.89x10^-9ms^- 1kpa^-1 and 1.02x10^-9ms^- 1kpa^-1.

These differences distance these results from those that the scientific community commonly obtains. However, the relationship between the potential obtained between salt water and cyanide water evaluated in the same system opens the possibility of finding great potential using cyanide water in more optimal systems like those found in previous studies.

**Acknowledgements**

The financial support by Colciencias to undertake this work are gratefully acknowledged. The authors would like to sincerely thank Dr. Mehrab Mehrvar, PhD, Titular professor of Chemistry Engineering Department of University of Ryerson and Environmental Modeling research group (GIMA) of University of Cartagena for them useful feedback and comments. Thanks also go to Javier Mouthon PhD, Ciro Bustillo Lepcamte

Rosario Muñoz, MSc, Calendaria Tejada Tovar, M.Sc. and Dalia Moreno Egel, M.Sc. for their valuable assistance with development and scale plants operation. We would also like to thank Shahid Hossaini, Diego Gonzales and Mary Mekhail, University of Guelph for their help.

## References

Achilli, A., & Childress, A. 2010. Pressure retarded osmosis: From the vision of Sidney Loeb to the first prototype installation - Review. *Desalination*, 205-211

Attarde, D. Jain, M. Chaudhary, K. Gupta, SK. 2015. Osmotically driven membrane processes by using a spiral wound module, *Desalination* **361** 81–94

De La Hoz, J. V. 2009. *Economía y conflicto en el Cono Sur de Bolivar.* Cartagena.

Elimelech, M., Yin Yip, N., Hoover, L., Phillip, W., & Tiraferri, A. 2011. Forward with Osmosis: Emerging Applications for Greater. *Environmental: Science & Technology*, 9824 - 9830.

Elimelech, M. 2006. The global challenge for adequate and safe water, J.Water Supply Res. Technol. *AQUA* **55** 3–10.

Gang, H. Sui, Z. Xue, L. Tai-Shung, C. 2015. Progress in pressure retarded osmosis (PRO) membranes for osmotic power generation . *Polimer Science, **51***, 1-27.

Helfer, F. Lemckert, C. & Anissimov, Y. 2014. Osmotic Power with Pressure Retarded Osmosis: Theory, Performance and Trends - A Review. *Journal of Membrane Science, **453***, 337-358.

Hernadez, A. Rios, G. Vicencio, A. 2010. *Generacion de energia a partir de la presion osmotica existente entre aguas con diferentes gradientes salino.* Tamaulipas, Mexico.

Hilson, G. & Monhemiusb, A. 2006. Alternatives to cyanide in the gold mining industry: what prospects for the future? *Journal of Cleaner Production*, 1158-1167.

Ministerio de Minas y energía. (2011- 2012). *Memorias al Congreso de la República.* Colombia: Ministerio de Minas y energía.

Murthy, Z., & Gupta, S. 1999. Sodium cyanide separation and parameter estimation for reverse osmosis thin film composite polyamide membrane. *Journal of Membrane Science*, 89-103.

Quiñones Bolaños, E. 2005. *Evaluación del grado de remoción de cianuro de sodio y cianato de sodio de aguas contaminadas sintéticas por el método de pervaporación utilizando membranas elastomericas bajo diferentes condiciones de operación.* University of Guelph.

Ramos, V. (2012). *Remoción de cianuro en relavés auríferos utilizando bichar reducido a partir de talles de Gliricidia sepium.* Cartagena: Universidad de Cartagena.

R.J. Aaberg, 2003. Osmotic power: a new and powerful renewable energy source? *Refocus* **4** 48–50.

Sharif, A. Merdaw, A. Aryafar, M. Nicoll, P. 2014. Theoretical and Experimental Investigations of the Potential of Osmotic Energy for Power Production Membranes, *Membranes*, **4**, 447-468;

Statkraft. 2009. *Osmotic power - A new, renewable energy source.* Alemania: Statkraft Publishmen.

Leadership in Sustainable Infrastructure
*Leadership en Infrastructures Durables*

Vancouver, Canada
May 31 – June 3, 2017/ *Mai 31 – Juin 3, 2017*

# REMEDIATION IN CHALLENGING ENVIRONMENTS: PRELIMINARY NUMERICAL SIMULATION AND MEASUREMENT OF UNFROZEN WATER CONTENT FOR AN OUTDOOR PILOT-SCALE BIOPILE IN COLD CLIMATE

Kim, Jihun[1], and Chang, Wonjae[1, 2]
[1] University of Saskatchewan, Canada
[2] wonjae.chang@usask.ca

**Abstract:** Cold-adapted bacteria are able to remain metabolically active in petroleum hydrocarbon-contaminated soils at low temperatures, and thus have been the basis for bioremediation efforts at cold-region sites. Recent studies have consistently reported significant microbial activity in partially frozen and frozen habitats, including in soils, sea ice cores and even at the interface between ice and water. Unfrozen liquid water in freezing and frozen soils is likely prerequisite for extended microbial activity. Predicting the presence and retention of unfrozen liquid water in cold site soils can be useful for planning, managing and implementing bioremediation for petroleum-contaminated soils at remote northern sites. However, available numerical simulation tools for predicting soil water content have not been extensively considered in bioremediation research for cold sites. For this research, a preliminary study was performed to calibrate a TEMP/W model for predicting soil temperatures and unfrozen water content in field-aged, petroleum-contaminated, clayey soils from outdoor pilot-scale biopiles exposed to winter temperatures of Saskatoon, Saskatchewan (Canada). The simulation of unfrozen water content in the biopiles subjected to representative winter temperatures was conducted using TEMP/W. The preliminary study predicted that significant quantities of unfrozen water remain available in the biopile during the seasonal transitional and winter periods. By comparing the simulated and measured unfrozen water data for the site soils, this study conservatively predicted an abundance of unfrozen water in the site soils during the winter season.

## 1 Introduction

Petroleum hydrocarbons are one of the most frequently identified contaminants in Canada, including at northern sites, mainly due to intensified anthropogenic activities and associated impacts. Bioremediation is often considered a cost-effective and less-destructive remediation technology for petroleum-contaminated soils at cold sites (Snape et al. 2008). The potential to enhance hydrocarbon biodegradation in cold site soils by stimulating cold-adapted hydrocarbon-degrading bacteria has been frequently reported over the last ten years (Whyte et al. 2001). More recently, microbial activity extended to sub-temperatures in freezing and frozen soils has been reported, implying that unfrozen liquid water content plays a critical role for maintaining metabolic activity of microbial survivors when temperatures are near or below the freezing-depression point of freezing and frozen soils (Panikov et al. 2006). Temperatures alone cannot explain the unique adaptations of microorganisms in freezing soils, where microbial habitats are subjected to multiple stresses. For example, Deming (2002) proposed an ecophysiological group, *eutectophiles*, to designate unique microorganisms that take advantage of soil phase changes at the water-ice interface. The microbial habitat of eutectophiles at the water-ice interface is influenced by a combination of factors including temperature, solute concentrations, and the physical state of the water.

It has been frequently observed that changes in unfrozen water availability is significantly correlated with the seasonal onset and inactivation of microbial activity in freezing and frozen soils (Chang et al. 2011). Rivkina et al. (2000) reported that microbial activity is correlated with the thickness of unfrozen water films in freezing soils. Similarly, Panikov et al. (2006) indicated that $CO_2$ evolution in freezing soils is related to changes in unfrozen water content. However, these prior studies are small lab-scale experiments using uncontaminated soils.

Some low-temperature studies have considered field-aged, petroleum hydrocarbon-contaminated sub-Arctic soils. Significant in-situ soil gas concentrations ($CO_2$ and $O_2$) were measured at a sub-Arctic site in Norway to estimate microbial respiration in the active and permafrost layers of oil-contaminated soils during a seasonal transition period and winter (Rike et al. 2003; 2005). Chang et al. (2011) observed a significant correlation between temporal changes in unfrozen water content and apparent microbial respiration activity ($CO_2$ and $O_2$) in soils subjected to varying site-representative freeze/thaw temperatures in an indoor, pilot-scale biodegradation experiment. However, that study was conducted in a laboratory by mimicking site temperatures using a temperature-programmable cold room.

On the other hand, the relationship between unfrozen water content and physical soil properties has been studied (Andersland and Ladanyi 2004). Empirical equations and predictive numerical soil thermal models are currently available to estimate unfrozen water content in various types of unsaturated soils (e.g., TEMP/W). Yet, a comparison of predicted unfrozen water content over time and measured unfrozen water data obtained for an outdoor biopile of field-aged petroleum-contaminated soils has rarely been reported.

The objective of this study is to compare simulated unfrozen water content with measured unfrozen water content in field-aged petroleum hydrocarbon-contaminated soils during seasonal freezing in an outdoor pilot-scale biopile installed in Saskatoon, Saskatchewan (SK), Canada. This preliminary study calibrated the numerical simulation tools and associated input parameters specifc to the site soils. This study is particularly important for planinng and managing practices for contaminated site remediation at remote cold sites, where only short summers are allowed for active site monitoring and soil treatment.

## 2 Materials and methods

### 2.1 Site soil

The site soils are field-aged, petroleum hydrocarbon-contaminated clayey soils. TPH concentrations were determined using the Canada-Wide Standard for Petroleum Hydrocarbons (CWS PHC) in Soil - Tier 1 Method. Briefly, wet soil samples (10 g) were placed in a cellulose extraction thimble (Gerhardt GmbH & Co. KG, Germany) with a surrogate standard (o-terphenyl; Sigma Aldrich Canada) for estimating hydrocarbon extraction efficiencies. For hydrocarbon analyses, approximately 150 mL of a 50:50 hexane/acetone (v/v) solvent was used for hydrocarbon extraction from the soil samples by automatic Soxhlet extraction (Gerhardt Soxtherm, Germany). Extract solutions were filtered through a column containing silica gel and sodium sulfate to remove polar compounds before 20 mL of a 50:50 hexane/ dichloromethane (DCM) solvent mixture was passed through the same column. The extracts obtained after the silica gel cleanup were concentrated by a nitrogen gas blow-down concentrator. The final volumes of concentrated extracts were measured. Hydrocarbons in the concentrated extract solutions were analyzed using gas chromatography (7890A, Agilent with J&W DB-1HT capillary column) with a flame ionization detector (FID).

### 2.2 Viable hydrocarbon degraders

For microbial enumeration, 10 g of site soils were added to 95 mL of phosphate buffer solution (PBS). Serial soil dilutions ($10^{-1}$ to $10^{-5}$) were prepared after a 30-min homogenization of the soils in the PBS solution. 0.1-mL aliquots of the soil dilutions were spread onto Bushnell Haas (BH) agar plates (0.2 g $MgSO_4$, 0.02 g $CaCl_2$, 1 g $KH_2PO_4$, 1 g $K_2HPO_4$, 1 g $NH_4NO_3$, 0.05 g $FeCl_3$ and 20 g agar in 1 L distilled water; Sigma Aldrich Canada). 0.1 mL of filter-sterilized diesel was added onto the plates as the sole source of carbon. Plates were incubated at 17 °C for 9 days. Colony forming units (CFU) were counted after the incubation.

### 2.3 Outdoor biopile construction

Prior to winter, a set of pilot-scale biopiles was constructed in September and October, 2015, at an outdoor soil remediation facility (PINTER & Associates) in Saskatoon. The size of the biopiles is approximately 2 m long, by 2 m wide and 1 m high. Approximately 3.2 tons of contaminated soils were used for each biopile. A drainage layer (gravel and collection pipes) was installed at the bottom of each biopile. Soil gas collection tubes were embedded in the biopile. Soil gases ($CO_2$ and $O_2$) were monitored using MX6i multi gas detectors (Industrial scientific, USA). Two sets of real-time soil temperature and water content sensors (5TM, Decagon devices, USA), along with a data logger (EM50, Decagon devices, USA), were installed in the top (at 90 cm) and bottom (at 10 cm) layers of the outdoor biopile. Soil temperature and water content in the biopile were monitored every six hours.

### 2.4 Preliminary simulation of soil temperatures and unfrozen water content

Soil temperature and unfrozen water content were simulated using TEMP/W (Geo-slope International Ltd.). An air temperature profile was constructed based on 10 years of historical air temperature data for Saskatoon (Environment Canada). The outside air temperature profile was then imported into TEMP/W to predict temperatures and water content in the soils subjected to seasonal freezing conditions. The simulated data were compared to the measured data.

## 3 Results and discussion

### 3.1 Biostimulation potential prior to the biopile experiment

Microcosm experiments were conducted with nutrient-amended and unamended soils, and viable hydrocarbon degraders were enumerated and compared. A C:N ratio of 100:1 was used for the site soil microcosms and the enumeration of viable hydrocarbon-degrading bacteria. As shown in Fig. 1, significant biostimulation potential was indicated in the nutrient-amended site soils by large populations of viable diesel-degrading bacteria ($2.2 \times 10^4$ CFU/ g) compared to in the unamended soils ($4.5 \times 10^3$ CFU/ g), based on the preliminary soil microcosm testing. The difference was statistically significant with $p < 0.05$ (t-test, GraphPad).

For further adjustment of the nutrient dosage, a range of C:N ratios from 100:0.5 to 100:2 were tested. These nutrients were applied in soil microcosms with 100 g soils, while soil respiration activity ($CO_2$ and $O_2$) was monitored. As shown in Fig. 2, the site soil readily responded to the different nutrient amendments. $CO_2$ production and $O_2$ consumption were significant greater in the nutrient-amended site soils compared to the unamended soils. Of the nutrient amendments tested, a C:N ratio of 100:0.8 exhibited the highest $CO_2$ production and $O_2$ consumption, based on the early response of the microcosm to the nutrient-amendment within 48 hours. The C:N ratio of 100:0.8 was used during the biopile setup for biostimulation.

It is important to indicate the presence of viable indigenous bacteria capable of degrading target hydrocarbon fractions (diesel fractions in this study), as well as determine favorable nutrient amendments, prior to installation of large-scale biopiles, in order to achieve successful bioremediation in practice. In particular, cold region sites have a short seasonal window for active soil treatment and field logistics are a concern. Characterizing the favourable soil conditions for stimulating indigenous hydrocarbon degraders is prerequisite. Changes in soil gas are a useful indicator, and soil gas monitoring ($CO_2$ production and $O_2$ consumption) has been widely employed to estimate microbial enhancement during bioremediation (Chang and Ghoshal 2014). Soil gases are monitored to demonstrate microbial response during biodegradation (Walworth et al. 2013; Akbari and Ghoshal 2015).

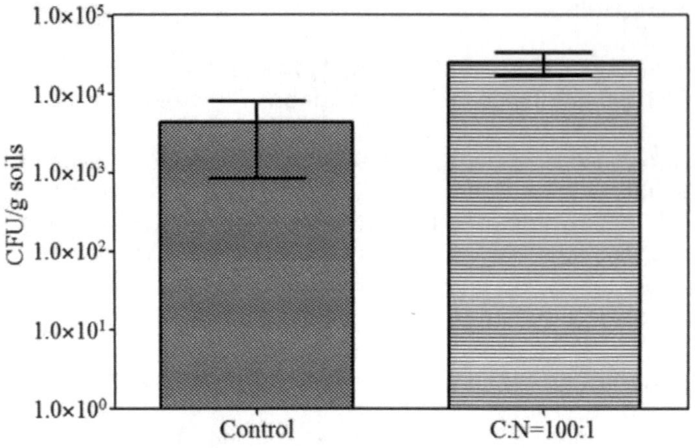


Figure 1: The microbial enumeration of hydrocarbon-degrading bacteria from the nutrient-amended soils with a C:N ratio of 100:1 and from unamended site soils. Enumeration was performed using diesel-spiked Bushnell Hass plates at 25 °C (Error bar: standard error)

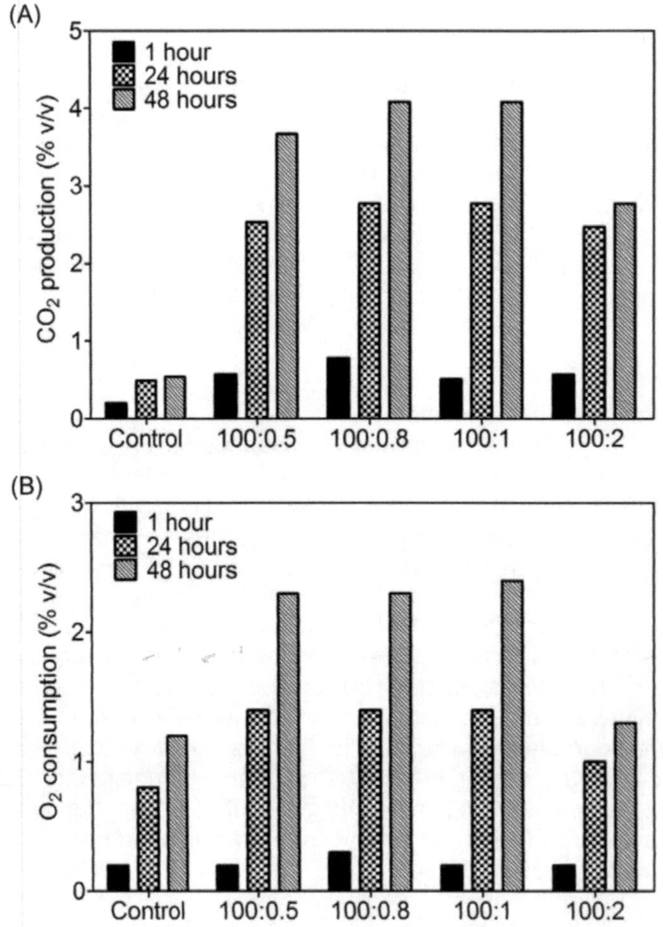


Figure 2: Soil $CO_2$ production (A) and $O_2$ consumption (B) in the site soils in response to different C:N ratios in the nutrient amendment.

### 3.2 Preliminary calibration of TEMP/W model

Prior to the field biopile experiment, temporal changes in the content and distribution of unfrozen water in the biopile were predicted using the historical temperatures (10-year data) and the TEMP/W model. The field biopile was directly exposed to ambient temperature conditions and snow cover. However, the simulation did not consider the effect of snow cover and wind. The highest, average and the lowest ambient temperature profiles constructed from historical data were used for the simulations, producing a range of unfrozen water content profiles as a result. In the present study, the unamended biopile was considered first for as a baseline study.

As shown in Fig. 3, the distributions of simulated unfrozen water content in the biopile during seasonal freezing is illustrated. Significant quantities of unfrozen water could be available in the bottom layers of the biopiles until early November, implying that soil conditions could accommodate microbial populations as long as other soil parameters, such as oxygen and nutrient diffusion, do not limit microbial activity. Although free unfrozen water is limited beginning in the middle of November, small amounts of unfrozen water were still measureable in microsites in the middle of November.

The predicted soil temperatures, based on the highest, average and lowest ambient temperature profiles constructed using historical climate data, were generated and compared with the measured soil temperatures from the biopiles in the field. As shown in Fig. 4, the measured soil temperatures are closer to the soil temperatures corresponding to the 10-year average soil temperature scenario. However, the measured soil temperatures were slightly higher than soil temperatures simulated using the 10-year average ambient temperatures of the site. On-site soil temperatures in the biopile (unamended) remained near 0 °C and fluctuated between 0 and -5 °C during winter. The soils were not completely frozen until January of the following year (2016). In reality, unfrozen liquid water content did not decrease during the period in which the biopile soil temperature was below 0 °C, but were not lower than the typical freezing-depression point of clayey soils (around -5 °C). This result implies that the biopile soils may exhibit the metabolic activity of cold-adapted microorganisms, even though the kinetics of microbial activity are slowed due to the low temperature. In the middle of November, the simulation using 10-year average temperatures indicated the depletion of unfrozen water content. However, measured unfrozen water content from the biopile is closer to the simulated unfrozen water content that used the highest soil temperature scenario.

The on-site climate conditions (e.g., snow cover) and soil properties (e.g., clay content) seem to influence the retention of unfrozen water. The simulation of soil temperatures and unfrozen water content prior to the field biopile implementation provided indications of temporal changes in unfrozen water availability over time during a freezing period. Currently, more sensitive analyses of the effects of a variety of parameters are being conducted to produce better results. In addition, the on-site data for microbial respiration activity and hydrocarbon biodegradation in the biopiles are being obtained, which will further highlight the implications of this study's findings for planning and management of bioremediation for petroleum-contaminated soils in cold climates. The present study provides a framework for investigating the feasibility extended bioremediation during seasonal freezing through the prediction of soil temperatures and unfrozen water content in field-aged petroleum hydrocarbon-contaminated soils.

## 4 Summary and future studies

This preliminary study indicated sufficient unfrozen water availability during freezing in both simulated and measured soil data. In general, existing numerical tools for predicting temporal changes in the distributions of soil temperatures and unfrozen water contents are available and applicable to contaminated soils in pilot-scale biopiles. The significant retention of unfrozen water, which is likely required for maintaining microbial activity in partially frozen and frozen soils, was observed in both the simulation and measurements of unfrozen water content in the testing biopile. With future investigations of microbial activity, soil properties and on-site climate conditions, the simulation study may produce more accurate results. The current study implies the feasibility of predicting soil temperatures and unfrozen water availability using available numerical tools, which has particularly important implications for the planning and management of contaminated site remediation in remote and cold regions.

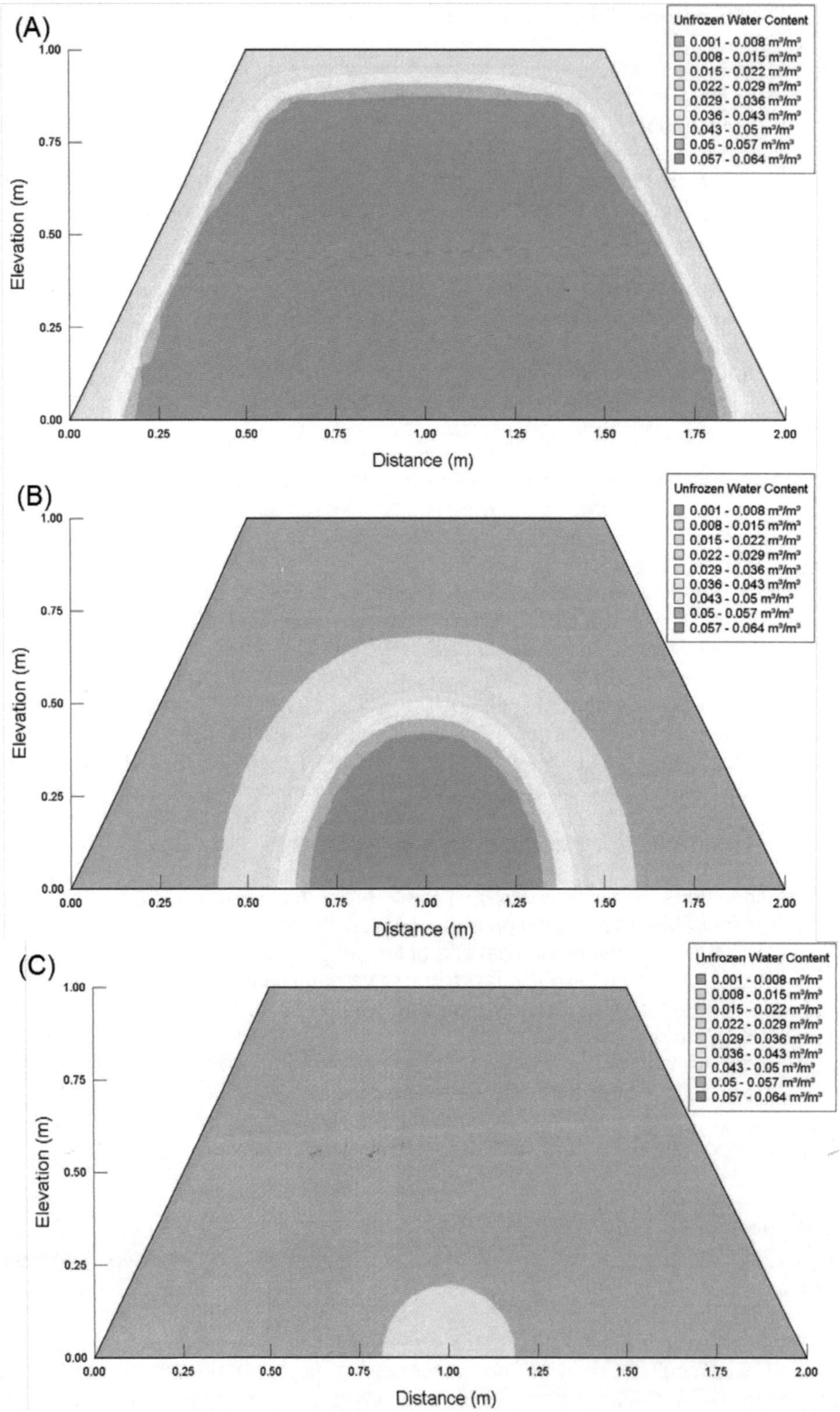


Figure 3: Results of the preliminary calibration of the TEMP/W model. The contour maps show simulated unfrozen water distributions at the end of October (A), in early November (B) and in the middle of November (C).

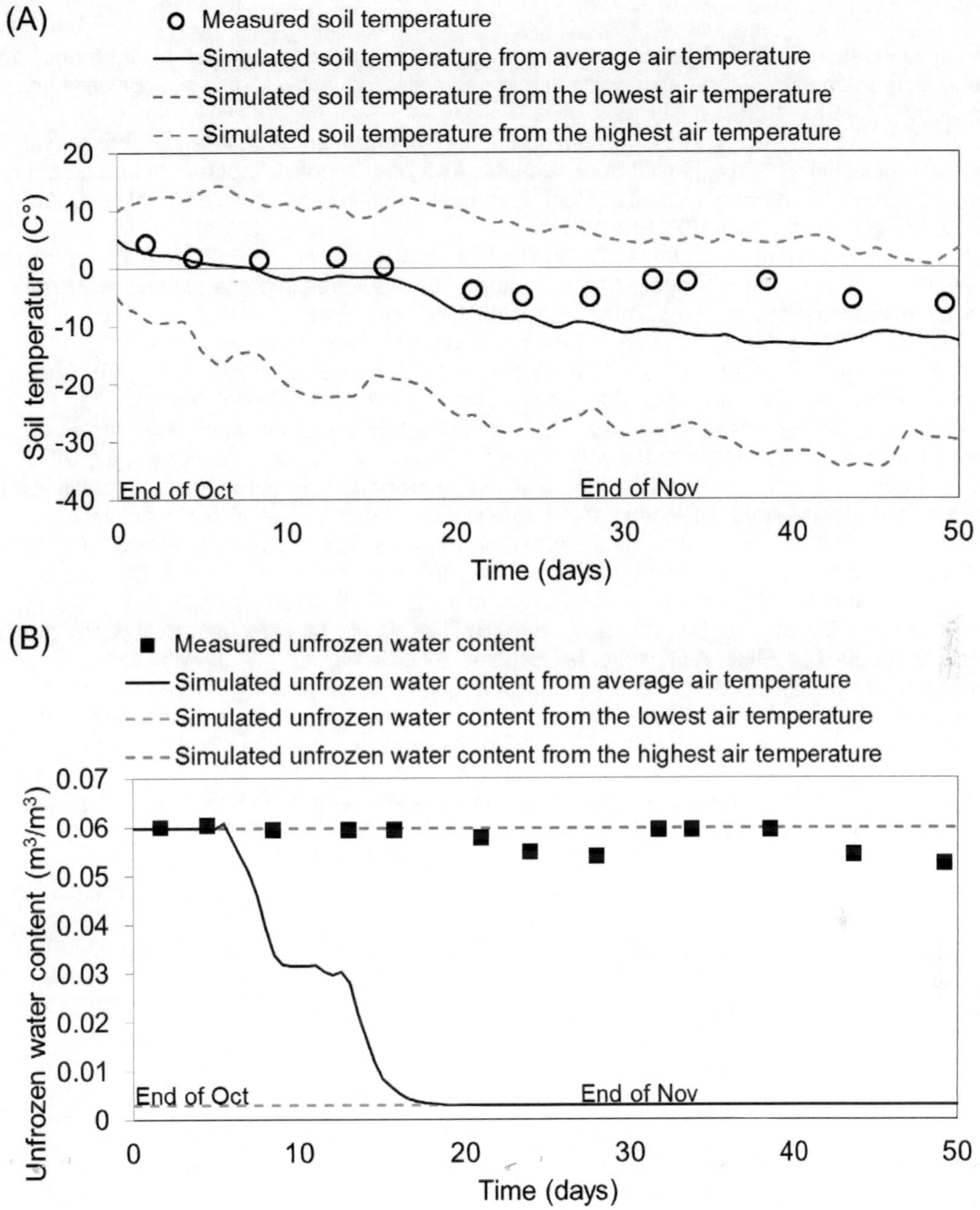


Figure 4: The preliminary comparison of the simulated data with the measured data (soil temperature and unfrozen water measurements from the end of October and November). The measured unfrozen water contents were obtained from a preliminary dummy biopile without the soil treatment and plastic cover. The plots show (A) soil temperatures and (B) unfrozen water content.


## Acknowledgements

We thank Ryan Riess, Dustin Hicke and Lawrence Pinter from PINTER & Associates Ltd. and Prairie Waste Management Ltd. in Saskatoon, as well as the Natural Sciences and Engineering Research Council of Canada (NSERC; NSERC EG 488995).

## References


Akbari, A. and Ghoshal, S. 2015. Effects of diurnal temperature variation on microbial community and petroleum hydrocarbon biodegradation in contaminated soils from a sub-Arctic site. *Environmental Microbiology*, **17**, 4916–4928.

Andersland, O.B. and Ladanyi, B. 2004. *Frozen ground engineering*, 2nd ed., Wiley, Hoboken, NJ, USA.

Chang, W. and Ghoshal, S. 2014. Respiratory quotients as a useful indicator of the enhancement of petroleum hydrocarbon biodegradation in fi eld-aged contaminated soils in cold climates. *Cold Regions Science and Technology*, **106–107**, 110–119.

Chang, W. Klemm, S. Beaulieu, C. Hawari, J. Whyte, L. and Ghoshal, S. 2011. Petroleum hydrocarbon biodegradation under seasonal freeze-thaw soil temperature regimes in contaminated soils from a sub-Arctic site. Environmental Science & Technology, **45**, 1061–6.

Deming, J.W. 2002. Psychrophiles and polar regions. *Current Opinion in Microbiology*, **5**, 301–9.

Panikov, N.S. Flanagan, P.W. Oechel, W.C. Mastepanov, M. a. and Christensen, T.R. 2006. Microbial activity in soils frozen to below −39°C. *Soil Biology and Biochemistry*, **38**, 785–794.

Rike, A.G. Haugen, K.B. Børresen, M. Engene, B. and Kolstad, P. 2003. In situ biodegradation of petroleum hydrocarbons in frozen arctic soils. *Cold Regions Science and Technology*, **37**, 97–120.

Rike, A.G. Haugen, K.B. and Engene, B. 2005. In situ biodegradation of hydrocarbons in arctic soil at sub-zero temperatures—field monitoring and theoretical simulation of the microbial activation temperature at a Spitsbergen contaminated site. *Cold Regions Science and Technology*, **41**, 189–209.

Rivkina, E.M. Friedmann, E.I. and Mckay, C.P. 2000. Metabolic Activity of Permafrost Bacteria below the Freezing Point. *Applied and Environmental Microbiology*, **66**, 3230–3233.

Snape, I. Acomb, L. Barnes, D.L. Bainbridge, S. Eno, R. Filler, D.M. Plato, N. Poland, J.S. Raymond, T.C. Rayner, J.L. Riddle, M.J. Rike, A.G. Rutter, A. Schafer, A.N. Siciliano, S.D. and Walworth, J.L. 2008. Contamination, regulation, and remediation: an introduction to bioremediation of petroleum hydrocarbons in cold regions, in: Filler, D.M., Snape, I., Barnes, D.L. (Eds.), *Bioremediation of Petroleum Hydrocarbons in Cold Regions*, Cambridge University Press, Cambridge, Cambridgeshire, England, pp. 1–37.

Walworth, J. Harvey, P. and Snape, I. 2013. Low temperature soil petroleum hydrocarbon degradation at various oxygen levels. *Cold Regions Science and Technology*, **96**, 117–121.

Whyte, L.G. Goalen, B. Hawari, J. Labbé, D. Greer, C.W. and Nahir, M. 2001. Bioremediation treatability assessment of hydrocarbon-contaminated soils from Eureka, Nunavut. *Cold Regions Science and Technology*, **32**, 121–132.

Leadership in Sustainable Infrastructure

*Leadership en Infrastructures Durables*

**Case Study**

Vancouver, Canada

May 31 – June 3, 2017/ *Mai 31 – Juin 3, 2017*

# REMEDIATION OF OIL SANDS PROCESS-AFFECTED WATER (OSPW) BY INDIGENOUS MICROORGANISMS BASED BIOFILTRATION

Lei Zhang [1], Yanyan Zhang [1], Mohamed Gamal El-Din [1,2]

[1] Department of Civil and Environmental Engineering, University of Alberta, Edmonton, Alberta, T6G 1H9, Canada.

[2] mgamalel-din@ualberta.ca

## 1 Introduction

Large amount of oil sands process-affected water (OSPW) has been generated by the extraction of bitumen from the oil sands using hot alkaline water (Hwang et al. 2013). Naphthenic acids (NAs), which attribute the major toxicity of OSPW, are resistant to conventional wastewater treatments. Effective NA degradation processes are eagerly needed for the remediation and safe discharge of the OSPW. Previous researchers found that the toxicity of the OSPW towards to selected test organisms can be reduced and the NAs can be partially degraded through biological treatment. Biofiltration shows a clear advantage in terms of low energy cost. Biofiltration has been considered as one of the most important biodegradation processes for the removal of organic pollutants from wastewater (Hollender et al. 2009). The indigenous microorganisms based biofiltration system shows high possibility to be used for the OSPW treatment.

## 2 Objective

The main objective of this research was to investigate the efficiency of indigenous microorganisms based fixed bed biofilm reactor to treat OSPW using sand as a media.

## 3 Methods

Pure sand was used as a media for the establishment of the biofiltration system. Microorganisms from the raw OSPW are supposed to attach and form biofilm on the sand media when the OSPW is going through the fix-bed bioreactor. Sand media sample was collected for DNA extraction and applied for the qPCR detection three times. Z stack technology from the confocal microscope was used to measure the thickness of the biofilm on the sand surface two times. Next generation sequencing technique was used for the mega genomic sequencing analysis of the microorganisms on the biofilter. After the mature of the biofiltration system, the effect of the biofiltration on the remediation of the OPSW was assessed through the OSPW recirculation. Biofiltered OSPW was assessed in terms of NAs and toxicity towards *Vibrio fischeri.*

## 4 Results

Under steady state condition, the number of total bacteria on the biofilter media was $10^9$ copies per gram of sand media and the thickness of the biofilm on the sand filter was 36.3 μm. Through the 12 times of circulation on the biofilter, classical NAs from raw OSPW were removed by 39.05% (Figure 1). Microtox® bioassay was used for the toxicity analysis, and it was found that the toxicity of the OSPW toward *Vibrio fischeri* decreased from 27.27% to 20.33% (5 min) and from 31.22% to 24.06% (15 min) after 12 times of circulation on the biofilter. The microbial community structure investigation showed that *Rhodococcus spp* was the dominating bacterial species on the biofilter, which indicated that *Rhodococcus spp* played a critical role during the biodegradation of NAs.

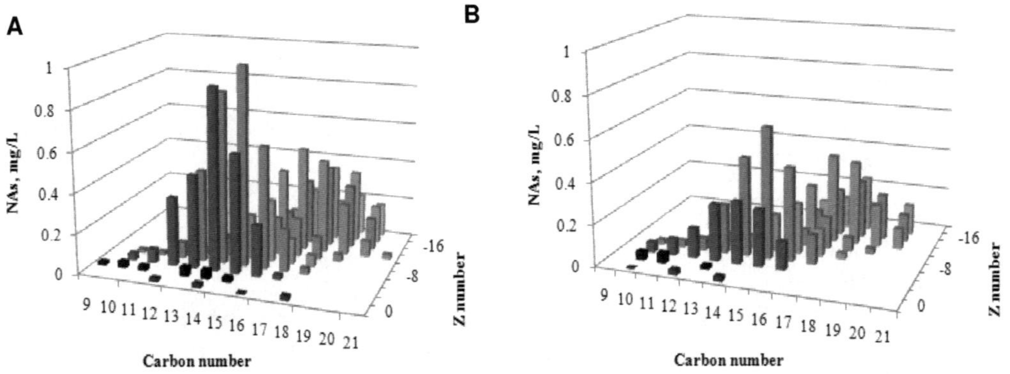


Figure 1: Removal of NAs from the OSPW by biofiltration. A: NAs concentration from raw OSPW; B: NAs concentration from biofiltered OSPW.

## 5 Conclusions

Indigenous microbes based fixed-bed biofiltration system can be established efficiently by using raw OSPW as influent wastewater. The mature biofiltration system can remove the NAs from OSPW effectively.

## Acknowledgements

This work was supported by grants from the Helmholtz-Alberta Initiative through Alberta Environment and Parks' ecoTrust Program, and a Natural Sciences and Engineering Research Council of Canada (NSERC) Senior Industrial Research Chair (IRC) in Oil Sands Tailings Water Treatment through the support by Syncrude Canada Ltd., Suncor Energy Inc., Shell Canada, Canadian Natural Resources Ltd., Total E&P Canada Ltd., EPCOR Water Services, IOWC Technologies Inc., Alberta Innovates Energy and Environment Solution, and Alberta Environment and Parks.

## References

Hollender, J., Zimmermann, S.G., Koepke, S., Krauss, M., McArdell, C.S., Ort, C., Singer, H., von Gunten, U. and Siegrist, H. 2009. Elimination of organic micropollutants in a municipal wastewater treatment plant upgraded with a full-scale post-ozonation followed by sand filtration. *Environ Sci Technol*, **43**(20): 7862-7869.

Hwang, G., Dong, T., Islam, M.S., Sheng, Z., Pérez-Estrada, L.A., Liu, Y. and Gamal El-Din, M. 2013. The impacts of ozonation on oil sands process-affected water biodegradability and biofilm formation characteristics in bioreactors. *Bioresour Technol*, **130**: 269-277.

Leadership in Sustainable Infrastructure
*Leadership en Infrastructures Durables*

**Case Study**

Vancouver, Canada
May 31 – June 3, 2017/ *Mai 31 – Juin 3, 2017*

# REMOVAL OF A MODEL NAPHTHENIC ACID – TRANS-4-PENTYLCYCLOHEXANE CARBOXYLIC ACID – USING MESOPOROUS CARBON XEROGEL ADSORBENT

Ibrahim, Mohamed D.[1], Messele, Selamawit A. [2], and Gamal El-Din, Mohamed [3,4]
[1,2,3] Department of Civil & Environmental Engineering, University of Alberta, Edmonton, AB, Canada T6G 1H9
[4] mgamalel-din@ualberta.ca

## 1 Research Overview

Due to Alberta's zero discharge approach, the oil sands process-affected water (OSPW) is stored in-situ for further treatment before its release to the surrounding environment (Alberta-Government 2014; Brown et al. 2013). OSPW is toxic to aquatic and terrestrial organisms due to the presence of organic matters, such as naphthenic acids (NAs), and non-organic constituents, such as heavy metals (Grewer et al. 2010; Jensen-Fontaine et al. 2014). NAs are a complex mixture of alicyclic and aliphatic carboxylic acids with general formulae of $C_nH_{2n}+ZO_x$, $C_nH_{2n+Z}O_xS$, and $C_nH2_{n+Z+1}O_xN$, where "n" is the carbon number ($7 \leq n \leq 26$), "Z" is zero or an even negative integer ($0 \leq -Z \leq 18$) that specifies the hydrogen deficiency resulting from rings or unsaturated bonding formation, and x represents the number of oxygen atoms ($2 \leq x \leq 6$) in the NA molecules (Grewer et al. 2010). Many recent studies have discussed the OSPW remediation using chemical, biological, physical, and a combination of two or more of these methods.

Adsorption is one of the promising treatment techniques that show a potential in removing the NAs from OSPW. Activated carbon (AC) is usually used as an adsorbent for OSPW treatment. AC shows high removal compared to other adsorbents such as petroleum coke, but the high cost of the AC makes it a less favourable solution for OSPW reclamation process. Carbon xerogel (CX) is a mesoporous carbon based adsorbent that can be used to remove the NAs from the OSPW for the safe discharge of the treated water to the environment.

## 2 Innovation

In this study, two CX adsorbents (i.e., CX-5.5 and CX-6.9), having different textural properties were prepared by the polycondensation of resorcinol and formaldehyde using two different solution pH values (pH 5.5 and pH 6.9). Trans-4-Pentylcyclohexane (TPCA) model compound was used to evaluate the effectiveness of the two adsorbents in removing the TPCA from aqueous solutions. TPCA stock solution was prepared in pH 8 phosphate buffer to simulate the OSPW water matrix. As a positive control, commercially available granular activated carbon (GAC) was used to assess the removal of TPCA from the water. Characterization result revealed that the GAC has the highest surface area (976 $m^2/g$) followed by CX-5.5 (573 $m^2/g$) then CX-6.9 (391 $m^2/g$). The TPCA isotherm was developed using all of the three adsorbents. Preliminary results show that the CX-5.5 removal percentage was close to that of the GAC for concentrations of TPCA up to 60 mg/L. The results highlight that the mesoporous CX 5.5 can be an alternative to the commercially available GAC in removing the TPCA.

## 3 Lessons Learned

Results show that the effect of the preparation pH on the CX was significant. This effect was reflected by the surface area and the pore sizes. GAC shows the highest TPCA removal (86%) after 24 hours, while CX-5.5 achieved 63% removal in the same time period. On the other hand, CX-6.9 shows almost no removal after 24 hours. This can be attributed to the adsorbent surface area and the internal pore volume for each adsorbent. The advantage of using the CX5.5 is evident when normalizing the adsorption capacity based on the adsorbent surface area. The CX5.5 removes about 0.11 $mg/m^2$, while the GAC only 0.08 $mg/m^2$ this shows the potential of the CX5.5 in removing the NAs model compound. The difference of removal can be explained by the mesoporousity of the CX-5.5. To fully understand the kinetics of the TPCA removal, more experiments need to be performed considering different temperatures; pHs as well as different solution ion strengths.

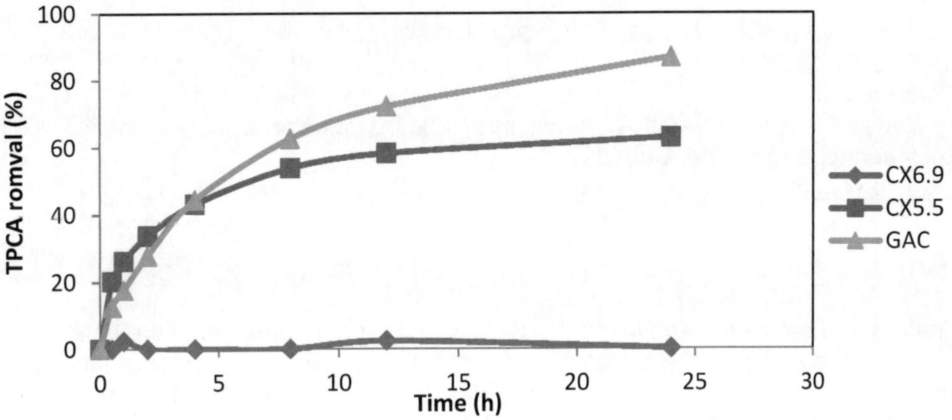


Figure 1: TPCA removal with time for three adsorbents (CX-6.9, CX-5.5, and GAC) at 20 °C.


### Acknowledgements

The financial supports provided by a Natural Sciences and Engineering Research Council of Canada (NSERC) Senior Industrial Research Chair (IRC) in Oil Sands Tailings Water Treatment, the Helmholtz-Alberta Initiative through Alberta Environment and Parks' ecoTrust Program are greatly appreciated. Dr. Pamela Chelme-Ayala helped in reviewing the abstract. Dr. Mingyu Li and Dr. Selamawit Messele helped in performing the LC-MS analysis.


### References


Alberta-Government. 2014. "Alberta oil sands Fact sheet ": Alberta Government.

Brown, L. D., L. Perez-Estrada, N. Wang, M. G. El-Din, J. W. Martin, P. M. Fedorak, and A. C. Ulrich. 2013. "Indigenous microbes survive in situ ozonation improving biodegradation of dissolved organic matter in aged oil sands process-affected waters." *Chemosphere* 93(11):2748-55.

Grewer, David M., Rozlyn F. Young, Randy M. Whittal, and Phillip M. Fedorak. 2010. "Naphthenic acids and other acid-extractables in water samples from Alberta: What is being measured?" *The Science of the Total Environment* (23):5997.

Jensen-Fontaine, M., W. P. Norwood, M. Brown, D. G. Dixon, and X. C. Le. 2014. "Uptake and Speciation of Vanadium in the Benthic Invertebrate Hyalella azteca." *Environmental Science & Technology* 48(1):731-38.

Leadership in Sustainable Infrastructure
*Leadership en Infrastructures Durables*

**Case Study**

Vancouver, Canada
May 31 – June 3, 2017/ *Mai 31 – Juin 3, 2017*

# SCALE-UP FACTORS WHEN APPLYING BIOVENTING RESULTS TO THE FIELD

Zytner, R.G.[1,5], Eyvazi, M.[2], Khan, A.A.[3] and Mosco, M.[4]
[1] University of Guelph, Canada
[2] UV Pure, Toronto, Canada
[3] Muhammad Nawaz Shareef University of Agriculture, Multan, Pakistan
[4] Meritech Engineering, Cambridge, Canada
[5] rzytner@uoguelph.ca (corresponding author)

## 1 BIOVENTING SCALE-UP

Soil contamination with petroleum and its by-products is an on-going problem in the entire the world. This contamination is caused frequently by leaking underground storage tank systems, accidental releases during transportation, or improper disposal of petroleum hydrocarbon wastes. Bioventing is an important technology that is used to remediate sites contaminated with petroleum hydrocarbons. It is a low cost non-destructive in-situ method that uses natural occurring microbiology to degrade the contaminants. The challenge in using bioventing is the determination or estimation of the biodegradation rate, as the remediation conditions vary greatly from site to site due to changes in environmental conditions, as well as site-specific conditions, like the type of soil, presence of degrading bacteria, physiochemical properties, and the concentration of contaminant. Consequently, crude estimations of the biodegradation rate are made, or expensive, time-consuming bench- or pilot-scale measurements for each site are completed.

Eyvazi and Zytner (2009) developed an empirical correlation using six different soils to provide a simple and reliable tool to estimate the degradation potential for different sites without undertaking expensive and time consuming laboratory studies. The correlation given in Equation 1 has a significance level of 95%:

$$\ln(k)=2.803\cdot PDP+0.210\cdot Sand+4.886\cdot OM-0.094\cdot Clay-0.004\cdot(SW\cdot Sand)-0.021\cdot(Sand\cdot PDP)-0.632\cdot(OM\cdot PDP)+0.004\cdot(SW\cdot Clay)-23.4997 \quad (1)$$

where, k is the biodegradation rate constant (1/d); PDP is the logarithm of initial population of hydrocarbon-degrading microorganisms in soil (log cfu/g); SW is the soil water content (%); Sand is the soil sand content (%); OM is the soil organic matter content (%); Clay is the soil clay content (%).

Development of Equation 1 showed that the soil organic matter and initial population of hydrocarbon-degrading microorganisms were the major factors affecting the biodegradation rate of synthetic gasoline in soil. It was also seen that the biodegradation rate of the gasoline was independent of the initial concentration, up to 8700 mg/kg soil for the study completed, indicating that this level of petroleum hydrocarbon contamination had no inhibitory effect on the activity of microorganisms.

Having the an estimated degradation rate for different site conditions is important as predictions on the time to closure could be completed, but the question arises as to whether the value determined from a soil sample of 200 g in size is appropriate for a site having tonnes of contamination. Accordingly it was decided to evaluate the scale-up factor to transfer the lab results to the field. Khan et al. (2015) completed work with 4 kg reactors to refine Equation 1 for a large scale and determine a scale-up factor.

Khan et al. (2015) monitored biodegradation in the 4 kg reactors with radial flow under similar environmental conditions to that studied by Eyvazi and Zytner (2009). It was determined that under ideal bioventing conditions, nutrients, temperature and oxygen, the degradation rate followed a two stage model, with the second stage of degradation representing degradation conditions that were similar to Equation 1. Equation 2 is the resulting stage-2 correlation developed by Khan et al. (2015):

$$\ln k_{Stg2} = 0.0231 \times Silt - 0.0355 \times Clay - 0.383 \times OM + 0.825 \times PDB_8 - 0.048 \times SW - 6.774 \quad (2)$$

where, $\ln k_{Stg2}$ is biodegradation rate constant (1/d); $PDP_8$ is the logarithm of initial population of hydrocarbon-degrading microorganisms in soil (log cfu/g) after 8 days; Silt is silt content (%).

Comparing the degradation rates determined from Equations 1 and 2 for the various soils tested gave an average scale-up factor (SF) of 1.8 ± 0.5. This scale-up factor showed that the increase in volume of soil from the 200 g micro-reactors to the 4 kg meso scale reactors increased the rate of hydrocarbon degradation by almost a factor of two. The increased degradation rate was mainly attributed to improved oxygen supply to the microorganisms.

Based on the findings of Khan et al. (2015), Mosco and Zytner (2017) undertook a bioventing study with a 80 kg soil reactor system to further explore scale-up conditions. The 80 kg system consisted of a climate chamber for water control in the soil, low flow venting system and an off gas capture device. Sandy and clayey soils were tested after being spiked with synthetic gasoline. Results showed that a two stage degradation system still existed, with the second stage degradation rate similar to that obtained by Khan et al. (2015) when comparing results from identical soils. Further review of the results showed that there was statistical difference between the degradation rates measured for the identical soils. However, when comparing all the scale-up factors determined for the 80 kg reactor, and comparing the results with the 200 g degradation results, an average scale-up factor of 2.3±0.4 was obtained. Completing a statistical analysis on the scale-up factors determined for the 80 kg and 4 kg reactor degradation rates when compared to the 200 g reactor showed no statistical difference.

#### 1.1.1 Significant Findings

Similar soils were tested under similar environmental conditions at various scales: 200 g, 4 kg and 80 kg. The 200 g bioventing tests showed a single stage degradation, while the 4 kg and 8 kg reactors demonstrated two stage degradation patterns. These degradation patterns were attributed to the ideal bioventing conditions induced from the radial flow induced in the larger reactors, with the second stage degradation rate matching the single stage degradation determined for the 200 g soil. Comparing all the degradation rates gave scale-up factors that averaged to 2.1±0.5, with no statistical difference. This indicates that there is no significant advantage in completing bioventing experiments at the larger 80 kg scale. As such, the Equation 1 can used as the base correlation for estimating the degradation rate, provided that a scale-up factor of approximately 2 is applied. Doing so can help remediation consultants estimate the length of time it will take to reach closure and evaluate if environmentally viable.

## REFERENCES

Eyvazi, M.J. and Zytner, R.G. (2009) A Correlation to Estimate The Bioventing Rate Constant, *Bioremediation Journal*, 13(3):141-153.

Khan, A.A., Feng, Z. and Zytner, R.G. (2015) Correlations and Scale-up Factor to Estimate Petroleum Biodegradation Rate in Soil, *Bioremediation Journal*, 19:32-46.

Mosco, M.J. and Zytner, R.G. (2017) Large Scale Bioventing Degradation Rates of Petroleum Hydrocarbons and Determination of Scale-up Factors, *Bioremediation Journal*, submitted.

# TOXICITY AND BIODEGRADABILITY STUDY ON ENHANCED PHOTOCATALYTIC OXIDATION OF POLYCYCLIC AROMATIC HYDROCARBONS IN OFFSHORE PRODUCED WATER

Liu, Bo[1,2], Chen, Bing[1], Zhang, Baiyu[1], Zheng, Jisi[1] and Liang, Jing[1]
[1] Northern Region Persistent Organic Pollution Control (NRPOP) Laboratory, Faculty of Engineering and Applied Science, Memorial University of Newfoundland, St. John's, NL A1B 3X5, Canada
[2] bo.liu@mun.ca

**Extended Abstract:**

Photocatalysis has showed excellent performance in oxidizing and mineralizing polycyclic aromatic hydrocarbons (PAHs). However, the efficacy of the sole technology could be significantly diminished in offshore produced water (OPW) due to its complicated composition. The prolonged retention time might not meet the requirements in the field because of the daily heavy discharge of OPW. It may further lead to the incomplete degradation of organics and the generation of intermediates with higher toxicity and persistence. Photocatalysis coupling with ozonation was proved to be a better solution in treating vary types of wastewater. Therefore, the integrated approach was applied in the treatment of PAHs in OPW. Glass slides coated with immobilized $TiO_2$ was used as catalysts in the process. The Microtox® toxicity and biodegradability of untreated, ozone treated and integrated process treated OPW effluents were evaluated.

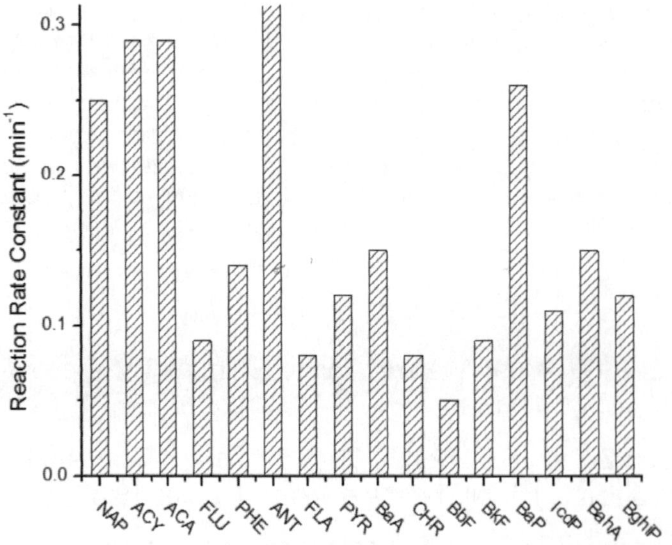


Figure 1: First order degradation rate constants of 16 PAHs in the integrated ozonation and photocatalysis process

The degradation of PAHs in the integrated ozonation and photocatalysis system was illustrated in Figure 1. NAP, ANY, ANA, PHE, ANT, BaA and BaP were completely removed within 15 minutes. 90% of 16 PAHs were degraded within 1 hour. Only BbF was more resistant but the concentration was reduced to a negligible level. The system showed a promising performance in degrading PAHs in produced water.

PAHs removal of integrated ozonation and photocatalysis was simulated with the apparent-first-order kinetics. In general, almost all the correlation coefficients ($R^2$) between the simulated model and experimental data in the integrated system were higher than 0.9, indicating a good fitness of the model. In comparison with ozonation, the reaction rate constants of almost all the PAHs were enhanced in the integrated system. Especially for the highly persistent and toxic compounds such as BaA, BaP and BghiP, their reaction rate constants were increased from 3 to 20 times. It was indicated that the significant enhancement by integrating photocatalysis and great potential of the developed method for produced water treatment.

The compounds with aromatic structure can be quickly degraded by the presence of ozone, UV light and $TiO_2$, as more than 90% of total phenols and PAHs were removed within 15-minute treatment time. However, aliphatic hydrocarbons, especially for those with higher molecular weight were relatively resistant to the system. The removal rate of aliphatic hydrocarbons was 23.8%.

Most of PAHs have the toxicity values lower than 1 mg/L. In addition, the concentration of phenols is much higher that their $EC_{50}$ values (15.1 mg/L), the toxicity of OPW is strongly related to their PAHs and phenols' concentrations. The depletion of PAHs during the treatment strongly increased the $EC_{50}$ values of treated OPW. The ozonated by products were proved as a key factor contributed to the toxicity of the effluent, especially the oxidative products of bromide (e.g. bromoform and bromate) was linearly correlated with the acute toxicity. In ozonation process, tribromomethane ($CHBr_3$) and bromodiiodomethane ($CHBrI_2$) was dramatically produced and persisted as high as 1000 ppb. The integration with photocatalysis reduced the concentration of these DBPs to a much lower level, as the $CHBr_3$ was declined to 8.3% of its peak concentration, which significantly reduced the toxicity of OPW effluent.

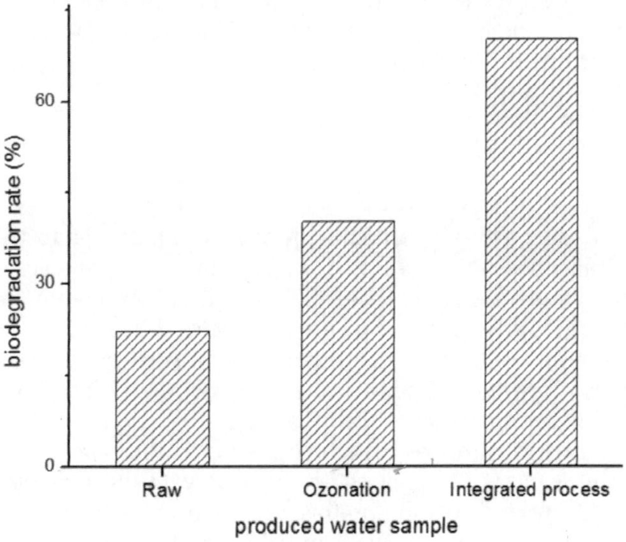


Figure 2: comparison of the biodegradation rates of untreated, ozonated, and integrated treated OPW samples

The advance of the integrated system was further proved by the enhanced biodegradability of OPW effluent (Figure 2). Compared with the degradation rates of untreated and ozonated samples, the effluent treated by the integrated system has the largest reduction rate of total organic carbon. The lower concentration of toxic compounds and the higher concentration of more biodegradable compounds produced by the integrated system thus led to the highest biodegradability of the treated OPW effluent. In this study, the integrated process showed a great potential in the effective treatment of OPW, the reduction of toxicity and the enhancement of biodegradability.

Leadership in Sustainable Infrastructure
*Leadership en Infrastructures Durables*

**Case Study**

Vancouver, Canada
May 31 – June 3, 2017/ *Mai 31 – Juin 3, 2017*

# TREATMENT OF OILY WASTERWATER USING ULTRAFILTRATION MEMBRANES MODIFIED BY NANO PARTICLES

Chen, Xiujuan[1,2] Huang, Gordon[1,3] An, Chunjiang[1,2]
[1] Institute for Energy, Environment and Sustainable Communities, University of Regina, Canada
[2] Faculty of Engineering and Applied Science, University of Regina, Canada
[3] chen240x@uregina.ca

**Abstract:** The development of membrane filtration provides a promising alternative for oily produced water treatment. Membrane fouling is a critical issue that affects treatment efficiency and service life of membrane. Membrane surface modification is a viable approach to improve its antifouling ability. In this study, Polyvinylidene Fluoride (PVDF) membrane was grafted with poly(acrylic acid) (PAA) by cold plasma induced graft copolymerization. PAA provided sufficient carboxyl groups as anchor sites for further binding of nano-$TiO_2$. The nano-$TiO_2$ binding was achieved through an effective self-assembled technique. The results indicated that both the PAA layer and nano-$TiO_2$ were uniformly and strongly attached onto the membrane surface. The surface hydrophilicity of modified membrane was significantly improved, which was proved by the decreased water contact angle. The irreversibly binding of super hydrophilic nano-$TiO_2$ endows the membranes with strong antifouling performance.

**Keywords:** Membrane; Filtration; Modification; Nano particles

## 1 BACKGROUND

Oily produced wastewater comes as a bi-product during recovery of natural gas and crude oil from onshore and offshore production operations (Zhao et al., 2014). Over the economic life of a producing field, the volume of produced water can be more than 10 times the volume of hydrocarbon produced. This volume is expected to increase from maturing conventional oil and gas fields (Zhao et al., 2014). Treating produced water from enhanced oil recovery efficiently for beneficial recycle and/or reuse, such as irrigation, reinjection, can mitigate the cost of water disposal, potentially adverse environmental impact and have a significant impact on sustainable development in Canada.

Various studies have been carried out on oily produced water treatment (An et al., 2016), membrane filtration is an alternative technology, which is considered to be with the advantages of simple operation and high treatment efficiency. Many kinds of membranes have been developed and used in produced water treatment (Padaki et al., 2015), in particularly PVDF membrane, it has been widely used because it has excellent thermal stability and chemical resistance to aggressive reagents like organic solvents and acids (Tao et al., 2014). However, it suffers from fouling, the accumulation of contaminants on or within the membrane, which eventually leads to irreversible decline in permeate flux and treatment efficiency (Mondal 2016). To improve the performance of PVDF membranes, membrane modification have been considered as one of potential methods to improve the membrane hydrophilicity and fouling resistance. Recently, the innovative use of $TiO_2$ nanoparticles (NPs) has gained considerable attention as a means of membrane modification to alleviate fouling (Ong et al., 2015). The challenge to assemble $TiO_2$ is the need to have key functional groups on the surface of the membrane that can stably hold $TiO_2$.

Therefore, in this study, PAA will be plasma-grafted on PVDF to facilitate the self-assembly of $TiO_2$ NPs. PAA is hydrophilic in nature, and its functional groups could strongly bind $TiO_2$ via ion coordination or hydrogen bonding. To maximize the number of active binding sites, the grafting efficiencies under different conditions of plasma treatment and will be examined. Afterwards, membranes will be immersed in varying concentrations of $TiO_2$ suspension to evaluate the effect of $TiO_2$ loading. The surface hydrophilicity of modified PVDF membranes will then be studied.

## 2 METHODS

Poly(acrylic acid) (PAA), PVDF ultrafiltration membrane (0.1 um pore size), $TiO_2$ NPs (P25) were used. To stimulate the assembly of $TiO_2$, four different concentrations of $TiO_2$ suspension were prepared. The plasma-grafted membranes were dipped in the aqueous $TiO_2$ suspension to attach $TiO_2$ NPs. The compositions of modified membrane surface were analyzed using XPS and FTIR. The concentration distribution of nano-$TiO_2$ on membrane surface was investigated through synchrotron based X-ray fluorescence spectroscopy, which were carried out at the beamline 07B2-1 (VESPERS) at the Canadian Light Source (Saskatoon, Canada). The morphology of membrane surface was tested by SEM. Water contact angles of neat and modified PVDF membranes were measured.

## 3 RESULTS

The present study investigated the modification of PVDF ultrafiltration membrane for produced water treatment. The activated species of low energy plasma can trigger polymerization reaction of AA monomer on or near the membrane surface with short bursts of plasma exposure. Extreme thin and even PAA coatings without compromising the bulk structure were formed at the membrane surface through cold plasma induced graft copolymerization. The $TiO_2$ NPs were self-assembled and evenly distributed in the PAA layer. Both the PAA layer and nano-$TiO_2$ were strongly attached onto the membrane surface. The surface hydrophilicity of modified membrane was significantly improved, the contact angle dropped from 83° before modification to 20° after modification. The irreversibly binding of super hydrophilic nano-$TiO_2$ endows the membranes with strong antifouling performance.

## 4 CONCLUSION

This study presented the potential of membrane surface modification for produced water treatment. The results can help to demonstrate the effects of surface modification on antifouling properties of PVDF membranes, and thus help to improve treatment efficiency and maximize membrane life. Further studies are also needed to obtain more theoretical foundation for the mechanisms of membrane antifouling. Different modification approaches for different membrane types will be also investigated for other membrane separation applications.

## Acknowledgements

This research was supported by the Natural Science and Engineering Research Council of Canada and the Canada Foundation for Innovation (CFI). The authors are also thankful to the Canadian Light Source for providing support in synchrotron infrared measurements and analysis

## References

An, C., G. Huang, Y. Yao and S. Zhao (2016). "Emerging usage of electrocoagulation technology for oil removal from wastewater: A review." *Science of The Total Environment.*

Mondal, S. (2016). "Polymeric membranes for produced water treatment: an overview of fouling behavior and its control." *Reviews in Chemical Engineering* 32(6): 611-628.

Ong, C., W. Lau, P. Goh, B. Ng and A. Ismail (2015). "Preparation and characterization of PVDF–PVP–TiO2 composite hollow fiber membranes for oily wastewater treatment using submerged membrane system." *Desalination and Water Treatment* 53(5): 1213-1223.

Padaki, M., R. S. Murali, M. Abdullah, N. Misdan, A. Moslehyani, M. Kassim, N. Hilal and A. Ismail (2015). "Membrane technology enhancement in oil–water separation. A review." *Desalination* 357: 197-207.

Tao, M., L. Xue, F. Liu and L. Jiang (2014). "An intelligent superwetting PVDF membrane showing switchable transport performance for oil/water separation." *Advanced Materials* 26(18): 2943-2948.

Zhao, S., G. Huang, G. Cheng, Y. Wang and H. Fu (2014). "Hardness, COD and turbidity removals from produced water by electrocoagulation pretreatment prior to reverse osmosis membranes." *Desalination* 344: 454-462.

Zhao, S., G. Huang, H. Fu and Y. Wang (2014). "Enhanced coagulation/flocculation by combining diatomite with synthetic polymers for oily wastewater treatment." *Separation Science and Technology* 49(7): 999-1007.

Leadership in Sustainable Infrastructure
*Leadership en Infrastructures Durables*

**Case Study**

Vancouver, Canada
May 31 – June 3, 2017/ *Mai 31 – Juin 3, 2017*

# WHEN MEMBRANE BIOREACTOR MEETS OZONATION: AN POTENTIAL SOLUTION TO THE OIL SANDS PROCESS-AFFECTED WATER (OSPW) ISSUE IN CANADA

Jinkai Xue[1,2], Yanyan Zhang[1,3], Yang Liu[1,4], Mohamed Gamal El-Din[1,4]

[1] Department of Civil and Environmental Engineering, University of Alberta, Edmonton, Alberta, T6G 1H9, Canada
[2] Department of Civil, Environmental and Geo- Engineering, University of Minnesota, Minneapolis, Minnesota 55404, United States
[3] Department of Civil Engineering, New Mexico State University, Las Cruces, New Mexico 88003, United States
[4] mgamalel-din@ualberta.ca

## 1 Introduction

In oil sands process-affected water (OSPW), naphthenic acids (NAs) are regarded as the most problematic organic compound group as they have acute toxicity to aquatic organisms. It is suggested that biodegradation is responsible for the natural, but slow dissipation of NAs in tailings ponds though the biodegradability (*i.e.*, $BOD_5$/COD ratio) of OSPW is usually low.[1-4] However, it is reported that the OSPW NAs' *in situ* biodegradation half-lives could be as long as ~13 years in tailings ponds.[5] To promote the biodegradation of OSPW NAs, it is essential to develop engineered biological processes that are efficient in mitigating OSPW organic compounds. Membrane bioreactor (MBR) has been widely used for municipal and industrial wastewater treatment because of its compactness, high treatment capacity, low sludge production and excellent effluent quality.[6, 7] To investigate the performance of MBR for OSPW treatment, two identical anoxic-aerobic MBRs were operated in parallel for raw and ozonated OSPW, respectively.[1-3] The report is focused on the biodegradation of NAs and membrane fouling of the two MBRs. The effects of various operating conditions and low-dose ozone pretreatment on the biodegradation of NAs and microbial community structure are investigated.

## 2 Methodology

The various operating conditions over the whole operation are tabulated in Table 1. Detailed description on system operation and analyses is provided elsewhere.[1-3]

**Table 1 Operating conditions examined**

| Stage | Supplemented nitrogen | | Hydraulic retention time (HRT) |
|---|---|---|---|
| | $NH_4$-N mg/L | $NO_3$-N mg/L | h |
| I | 25 | 50 | 48 |
| II | 75 | 0 | 48 |
| III | 50 | 25 | 48 |
| IV | 25 | 50 | 72 |
| V | 25 | 50 | 12 |
| VI | 25 | 50 | 24 |

## 3 Key results

The HRTs of 48-, 72-, 12-, and 24-h demonstrated classical NA degradation efficiency of 33.7%, 26.8%, 37.6%, and 36.0%, respectively, in the raw OSPW MBR; whereas the ozonated OSPW MBR showed NA degradation of 37.0%, 39.6%, 50.0%, and 32.6%, respectively. 12 h was considered as the optimal HRT based on the system's NA degradation performance and efficiency. The combined process of ozonation and MBR biological treatment achieved an NA removal rate of 94.0% with an HRT of 12 h. At Stage V (25 mg $NH_4$-N/L and 12-h HRT), the calculated half-lives of classical NAs in the raw OSPW and ozonated OSPW MBRs were 17.6 and 12.6 h, respectively. The half-lives of classical NAs in all the six operating conditions were substantially shorter than estimated half-life of 44 – 240 days in Mahdavi *et al.*[8] and Han *et al.*[9]'s suspended biological systems. The shortest half-life exhibited by our previous batch suspended reactors was 20 days.[4] Therefore, the performance of MBR for OSPW NA degradation is advantageous over conventional activated sludge systems. And the low-dose ozone pretreatment remarkably enhanced the MBR's performance in degrading OSPW NAs. Illumina MiSeq sequencing delineated the microbial community structures of the two MBR. Microbial species (e.g., *Rhodocyclales, Burkholderiales, Cytophagales*, etc.) known to be capable of degradation recalcitrant organic contaminants of petroleum origin[10-17] were found of numerical significance in the systems. Certain microorganisms might be closely correlated with the biodegradation of NAs with particular molecular structures. For instance, *Rhodocyclales* demonstrated strong positive correlations to the removal of NAs with 12 – 17 carbon atoms and 3 – 4 rings. Moreover, ozone pretreatment apparently affected the microbial community structures in the MBR by restricting the growth of certain bacteria (e.g., *Rhodocyclales* and *Burkholderiales*) known for facilitating biofilm (biofouling) formation.

## 4 Conclusions

The MBR system is efficient in degrading OSPW NAs. A half-life of 17.6 h was achieved without ozone pretreatment, which is enormously shorter than the *in situ* half-life in tailings ponds and calculated half-life in all previous engineered bioreactors. The low-dose ozone pretreatment not only substantially enhanced the MBR's NA biodegradation performance, but also significantly altered the microbial community structure. The combination of low-dose ozone pretreatment (i.e.,

30 mg $O_3$/L) and MBR is a competent candidate for OSPW treatment on a large scale in the future.

## References


1. J. Xue, Y. Zhang, Y. Liu and M. G. El-Din, *Water research*, 2016, **88**, 1-11.
2. Y. Zhang, J. Xue, Y. Liu and M. G. El-Din, *Chemical Engineering Journal*, 2016, **302**, 485-497.
3. J. Xue, Y. Zhang, Y. Liu and M. G. El-Din, *Water Research*, 2016, **105**, 444-455.
4. J. Xue, Y. Zhang, Y. Liu and M. G. El-Din, *Biodegradation*, 2016, **27**, 247-264.
5. X. Han, M. D. MacKinnon and J. W. Martin, *Chemosphere*, 2009, **76**, 63-70.
6. N. S. A. Mutamim, Z. Z. Noor, M. A. A. Hassan, A. Yuniarto and G. Olsson, *Chemical Engineering Journal*, 2013, **225**, 109-119.
7. Y. Tian, L. Chen, S. Zhang and S. Zhang, *Chemical Engineering Journal*, 2011, **168**, 1093-1102.
8. H. Mahdavi, V. Prasad, Y. Liu and A. C. Ulrich, *Bioresource technology*, 2015, **187**, 97-105.
9. X. Han, A. C. Scott, P. M. Fedorak, M. Bataineh and J. W. Martin, *Environmental science & technology*, 2008, **42**, 1290-1295.
10. A. Loy, C. Schulz, S. Lücker, K. Stoecker, C. Baranyi, A. Lehner, M. Wagner, S. Lu and A. Scho, *Applied and environmental microbiology*, 2005, **71**, 1373-1373.
11. T. H. Bell, E. Yergeau, C. Maynard, D. Juck, L. G. Whyte and C. W. Greer, *The ISME journal*, 2013, **7**, 1200-1210.
12. D. Pérez-Pantoja, R. Donoso, L. Agulló, M. Córdova, M. Seeger, D. H. Pieper and B. González, *Environmental microbiology*, 2012, **14**, 1091-1117.
13. W. F. Röling, M. G. Milner, D. M. Jones, K. Lee, F. Daniel, R. J. Swannell and I. M. Head, *Applied and environmental microbiology*, 2002, **68**, 5537-5548.
14. X. Wang, M. Chen, J. Xiao, L. Hao, D. E. Crowley, Z. Zhang, J. Yu, N. Huang, M. Huo and J. Wu, *PloS one*, 2015, **10**, e0132881.
15. W. Van den Tweel, J. Smits and J. De Bont, *Archives of microbiology*, 1988, **149**, 207-213.
16. B. Cao, K. Nagarajan and K.-C. Loh, *Applied microbiology and biotechnology*, 2009, **85**, 207-228.
17. C. Whitby, *Advances in applied microbiology*, 2010, **70**, 93-125.

# AUTHOR INDEX

# AUTHOR INDEX